W9-AMN-091

Biotechnology and Food Process Engineering

ift Basic Symposium Series

Edited by
INSTITUTE OF FOOD TECHNOLOGISTS
221 N. LaSalle St.
Chicago, Illinois

Foodborne Microorganisms and Their Toxins:
Developing Methodology *edited by Merle D.
Pierson and Norman J. Stern*

Water Activity: Theory and Applications to
Food *edited by Louis B. Rockland and
Larry R. Beuchat*

Nutrient Interactions *edited by C. E. Bodwell
and John W. Erdman, Jr.*

Food Toxicology: A Perspective on the Relative
Risks *edited by Steven L. Taylor and Richard
A. Scanlan*

Biotechnology and Food Process Engineering
*edited by Henry G. Schwartzberg and
M. A. Rao*

Biotechnology and Food Process Engineering

edited by

Henry G. Schwartzberg
University of Massachusetts
Amherst, Massachusetts

M. A. Rao
Cornell University—Geneva
Geneva, New York

Marcel Dekker, Inc.   New York and Basel

Library of Congress Cataloging-in-Publication Data
Biotechnology and food process engineering/edited by Henry G.
  Schwartzberg, M.A. Rao.
       p.   cm. -- (IFT basic symposium series)
    Papers presented at a symposium held June 23-24, 1989, sponsored
by the Institute of Food Technology and the International Union of
Food Science and Technology.
    Includes bibliographical references.
    ISBN 0-8247-8363-8 (alk. paper)
    1. Food--Biotechnology--Congresses. 2. Food industry and trade-
-Congresses. I. Schwartzberg, Henry G. II. Rao, M. A.
III. Institute of Food Technology. IV. International
Union of Food Science and Technology. V. Series.
TP248.65.F66B58 1990
664--dc20   90-3092
                                                   CIP

This book is printed on acid-free paper.

MARCEL DEKKER, INC.
270 Madison Avenue, New York, New York 10016

Current printing (last digit):
10   9   8   7   6   5   4   3   2   1

PRINTED IN THE UNITED STATES OF AMERICA

Preface

Each year the Institute of Food Technologists (IFT) and the International Union of Food Science and Technology sponsor a two-day symposium, held in conjunction with the IFT Annual Meeting. This symposium, Advances in Bio- and Food Process Engineering, the 13th in the series, was held June 23–24, 1989, immediately prior to IFT's 49th Annual Meeting in Chicago.

This volume deals with advances in bio- and food process engineering. Since we cannot hope to cover fully these fields, we have selected for coverage operations and areas that illustrate typical advances and problems. These include operations that are fairly traditional but extremely diverse in their range of application, for example, drying and freezing; operations and areas that are rapidly evolving at present, for example, microwave heating, aseptic processing, extrusion, supercritical extraction and use of computers in processing; and potentially important operations that are not widely used in food processing at present, for example, genetic manipulation, plant tissue culture, and use of membrane bioreactors.

Bioreactors have been used in producing fermented food products for thousands of years; recent advances have greatly clarified the nature of mass-transfer, heat-transfer, kinetic and biochemical processes involved and have provided bases for design and scaleup. Kinetic modeling, rheological complications, gas transfer, and power requirements for mixing are dealt with in the chapter by Harvey Blanch. Munir Cheryan and Mohamed Mehaia describe new techniques that provide enhanced selectivity and greater ease of product recovery in the chapter on membrane bioreactors. Mike Shuler and co-workers cover the emerging field of plant tissue culture. Genetic manipulation provides challenges and opportunities described by Carl Batt.

Use of microwave heating in food preparation at home has greatly expanded in recent years. Principles outlined in the chapter by Dick Mudgett provide a framework for designing both industrial processes and products for microwave-based home preparation. Marc Karel describes encapsulation and controlled release techniques that enhance flavor retention, improve storage stability, and provide controlled delivery of ingredients during preparation and consumption of foods. Solvent selectivity can be manipulated, and, in certain cases, food components can be extracted and fractionated more selectively and rapidly when supercritical extraction is used. Supercritical carbon dioxide has replaced solvents that leave potentially toxic residues. Mark McHugh discusses problems, opportunities, and process limitations in this area.

Freezing is examined from two points of view. R. Paul Singh and Jatal Mannapperuma describe techniques for predicting changes in physical and transport properties during freezing temperature vs. time behavior. Henry Schwartzberg examines how changes in ice crystal structure affect freeze concentration. Drying is used in processing many different food products, and many different methods of drying are used. Enrique Rotstein provides an approach that should prove useful in this extremely diverse field.

Twin-screw extrusion has greatly enhanced extrusion flexibility, and extrusion technology has evolved rapidly in recent years. Jud Harper examines changes in this area. Andy Rao discusses progress, new knowledge, regulatory hurdles, and problems relating to aseptic processing, one of the most actively investigated areas in food processing.

Use of computers in food processing is rapidly spreading, computational capability has expanded even more radically, and computer costs have dropped dramatically. Sam Saguy, Leon Levine, Steve Symes, and Enrique Rotstein examine relevant food processing case histories and new processing opportunities. Robotics involves intelligent decision making by machines. It involves computation and differs from simple automation, where decision-making options are limited. Peter Clark examines how robotics are being used in food processing.

Help provided by present and former Basic Symposium Committee members, Drs. Ron Josephson, Merle Pierson, Phil Crandall, V. N. Mohan Rao, Cavit Akin, and Barbara Klein, contributed greatly to setting up this symposium. A special debt of gratitude is owed to Mike Shuler for help with the biotechnology portion of the symposium.

The symposium organizers also want to thank Dr. T. P. Labuza, 1988–1989 IFT President, for his enthusiastic support; H. W. Mattson, IFT Executive Director; J. B. Klis, Director of Publications; Anna May Schenck, Associate Scientific Editor; and the other IFT staff members who provided support and coordination for this symposium.

But most especially, we gratefully acknowledge the speakers for their contribution to the success of the Symposium and this volume. Without their dedication, expertise, and hard work, publication of these proceedings in such a timely manner would not have been possible.

Henry G. Schwartzberg
M. A. Rao

Contributors

Carl A. Batt, Ph.D. Assistant Professor, Department of Food Science, Cornell University, Ithaca, New York

Harvey W. Blanch, Ph.D. Professor, Department of Chemical Engineering, University of California at Berkeley, Berkeley, California

V. Bringi, M.S.Ch.E. Graduate Student/Research Associate, School of Chemical Engineering, Cornell University, Ithaca, New York

Munir Cheryan, Ph.D. Professor, Food and Biochemical Engineering, Department of Food Science, University of Illinois, Urbana, Illinois

J. Peter Clark, Ph.D. President, Epstein Process Engineering, Inc., Chicago, Illinois

Judson M. Harper, Ph.D.* Vice President for Research, Research Office, Colorado State University, Fort Collins, Colorado

Thomas J. Hirasuna, M.S.Ch.E., P.E. Graduate Student/Research Associate, Department of Food Science, Cornell University, Ithaca, New York

Marcus Karel State of New Jersey Professor of Food Science, Department of Food Science, Rutgers University, New Brunswick, New Jersey

Leon Levine President, Leon Levine & Associates Inc., Plymouth, Minnesota

Jatal D. Mannapperuma, Ph.D. Post Graduate Researcher, Department of Agricultural Engineering, University of California at Davis, Davis, California

---

Current affiliation: Interim President, President's Office, Colorado State University, Fort Collins, Colorado

Mark A. McHugh, Ph.D.   Department of Chemical Engineering, The Johns Hopkins University, Baltimore, Maryland

Mohamed A. Mehaia, Ph.D.   Assistant Professor, Dairy Technology Laboratory, College of Agriculture & Veterinary Medicine, King Saud University, Qassim Branch, Buriedah, Saudi Arabia

Richard E. Mudgett, B.S.E.E., B.B.A., M.S., Ph.D.   Professor, Food Science Department, University of Massachusetts, Amherst, Massachusetts

Christopher L. Prince, M.S.Ch.E.   Graduate Student/Research Associate, Department of Food Science, Cornell University, Ithaca, New York

M.A. Rao, Ph.D.   Professor of Food Process Engineering, Department of Food Science and Technology, Cornell University—Geneva, Geneva, New York

Enrique Rotstein   The Pillsbury Company, Minneapolis, Minnesota

Israel Saguy   The Pillsbury Company, Minneapolis, Minnesota

Henry G. Schwartzberg, Ph.D.   Professor, Department of Food Engineering, University of Massachusetts, Amherst, Massachusetts

Michael L. Shuler, Ph.D.   Professor, School of Chemical Engineering, Cornell University, Ithaca, New York

R. Paul Singh, Ph.D.   Professor of Food Engineering, Department of Agricultural Engineering, University of California, Davis, California

Stephen T. Symes   Senior Systems Engineer II, Industrial and Systems Engineering, The Pillsbury Company, Minneapolis, Minnesota

Contents

1

Biotechnology and Its Implication for the Future Design and Production of Food Ingredients

Carl A. Batt

Cornell University
Ithaca, New York

INTRODUCTION

The physical attributes of any food are dictated by the complex behavior of its macromolecular constituents in their native form and as modified during processing and storage. The study of these macromolecules is the heart of food science and involves a wide variety of disciplines including chemistry, physics, biochemistry, and biology. In addition to this fundamental knowledge, the complex interactions that occur between different components, both intrinsic and extrinsic to the food, must be addressed. In the past few years, the term biotechnology has been used to cover a conglomeration of activities related to the genetic manipulation and design of processes to produce products including those destined for human consumption as foods. A paradigm developed by Tony Sinskey of M.I.T. illustrates the individual facets of biotechnology and how each individual process might be described and addressed (Fig. 1.1). Biotechnology, both as an academic pursuit and as an emerging industry, has seen an explosive growth in the past 10 years. A number of "older" disciplines, including food science, chemical

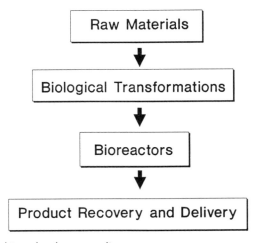

Figure 1.1 The biotechnology paradigm.

engineering, and biology, lay historic claim to its birth. Biotechnology has progressed primarily due to the development of techniques to manipulate in vitro the genetic structure of a biological system. This capability has facilitated the production and manipulation of proteins and hence their biological hosts to a degree previously unimaginable. To fully appreciate the impact that biotechnology has had and will continue to have on our society, a chronological review of its emergence is appropriate. To some in the food industry the initial predictions of the impact of molecular biological advances on foods have not proven correct. Unfortunately, food systems are extremely complex and cannot be addressed simply. Perhaps the greatest changes in the food industry as a result of molecular biology can be seen in diagnostic assays for food pathogens. Here the advent of DNA-based probes and monoclonal antibodies (another revolutionary development of the past 15 years) now enable the rapid detection of foodborne pathogens and their toxins, thus improving the safety of food products. This review will not cover these developments, and the reader is directed toward a number of recent examples in the literature (Flowers, 1985; Andrews, 1985; Todd et al., 1988; Klinger and Johnson, 1988).

THE BASIS OF GENETIC CONTROL

The attributes of any biological system, whether it be a simple bacterium or a human being, are dictated in part by the principle genetic material, deoxyribonucleic acid (DNA). This is of course not entirely true, as some viruses use ribonucleic acid (RNA) in place of DNA. DNA is organized in discreet units known as genes where each gene codes for a given protein. For example, the bacterium *Escherichia coli* contains approximately 1,000 genes; the number of genes increases with respect to the complexity of the system. The concept of genes was recognized over 100 years ago as the result of experiments conducted by an Austrian monk, Gregor Mendel. His "simple" breeding experiments with garden pea plants demonstrated that specific traits could be transferred to progeny and that they tended to segregate with a statistical probability. Curiously, although the concept of genes and their inheritance was well established by the beginning of this century, the nature of this genetic material was not known until the 1940s. The elegant work of Oswald Avery, Colin MacLeod, and Maclyn McCarty (Avery et al., 1944) of the Rockefeller Institute elucidated the role of DNA as the chemical element responsible for inheritance of traits in biological systems. Another milestone was the initial solution of the double helix structure of DNA by James Watson and Frances Crick using crystallographic data (Watson and Crick, 1953).

Since the realization that DNA was the factor that mediated the character of a biological system, great strides have been made in understanding the process. Genes for the most part (with the exception of ribosomal and transfer RNA) are transcribed and translated into proteins. The nucleotide sequence of a gene is almost always faithfully deciphered into a specific amino acid sequence, the result of which is a protein. The protein has a distinctive structure, a three-dimensional conformation, which is a function of the amino acid sequence and the environment in which it resides. The structure of the protein, in turn, dictates its function. Therefore the enzyme β-galactosidase cleaves lactose by virtue of its structure—a direct consequence of its amino acid sequence encoded by the nucleotides in the gene. In contrast, casein has an entirely different amino acid sequence and does not cleave lactose (in fact its true physiological function is not known).

The variety of biological genera found in nature is due to continual mutation and selection. Prokaryotic microorganisms, bacteria, have been a part of life on earth well before the appearance of man. Through a variety of environmental pressures, they have developed and adapted into a wide array of genera, only a fraction of which we have documented. Depending upon their given niche, these bacteria have evolved to utilize the available nutrients and to survive the environmental stresses they may encounter. The evolutionary process that has created the current phenotypes can be considered a form of genetic engineering, and it clearly has been successful in terms of bacterial survival and diversity. Most biological systems exhibit some sort of regulation which is designed in part to conserve energy and produce enzymes or other macromolecules only when necessary. There has been, and continues to be, a school of thought which contends that, given the vast array of environments on the earth, a microorganism can be found to carry out virtually any bioconversion. If one subscribes to the existence of extraterrestrial life, an even greater number of environments could be predicted. This approach, which could be described as ecological screening, has in fact resulted in the identification of a number of very important microorganisms, especially those that degrade complex carbon compounds including cellulose and hydrocarbons. An environment in which a microorganism with the desired trait can have a selective advantage over others and thus outproduce competitors and evolve must be available for ecological screening to be successful. Unfortunately, it is not always possible to identify such an environment and, more importantly, there are a lot of rocks in the world to look under.

The process by which the microorganisms have evolved requires a mutational event followed by some selective process favoring a given phenotype. Mutational frequencies for a given biological system vary; for example, a mutation occurs in the bacteriophage T2 once in every 1×10^8 gene replications, whereas a mutation occurs once in every 8×10^8 asexual spores of a common fungus. The mutation rate depends upon the integrity of the mechanisms that repair the damage inflicted on the DNA by environmental chemicals or ionizing irradiation or on the accuracy of DNA replication. The mutational rate can be enhanced by treatment with a variety of mutagens or by the selection of mutants which have some defect in the biological systems mentioned above. Exposure to a mutagen can increase the mutation frequency by at least 1,000 times. Selection of the desired phenotype still requires that it be identifiable within the context of the number of candidates that are feasibly screened. This number depends upon the complexity of the assay procedure and man-years that can be devoted to its identification.

The classical approaches toward strain improvement through mutagenesis and selection have resulted in the development of a great number of extremely valuable microorganisms. A variety of strains used in the food and pharmaceutical industries have been obtained via this approach. A recent discovery exemplifying this approach is the identification of *Bacillus brevis* HPD 31, which produces an amylase 100 times faster than those obtained through genetic engineering. It was selected after screening over 100,000 bacteria. Unfortunately, mutagenesis and selection cannot deliver a microorganism that does not have the latent genetic potential to code for the desired improvement. Given enough time, effort, and the appropriate biological selection, a desired microorganism may evolve with the desired traits. It is, however, often difficult to create the selectable environment necessary to effect the desired change. It is difficult to imagine the type of environment that would be advantageous for an insulin-producing microorganism to proliferate. Obviously, despite the enormous progress that has resulted from ecological screening and from mutagenesis and selection, a new technology was needed to facilitate the design of microorganisms with novel properties.

THE DEVELOPMENT OF RECOMBINANT DNA TECHNOLOGY

A number of scientific events during the late 1960s and early 1970s lay the foundation for establishing recombinant DNA technology. The highlight was the report by Stanley Cohen and colleagues from Stanford University, who demonstrated that DNA could be restricted and joined in vitro using enzymes isolated from bacteria. These simple biochemical tools could be used to combine DNA isolated from diverse sources, and hence genes could be "cloned." The nature of the manipulations that could be carried out were always possible in nature, but now we could perform them in vitro. The recombination of DNA from two dissimilar sources was now feasible and transgressed the normal boundaries of species. DNA from a mouse could be propagated in a bacterium (and now the reverse is possible). Furthermore DNA could be unequivocally analyzed in vitro in contrast to simply observing its effect on the phenotype of the organism. This in vitro analysis now can be as detailed as determining the precise sequence of nucleotides in the gene. The reader may be aware of the efforts now underway to initiate a program to sequence the human genome. A number of associated problems stand in the way of the successful completion of this enormous feat. Not only is there the technical aspect of obtaining the primary sequence information, but the processing and cataloguing of the 3 billion nucleotides in the human genome will require a new class of computer processors and algorithms. The absolute value of the information contained in a nucleotide sequence can be difficult to interpret. Although we have a basic understanding of the translation of a nucleotide sequence into an amino acid sequence, other meaningful details are not as obvious. What, for example, controls the expression of certain genes in specific organs? Furthermore, as alluded to previously, the expression of most genes is regulated, and the regulatory scheme can be a myriad of complex events. Differentiation in higher eukaroytes is extremely complex, and yet its control is buried somewhere in the genetic code of the organism.

The required tools for genetic engineering of a given organism are a vector and a transformation system. A vector is a DNA sequence which functions to maintain the desired gene in the host and is usually constructed from an extrachromosomal element. The only requirement is that it replicate whenever the host replicates to avoid its segregation and loss during cell division. If the vector integrates it must be replicated along with the chromosome; if it is extrachromosomal it must have a compatible and function replicon. Lastly, a vector must have a selectable marker in order to differentiate the transformed host from the untransformed host. Selectable markers usually either impart antibiotic resistance to the host or complement a nutritional deficiency.

The second tool for genetic engineering of a microorganism is a transformation system for the uptake of naked DNA by the host. Since most genetic manipulations are performed in vitro, it is essential to be able to introduce the engineered DNA into the host. A cell can be transformed using a variety of techniques to make it competent (capable of DNA uptake). Some organisms can achieve competence naturally or through manipulation of physiological conditions, while others require treatment with divalent cations. Alternatively, the cell wall can be removed, rendering the protoplasted cell capable of DNA uptake. Although a number of systems for transforming bacteria via protoplasts have been developed, the advent of electroporation has replaced many of the protocols. Electroporation is a process by which cells (animal, plant, or bacteria) are rendered permeable to DNA by exposure to high voltages. Through a mechanism not clearly understood, the cell membrane is temporarily altered allowing DNA to enter the cell. A number of parameters, including the time (on the order of milliseconds), voltage (approximately 5000 V), and buffer composition can affect the transformation efficiency.

A distinction should be made between transforming a single cell vs. multicellular organism. In the latter case, the issue becomes whether the individual cell once transformed can develop or regenerate into a multicellular organism. A mammalian cell in culture can be easily transformed using a variety of retroviral vectors but it cannot regenerate into a mammal. Similarly, only certain plant cells in culture can be regenerated into intact plants. The problem of regeneration is complex and relates to the genetic signals that control the developmental process. The developmental process from fertilized eggs into a mature organism is not understood, and to date a mammalian cell in culture has not been regenerated into a mammal. In plants, the developmental process may be simpler, and treatment with specific hormones seems to be able to initiate the process. In the absence of a system to regenerate a transformed cell in culture into an intact organism, other approaches to transformation were necessary.

Perhaps the most unusual approach to introducing DNA into a host was developed by John Sanford and Ed Wolf of Cornell University. They reasoned (in the face of a great number of unbelievers) that a small high-velocity particle coated with DNA could penetrate the cell membrane and deliver the DNA. The term "biolistic" was coined to describe the approach, and the system consisted of a .22 caliber driven projectile which accelerated 5-μm tungsten particles to speeds sufficient to pierce the cell. To date a number of yeast and plant cells have been transformed with a variety of vectors, including some hosts that have proven to be recalcitrant to other methods of transformation.

The scope of genetic manipulations extend to the animal kingdom as well. Animals cells grown in tissue culture can be transformed with exogenous DNA and the effect of the transforming DNA on the cell observed. It is not currently feasible to regenerate animals from cells in culture as can be achieved (in certain cases) with plants. In 1982, Ralph Brinster (University of Pennsylvania) and Richard Palmiter (University of Washington) produced transgenic mice by inserting a gene into the male pronucleus of a fertilized egg. The microinjected egg was then implanted into a foster mother and allowed to come to term. In a fraction of the progency, the system is not 100% efficient; the injected DNA is integrated into the chromosome, resulting in a transgenic animal. The result in these initial experiments was a transgenic mouse which grew faster than its control (nontransgenic) siblings due to the introduction of a gene coding for growth hormone. To date a number of transgenic animals have been produced including rabbits, sheep, pigs, fish, and cattle. The major obstacles are in part technical: how to isolate, inject, and reimplant fertilized eggs efficiently and in the end produce transgenic animals at a high frequency. A major breakthrough just reported that could revolutionize the production of transgenic animals is the use of sperm as a vehicle for delivering the DNA. Lavitrano and coworkers demonstrated that sperm incubated in the presence of exogenous DNA could be used to fertilize eggs in vitro and that these eggs when implanted into a foster mother gave rise to transgenic progeny at a rate of approximately 30% (Lavitrano et al., 1989). This circumvents one of the problems associated with the current methods of generating transgenic animals, namely the need to microinject the DNA into the fertilized egg and the problems associated with isolating fertilized eggs and damaging the egg during the injection process.

It would be virtually impossible to construct a list of all the microorganisms, plants, or animals that have been genetically transformed, since the number grows with each day. It is not clear what the limitations are and if any biological system cannot be transformed given a sufficient effort.

Once a vector/transformation system has been developed, it can serve as a tool for not only elucidating a gene function but also improving that function. One simple benefit from the use of plasmid vectors is the enhancing of the activity of the cloned gene product. In most, but not all cases, a gene present in many copies results in a higher level of encoded protein and hence a greater specific activity.

CURRENT AND FUTURE TARGETS

The current crop of biotechnology products includes a number of high-value pharmaceutical proteins which prior to their cloning and expression in either bacteria, yeast, or cultured mammalian cells were only available in limited quantities. Currently, interferon, insulin, human growth hormone, tissue plasmino- gen activator (TPA), and erythropoietin among others are being marketed both in the United States and throughout the world. The real economic future of these types of products is not clear, and market forecasts have invariably proven too optimistic. The reasons behind these inaccurate estimations are complicated but may involve the inability to estimate the market for a product whose existence and therefore actual potential could not be predicted in advance. How well, for example, would TPA sell in comparison to streptokinase (given the vast cost difference) when no actual comparative market studies were feasible? The generation of new biomedical proteins will obviously continue, especially as their true physiological roles are elucidated, given an adequate supply for analysis. It is reasonable to assume that the commercial interest and estimated time for entry into the marketplace for any given product is a function of the potential value of that product and the technical feasibility of producing it.

In agriculture, biotechnology has provided a means to develop disease, frost, herbicide, and insect-resistant plant varieties. For example, the gene coding for an insecticidal protein has been cloned from a bacterium (*Bacillus thurigiensis*) and expressed in a variety of plants. The result is a plant that can resist defoliation by certain insects. As with pharmaceutical proteins the true market value (and hence any future development) is difficult to predict.

One of the more interesting potentials of transgenic animals is to produce high-value pharmaceuticals in their milk. The approach is to genetically fuse the gene coding for the desired protein (i.e., TPA) to the expression sequences and secretion signal of a milk protein. Paul Simons and John Clark (AFRC, Edinburgh) have produced transgenic sheep carrying the gene coding for factor IX (a blood clotting protein) to the gene coding for ovine β-lactoglobulin. The result is a small amount of factor IX secreted in the milk of these animals. This approach takes advantage of the normal biosynthetic machinery in the animal and produces the product in a relatively simple medium such as milk which has only a small number of well-characterized proteins facilitating purification of the product. We have a long way to go in terms of understanding the rules of gene expression in higher animals. This is a prerequisite for intelligent genetic engineering of animals in order to modify important agricultural properties, for example, to make a cow produce more protein and less fat or a chicken that produces eggs devoid of cholesterol. The technology described only gives us the tools to investigate and hopefully produce wise answers.

The long-range impact of genetic engineering on the food industry is difficult to predict. The targets are less well defined than in other areas especially with respect to pharmaceuticals. Also, as mentioned previously, food systems are extremely complex, and their problems cannot be addressed simply by targeting a single gene or a single biochemical event. Too often the sirens of biotechnology will proclaim that a given problem will be solved by simply inserting a gene into a given host without a clear understanding of the consequences of the act. Genes can be expressed and in most cases the correct protein in terms of the amino acid sequence produced. The difficulty is, however, whether that protein will have the desired catalytic activity in the new host, and whether the required cofactors for the reaction will be in sufficient supply for the reaction to function. What will be the consequence of the enzyme's activity on the host? Over the course of the last billion years, nature has, through a complicated process of natural selection, fine tuned biological systems to perform in a manner that may not stand the demands of the "paper" biochemists redesign. My laboratory has suffered through a classical scenario where we initially thought that the yeast *Saccharomyces cerevisiae* could be made to ferment xylose by simply expressing a prokaryotic gene coding for the only missing metabolic step in the process, the isomerization of xylose to xylulose. After considerable work we were able to express the *E. coli* gene for xylose isomerase in *S. cerevisiae* using a promoter derived from the alcohol dehydrogenase gene of *S. cerevisiae*. The correct xylose isomerase protein was produced as demonstrated by immunological analysis, but it lacked any detectable activity. On paper, our strategy appeared sound, but in actuality the complexity of protein structure/function prevented its success. (We are not the only unfortunate research group; see Sarthy et al., 1987; Amore et al., 1989). Fortunately, not all prophecies prove incorrect. This is exemplified by the work of Lonnie Ingram and his engineering of *E. coli* to produce ethanol. Ingram and coworkers identified the genes essential for ethanol production (alcohol dehydrogenase II and pyruvate decarboxylase) from *Zymomonas mobilis* and introduced them into *E. coli* with the purpose of directing its fermentation products to ethanol. The result was an *E. coli* strain that had an enhanced growth yield on glucose and produced ethanol as a major end product (Ingram et al., 1987; Ingram, personal communication).

PROTEIN ENGINEERING

Despite the best efforts of mother nature, not all proteins have the appropriate functional properties desired by the biotechnologist. For enzymes, their pH optimum, temperature stability, or substrate range may not be optimal for their intended usage in a specified process. Once again a resourceful investigator may target a given environment that might contain an organism producing the protein with the desired properties. Alternatively, it is feasible to change the structure and therefore the function of a protein by effecting a change to the nucleotide sequence. The term "protein engineering" describes altering a protein's structure via its nucleotide sequence.

Within the past five years techniques for oligonucleotide-directed site-specific mutagenesis have been developed to selectively and precisely replace nucleotides in a targeted gene sequence. The obvious requirements for such a system include a knowledge of the complete nucleotide sequence (and by deduction the amino acid sequence) in addition to some information about the catalytic/functional properties of specific regions of the protein. It is not feasible, in the absence of luck, to intelligently direct an improvement on a random basis. The only exception is in the case where a powerful selection scheme is available to identify the desired change. This latter approach is exemplified by the isolation of thermostable derivatives of a thermolabile enzyme using an organism that can grow at elevated temperatures. Liao et al. (1986) outline a strategy using the thermophile *Bacillus stearothermophilus* as a host for isolating thermostable derivatives of a cloned thermolabile gene product. The concept was to select based upon the requirement that only those *B. stearothermophilus* cells that carried a thermostable derivative of the cloned gene would be able to grow under the selective conditions. Specifically they isolated thermostable derivatives of the kanamycin resistance gene (whose protein product does function above 45°C) by selecting kanamycin resistance at 55°C. The result was a series of random mutations which conferred thermostability on the gene product that occurred. If a scheme as described here to select a thermostable derivative can be devised by demanding functionality at elevated temperatures, then a random approach is possible. In most cases, however, such a scenario is not practical, and a detailed knowledge of the structure is a prerequisite.

An understanding of the tertiary structure of a protein is usually a prerequisite for any site-directed modification. In a protein such as glucose isomerase there are approximately 400 amino acid residues and, given 20 different amino acid substitutions, the possible derivatives are overwhelming. The structure and hence a correlation to the function for a protein can be determined in a number of ways (Fig 1.2). The most widely used is determining the x-ray crystal structure of the protein. Proteins under certain conditions can form crystals which represent a collection of upward of 10^{10} molecules arranged in an orderly lattice. Crystallization of proteins is not an exact science, and the conditions vary for each protein. The first protein to be crystallized, jack bean urease, did so almost spontaneously,

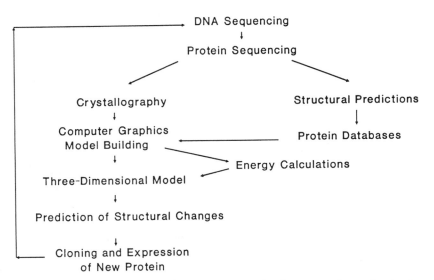

Figure 1.2 Routes toward the determination and refinement of protein structure. (*Adapted from Blundell and Sternberg, 1985.*)

and a researcher can suffer for years without success. Once a crystal is obtained, its structure can be determined by the diffraction pattern obtained using x-ray beams. The diffraction pattern is a reflection of the relative spacing between atoms within the protein and can be resolved to less than 1 Angstrom. Given sufficient data the position of amino acid side chains and potential intramolecular bonds can be mapped.

The data obtained from crystallographic analysis is not an absolute measure of the structure of the protein. Given the crystallographic coordinates indicating the positions of the atoms in the protein, the structure can be refined by bringing to bear the laws of thermodynamics. The energy of the structure can be minimized to yield the conformation that is most favorable. Molecular modeling is accomplished via computer analysis and high-resolution graphic terminals. The process is iterative, and the predictions must be tested against the original x-ray diffraction data. In the end any structural theory must be tested biologically since the laws of nature mediate the final structure.

It is not always necessary to determine de novo the structure of every new protein from primary crystallographic data. Indeed, since any given protein might carry out a similar if not identical function as a protein whose structure is known, it is possible to make structural predictions for the former based upon homology to the latter.

Analyzing the structure facilitates the prediction of changes that might improve the protein's function. The predicted changes are then made by altering the nucleotide sequence of the gene through site-directed mutagenesis, producing the protein and finally analyzing the effect. It is usual to then determine the structure of the modified protein as described above. This process is cyclical, and several rounds of structure determination, mutagenesis, and functional analysis are routine.

One of the more elegant examples of engineering a protein to improve its performance is the work of Jim Wells and coworkers at Genentech. Subtilisin is a protease isolated from the bacterium *Bacillus amyloliquifaciens*. It could potentially be used in laundry detergents, but it is sensitive to oxidation, a problem where the addition of oxidizers to detergents is desirable. Through a variety of biochemical and structural studies, a methionine group at position 222 was implicated as the residue which when oxidized inactivated the enzyme. The crystal structure of subtilisin suggested that this methionine residue was in the active site supporting the hypothesis that it was in the active site supporting the hypothesis that it was the target for oxidation. Through site-directed mutagenesis the methionine residue was replaced with a variety of other amino acids, some of which actually improved the catalytic activity of the enzyme. Of these replacements, serine and alanine substitutions resulted in an enzyme that was much more resistant to chemical oxidation. Recently, this modified enzyme was the first engineered protein to be patented in the United States.

MOLECULAR BIOLOGY OF FOODS

Foods are extremely complicated systems in terms of the different components and the interactions of these components. Add to this the effect of processing and storage and it becomes difficult to target a single constituent for molecular genetic manipulation. There are, however, a few good examples of how molecular biology and food science interface, and one from my own laboratory is presented.

Bovine β-lactoglobulin is a whey protein, found in the milk of many mammals including bovine species, sheep, deer, dogs, and pigs but not in humans or rodents (Pervaiz and Brew, 1985). The molecule occurs as several different variants due to genetic variations that exist in given animal species. β-Lactoglobulin is a single polypeptide of 18 kDa, which corresponds to 162 amino acids residues, and the bovine protein normally exists as a dimer. The complete amino acid sequence of bovine β-lactoglobulin has been reported and genetic variations in the amino acid sequence identified (Creamer et al., 1983). The tertiary structure of β-lactoglobulin has been resolved by x-ray crystallography by Lindsay Sawyer and colleagues in Edinburgh.

The utility of whey is limited in part by the thermally induced precipitation of β-lactoglobulin, a major whey protein. β-Lactoglobulin is a globular protein and the primary sequence reveals two disulfide bonds between amino acids residues 66–160 and 106–119 and a free cysteine group at 121. The free cysteine group buried within the molecule appears to be the reactive thiol group in β-lactoglobulin. Upon heating, β-lactoglobulin undergoes a conformational change leading to the formation of aggregates (Sawyer, 1967). It can react not only with itself but with other milk proteins, most notably κ-casein (McKenzie, 1971). Based upon the known disulfide linkages between the Cys-66/Cys-160 and Cys-106/Cys-119 residues, the free thiol Cys-121 appears to mediate the intermolecular disulfide linkage formed upon heating (Creamer et al., 1983; Papiz et al., 1986).

In 1987, at the suggestion of John Kinsella, we began to examine the problem of the thermal stability of β-lactoglobulin and quickly reasoned that if the gene could be cloned and sequenced the Cys-121 residue could be replaced. We isolated a gene coding for β-lactoglobulin A from a cDNA bank constructed from poly(A+)mRNA isolated from the bovine mammary gland (Jamieson et al., 1987). Its nucleotide sequence codes for the β-lactoglobulin A, from amino acid residues Leu-11 to Ile-162, as based on the amino acid sequence reported by Braunitzer et al., (1973). The first 10 amino acids not present in our cDNA clone were introduced using a 55 bp synthetic oligonucleotide coding not only for these amino acids but a translational start and a properly spaced Shine-Delgarno sequence (Fig. 1.3). The final construct has been expressed in the vector pTTQ18, a tac promoter vector. Recombinant β-lactoglobulin A accounts for up to 10% of the total cellular protein and is found in inclusion bodies. The insoluble r-β-lactoglobulin can be solubilized using guanidine HC1 and renatured (Batt et al., 1989). We have now replaced the

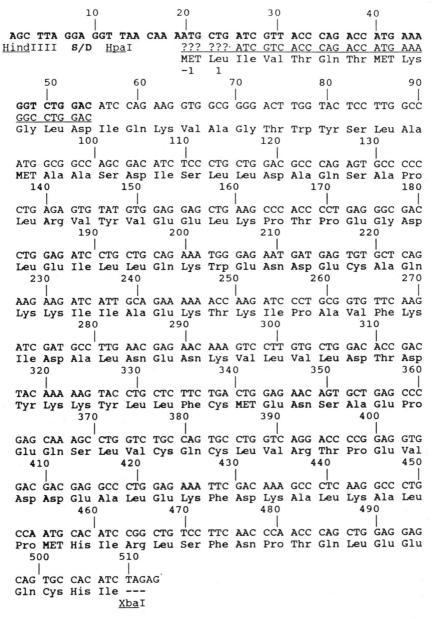

```
              10            20            30            40
               |             |             |             |
AGC TTA GGA GGT TAA CAA AATG CTG ATC GTT ACC CAG ACC ATG AAA
HindIIIII   S/D  HpaI   ??? ???· ATC GTC ACC CAG ACC ATG AAA
                        MET Leu Ile Val Thr Gln Thr MET Lys
                        -1   1

      50            60            70            80            90
       |             |             |             |             |
GGT CTG GAC ATC CAG AAG GTG GCG GGG ACT TGG TAC TCC TTG GCC
GGC CTG GAC
Gly Leu Asp Ile Gln Lys Val Ala Gly Thr Trp Tyr Ser Leu Ala
        100           110           120           130
         |             |             |             |
ATG GCG GCC AGC GAC ATC TCC CTG CTG GAC GCC CAG AGT GCC CCC
MET Ala Ala Ser Asp Ile Ser Leu Leu Asp Ala Gln Ser Ala Pro
    140           150           160           170           180
     |             |             |             |             |
CTG AGA GTG TAT GTG GAG GAG CTG AAG CCC ACC CCT GAG GGC GAC
Leu Arg Val Tyr Val Glu Glu Leu Lys Pro Thr Pro Glu Gly Asp
        190           200           210           220
         |             |             |             |
CTG GAG ATC CTG CTG CAG AAA TGG GAG AAT GAT GAG TGT GCT CAG
Leu Glu Ile Leu Leu Gln Lys Trp Glu Asn Asp Glu Cys Ala Gln
    230           240           250           260           270
     |             |             |             |             |
AAG AAG ATC ATT GCA GAA AAA ACC AAG ATC CCT GCG GTG TTC AAG
Lys Lys Ile Ile Ala Glu Lys Thr Lys Ile Pro Ala Val Phe Lys
        280           290           300           310
         |             |             |             |
ATC GAT GCC TTG AAC GAG AAC AAA GTC CTT GTG CTG GAC ACC GAC
Ile Asp Ala Leu Asn Glu Asn Lys Val Leu Val Leu Asp Thr Asp
    320           330           340           350           360
     |             |             |             |             |
TAC AAA AAG TAC CTG CTC TTC TGA CTG GAG AAC AGT GCT GAG CCC
Tyr Lys Lys Tyr Leu Leu Phe Cys MET Glu Asn Ser Ala Glu Pro
        370           380           390           400
         |             |             |             |
GAG CAA AGC CTG GTC TGC CAG TGC CTG GTC AGG ACC CCG GAG GTG
Glu Gln Ser Leu Val Cys Gln Cys Leu Val Arg Thr Pro Glu Val
    410           420           430           440           450
     |             |             |             |             |
GAC GAC GAG GCC CTG GAG AAA TTC GAC AAA GCC CTC AAG GCC CTG
Asp Asp Glu Ala Leu Glu Lys Phe Asp Lys Ala Leu Lys Ala Leu
        460           470           480           490
         |             |             |             |
CCA ATG CAC ATC CGG CTG TCC TTC AAC CCA ACC CAG CTG GAG GAG
Pro MET His Ile Arg Leu Ser Phe Asn Pro Thr Gln Leu Glu Glu
    500           510
     |             |
CAG TGC CAC ATC TAGAG·
Gln Cys His Ile ---
              XbaI
```

Figure 1.3. Nucleotide sequence of recombinant β-lactoglobulin gene. **Boldface** denotes the synthetic oligonucleotide sequence coding for the met (−1) and the first 12 amino acids of the β-lactoglobulin gene using an *E. coli* codon bias. Underlined nucleotides are the sequence for this region from the ovine β-lactoglobulin gene (Gaye et al., 1986). The final construct is cloned into the *Hind*III, *Xba*I sites of pTTQ18.

Cys-121 residue by site-directed mutagenesis and are evaluating the thermal polymerization of the amount r-β-lactoglobulin protein. Further structural modification is being planned in collaboration with Lindsay Sawyer and John Brady of Cornell's Department of Food Science. Our eventual plans include expression of the β-lactoglobulin gene in transgenic animals, an event already demonstrated in mice using the ovine β-lactoglobulin gene (Simons et al., 1987). In the end we see this program as a model for the modification of milk proteins to alter their functionality and hence utility. Kudos should go to the group at the University of California-Davis under the direction of Tom Richardson for their pioneering efforts in the area using casein as a model system.

WHAT'S NEXT?

Up to now, we have discussed the genetic engineering of proteins that have a known function, yet there are an unlimited number of potential reactions for which an enzyme capable of carrying out that bioconversion has not been isolated. Once again, the argument that if one searched long enough that enzyme might be found cannot be discounted. In the last two years, a novel approach to designing proteins capable of carrying out unique reactions was demonstrated by Peter Schultz (Univ. of California-Berkeley) and Richard Lerner (Scripps Institute). Remember that enzymes serve to catalyze a reaction by stabilizing the transition state between a reactant(s) and product. They reasoned that if an analog of a transition state compound could be made and monoclonal antibodies raised against it, those monoclonal antibodies might catalyze the reaction. (Monoclonal antibodies are proteins that specifically recognize a given molecule or antigen. They are produced by immunization of an animal with the antigen, then isolating the antibody-producing cells from the spleen of the animal. These antibody-producing cells are then fused in vitro to another cell line, which imparts on the antibody-producing cell the ability to grow in tissue culture almost indefinitely. The fusion product, a hybridoma, can then be cultured, and by dilution and selection, a hybridoma that secretes the antibody of interest can be isolated.) The monoclonal antibodies made against these transition state analogues could indeed catalyze, albeit slowly, the reaction of interest. The implication of this research is that a protein capable of mediating a novel reaction can be designed using the structure blueprint of the monoclonal antibody. Once the basic structure has been resolved, improvements can be made using site-directed mutagenesis and expression in a host, for example, a bacterium.

A long-term goal of biotechnologists is to be able to design structures based upon a desired function. This ambitious aim requires an understanding of the rules governing structure/function, the basis for a given reaction mechanism, and the important attributes of the protein. It is not unlike the alchemist trying to decipher the required steps to change lead into gold. The technical achievements over the last 15 years have only provided the basic machinery to study the complex nature of the problem, and we are a long way from unlocking the enigmatic relationships between structure and function.

The power of recombinant genetics is the ability to redesign biological systems and processes beyond the scope of natural evolution and selection. The fact that given the appropriate selective environment, any biologically and energetically feasible reaction can evolve cannot be easily disputed. It is difficult to imagine circumstances where, for example, a biological system would evolve that devotes much of its cellular energy to a given enzyme resulting in up to 30% of its total cellular protein. With the continuing success of protein engineering we are rapidly approaching the capability of not only overproducing proteins but redesigning them to function in the desired environment and to utilize the desired substrates. As our understanding of structure-function relationships of enzymes continues, de novo synthesis of completely unique enzymes will be feasible. Eventually our ability to produce proteins, hence enzymes in vitro obviate the need for bacteria or other biological hosts as production vehicles.

REFERENCES

Amore, R., Wilhelm, M., and Hollenberg, C. P. 1989. The fermentation of xylose—an analysis of expression of *Bacillus* and *Actinoplanes* xylose isomerase genes in yeast. *Appl. Microbiol. Biotechnol.* 30: 351–357.

Andrews, W. H. 1985. A review of culture methods and their relation to rapid methods for the detection of salmonella in foods. *Food Technol.* 39: 77–82.

Avery, O., MacLeod, C., and McCarty, M. 1944. Studies on the chemical nature of the substance inducing transformation of Pneumoccal types. *J. Exp. Med.* 79: 137–154.

Batt C.A., Rabson, L.D., Wong, D.W.J. and Kinsella, J.E. 1990. Expression of recombinant bovine β-lactoglobulin in *Escherichia coli. Agr. Biol. Chem.* (in press).

Blundell, T., and Sternberg, M. J. E. 1985. Computer-aided design in protein engineering. *Trends Biotechnol.* 3: 228–235.

Braunitzer, G., Chen, R., Schrank, B., and Stangl, A. 1973. Die sequenzalayse der β-lactoglobulins. *Hoppe-Seyler's Z. Physiol. Chem.* 354: 867–878.

Carter, P., and Wells, J. 1987. Engineering enzyme specificity by "substrate-assisted catalysis." *Science* 237: 394–399.

Creamer, L., Parry, D., and Malcolm, G. 1983. Secondary structure of bovine β-lactoglobulin B. *Arch. Biochem. Biophys.* 227: 98–105.

Estell, D., Graycar, T., Miller, J., Powers, D., Burnier, J., Ng, P., and Wells, J. 1986.

Probing steric and hydrophobic effects on enzyme-substrate interactions by protein engineering. *Science* 233: 659–663.

Estell, D. A., Graycar, T. P., and Wells, J. A. 1985. Engineering an enzyme by site-directed mutagenesis to be resistant to chemical oxidation. *J. Biol. Chem.* 260: 6518–6521.

Flowers, R. S. 1985. Comparison of rapid salmonella screening methods and the conventional culture method. *Food Technol.* 39: 103–108.

Gaye, P., Hue-Delahaie, D., Mercier, J.-C., Soulier, S., Vilotte, J., and Furet, J.-P. 1986. Ovine β-lactoglobulin messenger RNA: Nucleotide sequence and mRNA levels during functional differentiation of the mammary gland. *Biochimie.* 68: 1097–1107.

Hammer, R., Pursel, V., Rexroad, C., Wall, R., Bolt, D., Evert, K., Palmiter, R., and Brinster, R. 1985. Production of transgenic rabbits, sheep, and pigs by microinjection. *Nature* 315: 680–683.

Ingram, L. O., Conway, T., Clark, D. P., Sewell, G. W., and Preston, J. F. 1987. Genetic engineering of ethanol production in *Escherichia coli. Appl. Environ. Micro.* 53: 2420–2425.

Iverson, B. L., and Lerner, R. A. 1989. Sequence-speciic peptide cleavage catalyzed by an antibody. *Science* 243: 1184–1188.

Jamieson, A. C., Vandeyar, M. A., Kang, Y. C., Kinsella, J. E., and Batt, C. A. 1987. Cloning and nucleotide sequence of the bovine β-lactoglobulin gene. *Gene* 61: 85–88.

Janda, K. D., Benkovic, S. J., and Lerner, R. A. 1989. Catalytic antibodies with lipase activity and R or S substrate selectivity. *Science* 244: 437–440.

Janda, K. D., Schloeder, D., Benkovic, S. J., and Lerner, R. A. 1988. Induction of an antibody that catalyzes the hydrolysis of an amide bond. *Science* 241: 1188–1191.

Klein, T. M., Wolfe, E. D., Wu, R., and Sanford, J. C. 1987. High-base microprojectiles for delivering nucleic acids into living cells, *Nature* 327: 70–73.

Klinger, J. D., and Johnson, A. R. 1988. A rapid nucleic acid hybridization assay for Listeria in foods. *Food Technol.* 42: 66–70.

Lavitrano, M., Camaioni, A., Fazio, V. M., Dolci, S., Farace, M. G., and Spadafora, C. 1989. Sperm cells as vectors for introducing foreign DNA into eggs: Genetic transformation of mice. *Cell* 57: 717–723.

Liao, H., McKenzie, T., and Hageman, R. 1986. Isolation of a thermostable enzyme variant by cloning and selection in a thermophile. *Proc. Natl. Acad. Sci. USA* 83: 576–580.

McKenzie, H. A. 1971. β-Lactoglobulins. In *Milk Proteins*, Vol. II, pp. 257–330, Academic Press, N.Y.

Napper, A., Benkovic, S., Tramontano, A., and Lerner, R. 1987. A stereospecific cyclization catalyzed by an antibody. *Science* 237: 1041–1045.

Palmiter, R., Brinster, R., Hammer, R., Trumbauer, M., Rosenfeld, M., Birnberg, N., and Evans, R. 1982. Dramatic growth of mice that develop from eggs microinjected with metallothionein-growth hormone fusion genes. *Nature* 300: 611–613.

Papiz, M., Sawyer, L., Eliopoulos, E., North, A., Findlay, J., Sivaprasadarao, R., Jones, T., Newcomer, M., and Kraulis, P. 1986. The structure of β-lactoglobulin and its similarity to plasma retinol-binding protein. *Nature* 324: 383–385.

Pervaiz, S., and Brew, K. 1985. Homology of β-lactoglobulin, serum retinol-binding protein, and protein HC. *Science* 228: 335–337.

Sarthy, A. V., McConaugh, B. L., Lobo, Z., Sundstrom, J. A., Furlong, C. E., and Hall, B. D. 1987. Expression of the *Escherichia coli* xylose isomerase gene in *Saccharomyces cerevisiae*. *Appl. Environ. Micro.* 53: 1996–2000.

Sawyer, W. 1967. Heat denaturation of bovine β-lactoglobulins and relevance of disulfide aggregation. *J. Dairy Sci.* 51: 323–329.

Simons, J. P., McClenaghan, M., and Clark, A. J. 1987. Alteration of the quality of milk by expression of sheep β-lactoglobulin in transgenic mice. *Nature* 328: 530–532.

Simons, J., Wilmut, I., Clark, A., Archibald, A., Bishop, J., and Lathe, R. 1988. Gene transfer into sheep. *Biotechnology* 6: 179–183.

Todd, E. C. D., Szabo, R. A., Peterkin, P., Sharpe, A. N., Parrington, L., Bundle, D., Gidney, M. A. J., and Perry, M. B. 1988. Rapid hydrophobic grid membrane filter-enzyme-labeled antibody procedure for identification and enumeration of *Escherichia coli* O157 in foods. *Appl. Environ. Micro.* 54: 2536–2540.

Tramontano, A., Janda, K., and Lerner, R. 1986. Catalytic antibodies. *Science* 234: 1566–1569.

Watson, J. D., and Crick, F. H. C. 1953. A structure for deoxyribose nucleic acid. *Nature* 171: 737–738.

2

Introduction to Bioreactor Engineering

Harvey W. Blanch

University of California at Berkeley
Berkeley, California

INTRODUCTION

A biochemical reactor is a device in which materials are treated to promote biochemical transformation of matter by the action of living cells or cell-free enzyme systems. Biochemical reactors are widely employed in the food the pharmaceutical industries, to a limited extent in the chemical industry, in waste treatment, and in biomedical applications. In industrial processes, they are invariably at the heart of the process (see Fig. 2.1). Broadly speaking, there are two types of biochemical reactors: microbial fermenters and enzyme (cell-free) reactors. Depending on the process requirements (aerobic, anaerobic, solid state, immobilized), numerous subdivisions of this classification are possible.

In fermenters, cell growth is promoted or maintained to allow formation of products such as metabolites (e.g., antibiotics, alcohols, citric acid), biomass (e.g., baker's yeasts, SCP), transformed substrates (e.g., physiologically active steroids), or purified solvents (e.g., in water reclamation). Systems based on macroorganism cultures (consisting of mammalian or plant cells) are usually referred to as "tissue cultures," whereas those based on dispersed non–tissue-forming cultures of microorganisms (bacteria, yeasts, fungi) are loosely referred to as "microbial" reactors.

In enzyme reactors, substrate transformation is promoted without the life-support system of whole cells (e.g., enzymic saccharification of polysaccharides to make syrup). Frequently these reactors employ immobilized enzymes, where solid or semisolid supports are used to internally entrap or externally attach the biocatalyst so that it is not lost as in "free-enzyme" systems, and may be reused in a process. Virtually all biochemical reactors of technological importance deal with heterogeneous systems involving one or more phases. Thus, to be effective in achieving the required biochemical changes, interphase mass (and heat) transfer must occur in these systems. Depending on the primary interphase-contacting requirement, there are two basic types of biochemical reactors: fluid-fluid contactors and solid-fluid contactors. Numerous subdivisions of this classification are possible, e.g., bubble columns, stirred tanks, trickle-bed filters, fluidized beds, air-lift towers, etc.

We will examine various aspects of physical mass transfer (and by analogy heat transfer) which form the basis of the rational engineering design and operation of all types of biochemical reactors. The general aims of bioreactor design are illustrated in Table 2.1.

Our starting point in the design or analysis of a bioreactor is the coupled set of mass balance equations describing the conversion of reactants to products, whose product may be microbial cells in some cases. To illustrate the coupling of biochemical kinetics with mass transfer, we shall consider a simple, well-mixed

Table 2.1 Aims of Bioreactor Design

Overall:

To provide the most economical means (capital and operating costs) to optimize bioreactor productivity within the context of an integrated process

Biological/biochemical specifics:

(1) Design or select most appropriate biocatalyst (living cells, nonviable enzyme systems, etc.) for bioreaction(s).
(2) Design hardware to carry out (1)

Constraints:
    (1) Technical feasibility (e.g., in construction and installation)
    (2) Reliability
    (3) Biological (sterile operating, materials compatibility, etc.)

---

batch reactor in which substrate S is converted to cells (X) and product (P). Generalized mass balances have the form

(i) growth of cells

$$\frac{dXV}{dt} = r_X \cdot V$$

(ii) consumption of substrate

$$\frac{dSV}{dt} = r_S \cdot V$$

(iii) product formation

$$\frac{dPV}{dt} = r_P \cdot V$$

The constitutive rate expressions r_X, r_S, and r_P may be complex functions of (S,X,P) and in addition may depend on other metabolites, temperature, and pH. In aerobic systems, all these rate expressions depend on the liquid-phase oxygen concentration. Thus, the transport of, for example, oxygen from a gaseous phase to the liquid phase and finally to the biological catalyst is important in determining the performance of the reactor system.

Systems and Operating Constraints

The application of chemical engineering principles is useful in the analysis of the design and operation of biochemical reactors. However, classical approaches to the analysis are limited by the following special constraints:

1. The bulk densities of suspended microbial cells and substrate particles generally approach those of their liquid environments so that relative flow between the dispersed and continous phases is normally low. This situation may be contrasted with the relatively heavy metallic catalyst particles generally used in chemical reactors.

2. The sizes of single microbial cells are very small (in the range of a few μm) compared to catalyst particles; coupled with the above constraints, it is generally difficult to attain turbulent-flow mass transfer conditions.

3. Polymeric substrates or metabolites and mycelial growth often produce very viscous reaction mixtures which are generally pseudoplastic. These conditions tend to limit desirably high flow dynamics in biochemical reactors.

4. Multicellular microbial growth, especially fungal growth, generally forms relatively large cell aggregates such as mycelia, clumps, or pellets. Intraparticle diffusional resistances are often pronounced in these systems, e.g., leading to anaerobiosis.

5. Biochemical reactors frequently require critically close control of solute concentrations, pH, temperature, and local pressures in order to avoid damage or destruction of live or labile components which are essential to the process.

6. Very low concentrations of reactants and/or products are normally involved in biochemical reactors so that the concentration-deriving forces for mass transfer are often severely limited.

7. Microbial growth rates are substantially lower than chemical reaction rates so that relatively large reactor volumes and residence times are required.

As an illustration of some of the problems imposed by the above constraints, we noted that an adequate oxygen supply rate to growing cells is often critical in aerobic processes. Because of its solubility in water, gaseous oxygen, usually in the form of air, must be supplied continuously to the medium in such a way that the oxygen absorption rate at least equals the oxygen consumption rate of the cells. Even temporary depletion of dissolved oxygen could mean irreversible cell damage. It is worth noting, however, that the microbial species may show large variations in oxygen requirements, depending on the oxygen concentration to which they have been adapted.

We shall first examine the kinetic rate expressions characteristic of microbial and enzymatic reactions as these expressions determine the reactor volume for a specified conversion of reactant to product. The equations developed for microbial kinetics can be employed to describe many enzyme-catalyzed reactions.

CONSTITUTIVE RATE EXPRESSIONS FOR BIOLOGICAL PROCESSES

Model of Microbial Growth

The growth of microbial cells can be viewed from various perspectives and with varying degrees of complexity, depending on the extent to which we wish to distinguish between individual cells in a reactor and the degree to which we examine the metablic reactions occurring within the cell. While the most realistic model of the growth of a microbial population would consider all reactions occurring within each cell and the variations form cell to cell in a population, such a model would be very unwieldy. We must therefore make some simplifications; the extent of these will depend on the proposed use of the model. We shall make the following distinctions in models describing cell growth.

When the population is segregated into individual cells that are different from one another in terms of some distinguished characteristics, the models are *segregated*. Nonsegregated models consider the population as lumped into one "biophase" which interacts with the external environment, and can be viewed as one "species" in solution; the cell concentration can be described by one variable alone. Nonsegregated models have the advantage that they are relatively simple from a mathematical perspective. The usefulness of segregated models depends on our ability to distinguish between cells in a population. Often this is difficult.

The introduction of the nature of reactions occurring within a cell gives rise to the concept of *structure*. Structured models do not consider the cell as an entity which simply interacts with the environment; they consider individual reactions or groups of reactions occurring within the cell. The biomass is subdivided into a number of components, e.g. DNA, RNA, protein, etc., and the reactions occurring between these components are described. The external environment can influence the response of the cell in a variety of ways, e.g. lactose may induce the production of the enzyme β-galactosidase, conferring new growth characteristics on the cell following the induction period. Structured models attempt to describe this behavior, e.g. from the description of how the new enzyme is induced (transcription and translation) to the kinetics of breakdown of lactose to its component saccharides and their subsequent utilization.

Unstructured Growth Models

The simplest relationships describing exponential growth are unstructured models. We view the cell as a single species in solution and attempt to describe the kinetics of cell growth based on observations of cell and nutrient concentration profiles. Early models of cell growth did not account for the dependency of the exponential growth rate on nutrient concentration; they were devised to have a maximum achievable population built into the constitutive expressions employed. These models find applicability when the nature of the growth-limiting substrate is not understood. The simplest models is that of Malthus:

$$r_X = \mu X$$

where r_X is the volumetric rate of increase in dry cell weight (e.g., g DCW/liter·hr) and $\mu(hr^{-1})$ is a constant. V(liters) is the reactor volume. This predicts unlimited growth. To provide a means to limit growth, Verhulst (1844) and later Pearl and Reed (1920) proposed a cell concentration–dependent inhibition term:

$$\frac{dX}{dt} = kX(1 - \beta X)$$

and thus

$$X = \frac{X_o e^{kt}}{1 - \beta X_o(1 - e^{kt})}$$

where $X = X_o$ at $t = 0$. This result is known as the logistic equation. The maximum cell concentration is $1/\beta$, and the initial rate of growth is approximately exponential, as k is usually greater than βX_o.

Monod Model

One of the simplest models that includes the effect of nutrient concentration is the model developed by Jaques Monod based on observations of the growth of *E. coli* at various glucose concentrations. It is assumed that only one substrate (the growth-limiting substrate, S) is important in determining the rate of cell proliferation. The form of the Monod equation is similar to that of Michaelis-Menten enzyme kinetics; in fact, if substrate transport to the cell is limited by the activity of an enzyme, for example, a permease, cell growth might well be expected to follow this form:

$$r_X = \mu X = \frac{\mu_{max} SX}{K_S + S}$$

and thus for batch growth at constant volume:

$$\frac{dX}{dt} = \frac{\mu_{max} SX}{K_S + S}$$

μ_{max} is the maximum specific growth rate of the cells, and K_S is the value of the limiting nutrient concentration which results in a growth rate of half the maximum value. This equation has two limiting forms. At high substrate concentrations $S \gg K_S$, and it reduces to a zeroth order dependence on substrate concentration. At low substrate concentrations $S \ll K_S$ and a first-order dependence results.

$$\mu = \mu_{max} \qquad for\ S \gg K_S$$

and

$$\mu = \frac{\mu_{max}}{K_S} \cdot S \qquad for\ S \ll K_S$$

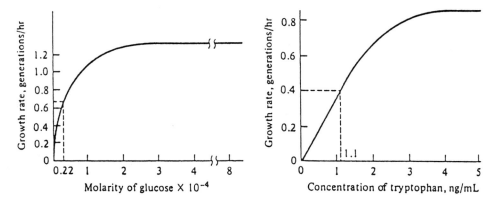

Figure 2.1 Dependence of the specific microbial growth rate substrate concentration, illustrating Monod model. (*Reproduced by permission from Stainer et al., 1970.*)

Typical results for the dependence of specific growth rate on substrate concentration are shown in Fig. 2.1.

Inhibition

Most substrates, if present at sufficiently high concentrations, can inhibit microbial growth by one or more mechanisms. Any changes in the properties of the medium induced by high substrate concentrations might be expected to have some effect on cell growth. These changes include the osmotic pressure, ionic strength, dielectric constant, and activity of solutes present, all of which result in an alteration of membrane fluidity. In addition, overproduction of a metabolite by one pathway may result in inhibition of another pathway. The Pasteur effect is one example of this phenomenon where high levels of oxygen inhibit glycolysis and cause respiration to be the predominant form of metabolism. Table 2.2 summarizes some of the effects an inhibitor may have on microbial growth.

The effect of inhibitory substrates can often be modeled by analogy with enzyme kinetics. We consider growth to follow the equation

$$X + S \longleftrightarrow SX \rightarrow \text{new cells}$$

where K_S is $[X][S]/[XS]$. The effect of an inhibitory substrate can be considered to result from the formation of a complex XS_2, which can no longer form new cells:

Table 2.2   Possible Types of Inhibitor Action

modify chemical potential of substrates, intermediates, or products
alter cell permeability
alter enzyme activity
dissociate enzyme aggregates
affect enzyme synthesis
influence functional activity of the cell

$$XS + S \longleftrightarrow XS_2$$

Here K_i is $[XS][S]/[XS_2]$. The overall rate of growth is given by $\mu = k[XS]$, and thus the resulting expression for the specific growth rate is

$$\mu = \frac{\mu_{max}S}{K_S + S + (S^2/K_1)}$$

The rate of growth increases with substrate concentration up to maximum value (S_{crit}), and beyond this value it declines. From the equation above, we see that the maximum growth rate occurs when

$$\frac{d\mu}{dt} = 0, \qquad S_{crit} = \sqrt{K_s K_1}$$

A variety of constitutive rate expressions have been proposed for inhibition as a result of products of metabolism. Many of these are given in the general references.

Models of Growth- and Non–Growth-Associated Product Formation

There are a wide variety of products that can be synthesized by the cell. Several schemes have been proposed to classify these products, with the objective of developing kinetic models of product formation that can be broadly applied to the resulting classes. One of the most widely applied schemes is based on whether the product is formed as a result of the primary metabolic functions of the cell or whether it is formed from secondary metabolism. End products of energy and carbon metabolism fit into the former ca'egory. Typical products include ethanol produced by the anaerobic fermentation of glucose by yeast and the production of gluconic acid from glucose by *Gluconobacter*. These products are referred to as growth-associated products, as their rate of production parallels the growth of the cell population. On the other hand, products such as antibiotics and vitamins are generally produced in batch cultures at the end of the exponential phase. Such secondary metabolites are termed non–growth-associated products, and their kinetics do not depend on the rate of growth of the culture. An intermediate class of products can also be identified, where product formation kinetics lies between the two classes above. Such products are partially growth associated and include amino acids, lactic acid, intermediates from the citric acid cycle (including citric acid), extracellular polysaccharides (such as xanthan and pullulan), and solvents such as acetone and butanol.

The development of constitutive rate expressions for these classes of product formation arise from the studies of Luedeking and Piret on the formation of lactic acid by *Lactobacillus delbrueckii*. Lactic acid production was found to depend on both the concentration of cells present and their growth rate. The resulting expression was proposed:

$$r_p = \alpha r_x + \beta X$$

$$= \alpha \mu X + \beta X$$

We see that this expression can be divided into growth and non–growth-associated parts, $\alpha \mu X$ and βX. The three classes of products can be characterized by the relationship between production kinetics and cell growth:

growth associated: $r_p = \alpha \mu X$
non–growth-associated: $r_p = \beta X$
partially growth associated: $r_p = \alpha \mu X + \beta X$

MASS TRANSFER IN BIOREACTORS

Rate-Controlling Steps

Figure 2.2 schematically describes a biochemical reactor subsystem involving two or more phases. An important example of this representation is an aerobic process in which a microbe utilizes oxygen (supplied by air bubbles which also desorb toxic carbon dioxide) and other dissolved nutrients (sugars, etc.) to grow and produce soluble extracellular metabolites. Eight resistances in the mass transfer pathways for the nutrient supply and utilization and for metabolite excretion and removal are possible at the following locations:

(1) in a gas film, (2) at the gas–liquid interface, (3) in a liquid at the gas–liquid interface, (4) in the bulk liquid, (5) in a liquid film surrounding the solid, (6) at

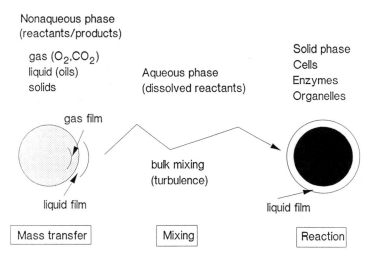

Figure 2.2 Generalized kinetic and transport processes occurring in a biochemical reactor.

the liquid–solid interface, (7) in the solid phase containing the cells, and (8) at the sites of the biochemical reactions. It should be noted that all the pathways except the last one are purely physical.

Figure 2.2 can depict a wide range of other practical situations. The continuous phase may be liquid or gas, the latter including special cases such as "solid-state" (e.g., composting, trickle-bed reactors, and "Koji") processes, while the dispersed phase may be one or more of the following phases: solid (e.g., microbial cells, immobilized enzyme particles, solid substrates), liquid (e.g., insoluble or slightly soluble substrates), or gas (e.g., air, carbon dioxide, methane).

In addition to the physical constraints considered above, the mass transfer pathways may be complicated by the biochemical reaction steps which occur concurrently inside the cellular or enzymic materials. For example, Fig. 2.3 illustrates various feedback mechanisms that may be generated by the biochemical changes in the reactor.

In biochemical reactors, because of the relatively high mass diffusivities in the gases compared to the liquid media (about 10^5-fold) and their low solubilities, and aqueous liquid-phase resistance invariably controls the overall physical mass transfer rate. For example, in Fig. 2.2, one of the following four liquid-phase resistances is rate-controlling.

1. A combined liquid phase resistance near and at a gas–liquid interface: This resistance is often rate-controlling in aerobic reactors because of the

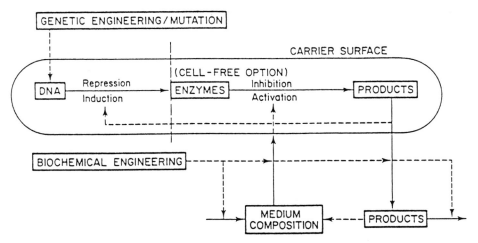

Figure 2.3 Factors affecting bioreactor performance. (*Reproduced with permission from MooYoung and Blanch, 1987.*)

relatively low solubility of oxygen in aqueous solutions and the retardation effects of adsorbed materials (e.g., surfactants) and electrolytes at the interface.

2. A liquid-phase resistance in the aqueous bulk medium separating the dispersed phases: This resistance if often insignificant because of the good liquid mixing promoted in practical reactor systems.

3. A liquid-phase resistance near and at the solid–liquid interface: This resistance can be significant because of the low density differences between the continuous aqueous medium and some dispersed phase (e.g., microbes, gel-entrapped enzymes liquid drops, polysaccharides).

4. A liquid-phase resistance inside a dispersed "solid" phase: This resistance can be significant in cell flocs, mold pellets, immobilized enzyme carriers, and insoluble substrate particles.

The complex interaction of these mass transfer steps and the biological reactions is illustrated in Fig. 2.3. This figure also indicates a relationship between aspects of "genetic engineering" and biochemical engineering in bioreactor design for whole-cell or cell-free systems. Table 2.3 lists a range of possible systems and operational variables.

Table 2.3 Examples of Bioreactor Systems

Biological/biochemical specifications:

Viable vs. nonliving enzymic systems
Aerobic vs. anaerobic systems
Aseptic vs. nonaseptic conditions
Mono vs. mixed cultures
Single cells vs. multicellular aggregates
Single vs. multisubstrate media

Physical operational modes:

Suspended vs. immobilized cells
Dissolved vs. immobilized enzymes
Soluble vs. insoluble substrates
Monomeric vs. polymeric substrates
Batch vs. continuous operations
Stirred vs. plug flow processing
Single vs. multistate systems
Single stream vs. recycle operation

Definition of Mass Transfer Coefficient

A mass transfer coefficient relates transfer rates to concentration terms and can be defined by a mass balance for a given reactant or product species in the bioreactor. For example, consider oxygen in air bubbles passing through a fermenter; the oxygen transfer rate N_a will be given by:

$$N_a = k_L a(C_i \cdot g - C_L)$$

where C_L is the local dissolved oxygen concentration in the bulk liquid at any time t, C^*_g is the oxygen concentration in the liquid at the gas–liquid interface at infinite time (equivalent to the saturation concentration), and a is the interfacial area; k_L is the local liquid phase mass transfer coefficient. Where necessary, methods of evaluating heat transfer rates between the dispersed and continuous phases in bioreactors can be calculated in the same way.

In the above equation, the mass transfer rate is dependent on the transfer coefficient, the interfacial area, and the concentration driving force. Interfacial area is controlled by bubble size, while the concentration driving force is affected by partial pressures (Henry's law). The mass transfer coefficient (k_L) is dependent on several factors, including the rheological properties of the fermentation broth and the type of bioreactor employed. We shall examine these in turn.

RHEOLOGICAL PROPERTIES OF BIOREACTOR MEDIA

The rheological properties of the materials being processed in bioreactors will influence the power consumption and, hence, the heat and mass transfer rates. These properties are particularly important in some antibiotic and related fermentations involving filamentous organisms and semisolid media because of the very viscous and frequently non-Newtonian behavior of the materials employed.

Three classes of fluids have been encountered in bioreactors: Newtonian, non-Newtonian, and (rarely) viscoelastic. The rheological characteristics of fluids can be described by the following general equation:

$$\tau = \tau_o + K\gamma^n$$

(a) Newtonian fluids exhibit no yield stress, so $n = 1$, $\tau_o = 0$, and K becomes the dynamic viscosity μ. Thus,

$$\tau = \mu\gamma$$

Newtonian fluids, with a constant viscosity and not affected by shear rate, will have constant rheological properties. This will apply to most bacterial and yeast fermentation fluids.

There are three types of time-dependent non-Newtonian fluids.

(b) Pseudoplastics occur most often and these fluids follow a power law model:

$$\tau = K\gamma^n$$

where K and n change during the course of the fermentation. With no linear relationship between shear rate and shear stress, as for Newtonian fluids, the viscosity of pseudoplastics will not be constant and we can define an apparent viscosity by the equation:

$$\mu_{app} = \frac{\tau}{\gamma} = K\gamma^{n-1}$$

in which the viscosity decreases with increasing shear rate (i.e., $n < 1$).

(c) Dilatants will also obey the power law model. In this case $n > 1$.

(d) Viscoplastic or Bingham plastic fluids are recognized by their relatively high yield stress which must be exceed to make the fluid flow. The fluid will then exhibit inherently Newtonian behavior.

(e) Casson behavior is exhibited by non-Newtonian fluids, which demonstrate an apparent yield stress followed by pseudoplastic behavior.

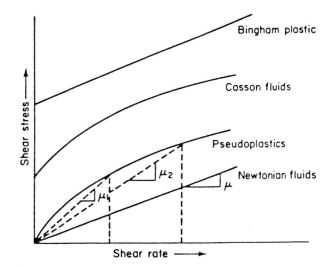

Figure 2.4   Flow behavior of biological fluids μ_1 and μ_2 are the apparent viscosities of the pseudoplastic fluid determined at different shear rates. (*Reprinted with permission from MooYoung and Blanch, 1987.*)

Filamentous fermentation broths and complex media with a high solids concentration will invariably exhibit non-Newtonian behavior. The rheological characteristics of different fluids are demonstrated on shear stress–shear rate diagrams, as shown in Fig. 2.4.

BIOREACTOR TYPES AND MODES OF OPERATION

We shall briefly review some of the principal types of bioreactors commonly employed for both microbial and enzyme systems.

Bubble Columns

Because of the very different bulk flow patterns that are induced, pneumatically agitated gas–liquid reactors may show wide variations in performance with the height-to-diameter ratio. In the production of baker's yeast, a tank-type configuration with a ratio of 3 to 1 is common industrially, whereas tower-type systems have height to diameter ratios of 6 to 1 or more (Fig. 2.5). As would be expected, the behavior of both gas and liquid phases may be quite different in these cases. In

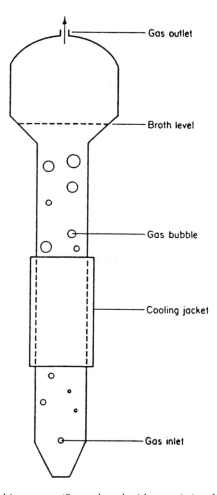

Figure 2.5 The tower bioreactor. (*Reproduced with permission from MooYoung and Blanch, 1987.*)

general, the gas phase rises through the liquid phase in plug-flow, under the action of gravity, in both types of system.

A variety of correlations for k_L are available for bubble columns. Generally they are of the form

$$k_L a = \text{constant}.V_s^n$$

when n is usually in the range 0.9–1.0 in the bubble flow regime. Many of the available correlations for $k_L a$ have been obtained on small-scale equipment and have not taken cognizance of the underlying liquid hydrodynamics. Thus, correlations developed on small-scale apparatus need to be reviewed in light of the varying interfacial area with column height.

The constant in the above equation can be rewritten to incorporate the power input from the gas phase to the liquid phase. As we shall see later, this is convenient as it permits a composition of bioreactor performance to be made.

Systems with Stationary Internals

Several reactors that include internal elements to enhance mass transfer rates have appeared. These include draught tubes, multiple sieve plates staged along the length of the column, and static mixing elements (see Fig. 2.6).

A considerable literature exists on draught tube columns, where liquid is circulated due to a bulk density difference between the inner core and the surrounding annular space. The downcoming liquid in the annular space entraps

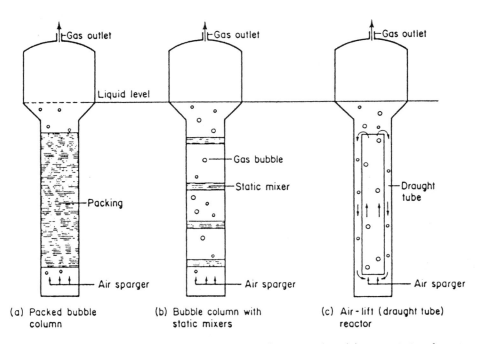

Figure 2.6 Bioreactors with stationary internals. (*Reproduced by permission from MooYoung and Blanch, 1987.*)

air bubbles, and thus hold-up in the central core and annular region will be different. Several reports on small-scale air-lift columns of varied design have appeared, while industrial-scale air-lift devices have been used mainly for SCP production and waste treatment.

The overall mass transfer correlations for air-lift devices are usually expressed as:

$$k_L a = \text{constant.} V_s^n$$

Static mixing elements have been incorporated into air-lift devices in order to provide additional mixing and hence greater transfer capabilities. Static mixers are becoming increasingly more common in oxidation ponds for biological waste water treatment. Here, fine bubbles may be produced as the gas-liquid mixture passes through the mixing elements. These mixing elements are usually 45–60 cm in diameter and are placed over the sparger pipes. A fairly intense liquid circulation can be developed by such mixers due to entrainment by the gas–liquid jet rising from the mixing element.

Stirred-Tank Reactors

To obtain better gas–liquid contact, mechanical agitation is often required. A typical geometric arrangement of a fully baffled sparged stirred tank reactor with submerged impellers is shown in Fig. 2.7. Several investigators have developed overall correlations for bubble column air-lift systems and stirred tank reactors. They usually take the form:

$$k_L a = \left(\frac{P_g}{V}\right)^n V_s^r$$

and their ranges of applicability for bubble columns, air-lift systems, and stirred tanks are shown in comparable form in Fig. 2.8.

From Fig. 2.8 it is seen that for a given power input, the magnitude of $k_L a$ obtained is about the same whether mixing is done mechanically in stirred tanks or pneumatically in bubble columns or air-lift devices. However, mechanically agitated systems offer the most commonly used means of providing the power inputs needed to secure really high $k_L a$ values such as are required for some antibiotic fermentations and the activated sludge method of treating waste water.

One criticism of the general equation for $k_L a$ given above is that no account is taken of the properties of the fluid. The rheological behavior of fermentation broths is time dependent and it would be desirable to include rheological properties into the correlation for $k_L a$. There have been attempts to develop more complete relationships between oxygen transfer coefficient and system parameter, using

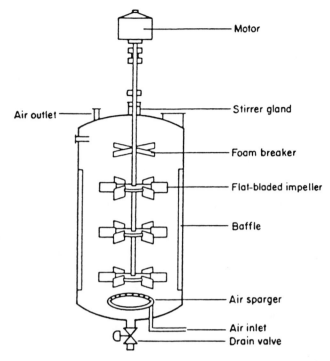

Motor

Stirrer gland

Air outlet

Foam breaker

Flat-bladed impeller

Baffle

Air sparger

Air inlet
Drain valve

Figure 2.7 A typical stirred tank bioreactor. (*Reproduced with permission from MooYoung and Blanch, 1987.*)

dimensional analysis and model solutions. However, the complex correlations that have emerged have found few applications.

Power Requirements

General concepts.   Mixing is used to promote or enhance mass and heat transfer rates in a bioreactor. The productivity of most fermentation processes is limited by the aeration capacity of the fermenter. Power is required, therefore, for both the mixing functions, which are to ensure good bulk mixing of the reactor content and to achieve high mass transfer rates. When the mixing is induced pneumatically or by pumping, it is fairly easy to evaluate the power consumption from pressure-drop considerations. For mechanically induced mixing, power consumption is more difficult to evaluate from operation variables. Since 100–400 kW m^{-1} (0.5–2.0

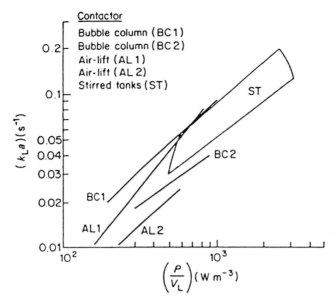

Figure 2.8 Oxygen transfer coefficients in cell-free systems for various bioreactor configurations. (*Reproduced with permission from MooYoung and Blanch, 1987.*)

HP/100 USG) is normally required and the usual range of capacities of industrial fermenters is 50–200 m³, the power requirement is often 40–600 kW (50–800 HP) per unit, which is an important consideration in the process economics.

Air-lift systems. In these systems, gas sparging must supply all the energy for the required bulk mixing and mass transfer. The power input from sparging air through a reactor can be quite substantial in large bioreactors. Assuming the kinetic energy change of the gas (due to velocity differences between inlet pipeline and reactor) can be ignored and isothermal expansion of the gas (from inlet pressure of P_0 to headspace pressure P_1) is the source of all power, the power input can be estimated as:

$$P_g = GRT \ln \left(\frac{P_1}{P_0} \right)$$

where G is the inlet gas volumetric flow rate.

Agitated ungassed systems. Standard impeller configurations generally employ impeller diameters which are one third of the tank diameter. In general, the agitator shaft is positioned centrally in an upright vertical cylindrical tank fitted with baffles. The flat-bladed turbine (Rushton turbine) is used extensively, and it provides adequate mixing for most bioreactors, and most published results are based on this configuration. Applying dimension analysis, it can be shown for Newtonian liquids that

$$\frac{P}{\rho N^3 D^5} = \left(\frac{D^5 N \rho}{\mu} , \frac{\Delta N^2}{g} , \text{geometric factors} \right)$$

The first term involves the inertia forces and is called the Power number, Po, the second term is Re, the Reynolds number, the third term is the Froude number, Fr, which takes into account gravity forces.

Although the complete functional relationship in the equation above is complex and can only be represented graphically, simple analytical expressions can be derived for certain ranges. (a) In the turbulent flow regime,

$$\frac{P}{\rho N^3 D^5} = Po = \text{constant}$$

$$P \propto \rho n^3 D^5$$

Thus, P is strongly dependent on impeller diameter but is independent of liquid viscosity. (b) In the laminar flow regime,

$$P_0 \propto \frac{1}{Re_1}$$

$$P \propto \mu N^2 D^3$$

Here, P is proportional to viscosity but independent of density.

Prediction of power consumption in non-Newtonian fluids can be more complex. The time-dependent rheological properties makes it difficult to estimate the Reynolds number. For power-law fluids the average shear rate around the impeller is related to its speed and an apparent viscosity can be defined as shown earlier. A generalized impeller Reynolds number for power-law fluids can thus be derived:

$$Re_l = \frac{d^2 N \rho}{\mu_{app}}$$

This procedure allows the use of conventional power curve formerly developed for Newtonian fluids.

Gassed systems.   The power required to agitate gassed liquid systems is less than for ungassed liquids since the apparent density and viscosity of the liquid phase decrease upon gassing. For Newtonian liquids this decrease may be as much as two thirds of the ungassed power. The reduction in gassed power, P_g/P, is generally given as a function of the ratio of the superficial gas velocity to the impeller tip speed (the aeration number, N_A):

$$P_g P = f(N_A)$$

where

$$N_A = Q/(ND^3).$$

The precision of this equation is poor and the correlation

$$P_g = \lambda X \left(\frac{P^2 \mu D^3}{Q^{0.56}} \right)^{0.45}$$

has been applied to both Newtonian and non-Newtonian systems. It should be noted that the equation is not dimensionally sound and predicts unrealistic results for very small gas flow rates.

A problem in the experimental determination of the effect of pseudoplasticity on gassed power consumption results when the size of the bubbles formed at the sparger is comparable to that of the impeller blade. The impeller may then spin within a "gas doughnut," leaving the bulk of the fluid unmixed. This can often occur in laboratory scale equipment, but not on an industrial scale.

SCALE-UP

Scale-Up Parameters

When the optimum process conditions are found at the laboratory scale, there is a need to translate these findings for use in large bioreactors. The scale-up methods that have been most often proposed are as follows:

1. Scale-up based on fixed power input
2. Scale-up based on fixed mixing time
3. Scale-up based on fixed oxygen transfer coefficient
4. Scale-up based on fixed environment (e.g., dissolved oxygen)
5. Scale-up based on fixed impeller tip speed

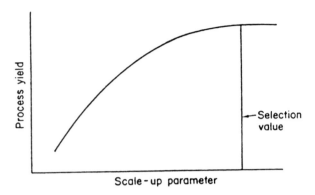

Figure 2.9 The principle of scale-up and selection of the value of the scale-up parameter. (*Reproduced with permission from MooYoung and Blanch, 1987.*)

The list is not necessarily exclusive. The principle involved in using one of the parameters listed above in scaling-up bioconversion processes is illustrated in Fig. 2.9. Assume it has been established that for a certain value of (say) the mixing time in a bioreactor the production of (say) enzyme is high and not affected by small alterations in the value of the mixing time. The process can then be scaled-up, using this particular value of the mixing time as a scale-up parameter. This assumes that everything else is constant or of little importance for process yield, which puts undue demand on the need for geometric similarities between bioreactors of different scale.

Scale-Up at Fixed k_La

We have already implied that if physical mass transfer rates drop below certain values, growth is hampered or destroyed. Thus, k_L is frequently used as a basis for scaling-up, especially anaerobic systems. Table 2.4 gives an example of scale-up based on constant k_La in both reactors. From this table, it is seen that it may not be possible to maintain equal volumetric gas flow rates (*vvm*) since the linear gas velocity (V_s) through the vessel will increase differently with the scale and in fact would be impractical if the upper limit of liquid blow-out action is reached. However, it may be possible to reduce the volume of gas per volume of liquid per minute (*vvm*) on scale-up while increasing power input by changing the reactor geometry and/or power input per unit volume (P/V) as shown in the table. k_L will remain constant in both cases.

It appears that k_L is often, but not always, a reasonable design approach. An

Table 2.4 Scaling-Up Based on Constant $k_L a$ for Gas–Liquid Contacting in a Sparged Stirred-Tank Reactor. Effect of Scale-up Parameters on Common Operating Parameters

Operating parameter	Laboratory reactor 751		Plant reactor 10 m³	
H_1/T	1	1	1	2.8
P/V	1	1	≥	1
vvm	1	1	0.2	0.1
V_s	0.1	0.5ᵃ	0.1	1
$k_L a$	1	—	1	1

ᵃImpractical liquid "blow-out" conditions.

increase in $k_L a$ can sometimes have an adverse effect because of damage to organisms in highly turbulent fermentation broth and/or oxygen poisoning. Other problems, such as gross bubble coalescence, are also important in nonmechanically stirred fermenters.

For scale-up maintaining geometric similarities, the general $k_L a$ equation may be used to predict the oxygen transfer coefficient. It must be noted, however, that the values for the exponents m and n are affected by scale as illustrated in Table 2.5.

It is important to realize that $k_L a$ is not necessarily constant during a fermentation. In highly viscous fermentations, or in fermentations producing surfactants, large variations in $k_L a$ can occur. Thus, using the oxygen uptake rate throughout the fermentation as a scale-up parameter is often more successful. In

Table 2.5 Effect of Scale on Exponent Values

Vessel size (liters)	Exponent	
	m	n
5	0.95	0.7
50	0.6	0.7
50000	0.4	0.5

this case, the oxygen uptake rate obtained in the small reactor is mirrored in the larger vessel by constantly manipulating the agitation and aeration. This allows for a much greater degree of freedom in geometric dissimilarities between the two vessels.

Scale-Up on Flow Basis (Constant Power Input)

Another common design approach is based on constant input of agitator power per unit reactor volume, i.e.,

$$P_1/V_1 = P_2/V_2$$

where subscripts 1 and 2 refer to small- and large-scale vessels, respectively. Under turbulent conditions, the agitator power input is given by

$$P \propto \rho N^3 D^5$$

For constant power input in geometrically similar vessels, we can therefore write

$$\frac{\rho N^3_1 D^5_1}{V_1} = \frac{\rho N^3_2 D^5_2}{V_2}$$

and this becomes

$$N_2 = N_1 = \left(\frac{V_2}{V_1}\right)^{1/3}\left(\frac{d_1}{D_2}\right)^{5/3}$$

giving the impeller speed in the large fermenter for successful scale-up.

It must be noted that this scale-up method relies on correlation for power input in ungassed systems and the demand for geometric similarity also limits its

Table 2.6 Examples of Incompatible Flow Parameters on Scaling-Up a Geometrically Similar Ungassed Stirred-Tank Reactor

Parameter	Laboratory reactor 201			Plant reactor 2.5m³	
P/V	1	1	25	0.2	0.0016
N	1	0.34	1	0.2	0.04
ND	1	1.7	5	1	0.2
Re	1	8.5	25	5	1

applicability. For constant power per unit volume in turbulent flow it is to be noted that impeller tip speed and, hence, shear increase with the cube root of the diameter, so that many other flow parameters cannot be maintained constant on scale-up (see Table 2.6). This table also demonstrates how different flow parameters will affect flow conditions in the reactor when used as a scale parameter.

SYMBOLS

Roman Letters

a interfacial area per unit liquid volume
C_L concentration of solute in bulk liquid (generally)
C^*g saturation (equilibrium) concentration of solute
D impeller diameter
G molar gas flow rate
g acceleration due to gravity
H Henry's constants
K consistency coefficient of power-law fluids
k thermal conductivity
k_L local liquid phase mass transfer coefficient
$k_L a$ volumetric liquid phase mass transfer coefficient
m an exponent
N speed of agitator
N_a oxygen transfer rate
n fluid behavior index of power-law fluids; an exponent
P agitator power input in ungassed systems
P_g power input in gassed systems
P_1 pressure in liquid at sparger
P_0 pressure at surface of liquid
R universal gas constant
T temperature; tank diameter
t time
V volume of reactor contents
V_s superficial gas velocity
vvm volume of air per unit volume of median per minute
X concentration of biomass

Greek Letters
β coefficient of thermal expansion (volumetric)
γ shear rate

μ viscosity (dynamic) of continuous phase
μ_{app} apparent viscosity (dynamic)
$\tau_{o\gg0}$ yield stress
τ shear stress

Abbreviations for Dimensionless Groups

Fr   Froude number
N_A aeration number
Po   power number
Re   Reynolds number for particles
Re_I Reynolds number for impellers

REFERENCES

Bailey, J. E., and Ollis, D. F. 1986. *Biochemical Engineering Fundamentals.* McGraw Hill, New York.

Leudeking, R., and Piret, E. Z. 1959. *Biotech. Bioeng.* 1: 393.

MooYoung, M. M., and Blanch, H. W. 1987. Transport phenomenon and bioreactor design. In *Basic Biotechnology.* (Ed.) Bu'Lock, J. and Kristiasen, P., p. 133. Academic Press, New York.

Pearl, R. 1940. *Introduction to Medical Biometry and Statistics,* 3rd ed., Saunders, Philadelphia.

Pirt, S. J. 1975. *Principles of Microbe and Cell Cultivation.* Blackwell Scientific Publ., Oxford.

Prokop, A. 1983. Reactor design fundamentals: Hydrodynamics, mass transfer heat exchange, control and scale-up. In *Foundations of Biochemical Engineering.* (Ed.) Blanch, H. W., Papoutsakis, E. T. and Stephanopolous, G., p. 355. ACS Symposium Series 207.

Schugert, K. 1987. *Bioreactor Engineering,* Vol. 1, *Fundamentals, Thermodynamics, Formal Kinetics, Idealized Reactor Types and Operation Modes.* John Wiley, New York.

Stanier, R. Y., Duodoroff, A., and Adelberg, E. A. 1970. *The Microbial World,* p. 315. Prentice-Hall, Englewood Cliffs, NJ.

Verhulst, 1844.

3

Bioreactor Considerations for Producing Flavors and Pigments from Plant Tissue Culture

Michael L. Shuler, Thomas J. Hirasuna, Christopher L. Prince, and V. Bringi

Cornell University
Ithaca, New York

INTRODUCTION

Many important food flavors and colors derive from plants. In most cases the compound can be obtained from field-grown plants. However, field-grown plants are seasonal and the quality of a flavor or color varies due to uncontrolled fluctuations in growing conditions. Plant tissue culture techniques could be used to ensure a continuous supply of a high-quality, uniform, predictable product (flavor or color) that could be used in food product formulation. The product would have the further advantage of being a natural product. The production of flavors and colors by plant cell tissue culture offers the advantage of control of quality and availability. The control also allows the possibility of specifically altering a flavor spectrum to allow the "tailoring" of a flavor to meet a specific objective in food formulation.

Plant cell tissue culture can also be used in the rapid clonal propagation of elite plant lines — particularly cell lines formed by asexual techniques, by protoplast fusion, or by genetic engineering. Plants having desirable flavor or color characteristics can be obtained from these techniques and would provide potentially superior sources for flavors or colors. Traditional techniques of micropropagation rely on the use of solid medium and are labor intensive. The use of bioreactors and liquid culture can greatly reduce the labor component. In Israel, bioreactor-based micropropagation has apparently shown significant commercial promise (Levin et al., 1988). A related technology is the production of "artificial seeds" (see Redenbaugh et al., 1987; Giles and Morgan, 1987), which relies heavily on effective bioreactor performance.

The emphasis of this review will be on the use of plant cell tissue culture for the direct production of flavors or colors. However, the techniques and bioreactor considerations used in metabolite production also provide important information applicable to micropropagation.

Before reviewing examples of how plant cell culture can be used for flavor and pigment production, we need to review basic aspects of plant cell culture.

WHAT ARE PLANT CELL CULTURES?

"Undifferentiated" Tissues

Plant cell tissue cultures have been established from several hundreds of plant species. The procedure for establishing such a culture makes use of any portion of a plant containing dividing cells. The plant material is first surface sterilized and the excised plant material is placed on a medium promoting rapid growth (inorganic nutrients, hormones, an exogenous carbon/energy source such as sucrose, and solidified with a low level of agar (0.6–1% by weight)). From the excised plant material a *callus* will form. The callus generally grows rapidly and the cells are "dedifferentiated." Unlike normal plant material there is minimal organization within the callus. The callus can be quite large — more than 1 cm in width. The concept of dedifferentiation must be treated carefully since the callus usually contains a mix of cell types. The plant part providing the source of callus (root, shoot, or embryo) can result in some subtle differences in the biosynthetic characteristics of the callus.

Suspension cultures are established by placing callus in liquid medium in a shake flask. Upon gentle to moderate agitation, cells slough off of the callus. The suspended cells or small aggregates grow and the remaining callus is removed when the culture is transferred. Suspension cultures contain aggregates predominantly rather than single cells.

The aggregates are due primarily to failure of progeny to separate. Cell-to-cell communication is important in plants. Normally cells in a whole plant are connected to one another by small pores called *plasmodesmata*. The plasmodesmata allow the interchange of lower molecular weight compounds (<800) and are formed primarily during cell division. Cells in aggregates can communicate through plasmodesmata and also through diffusable species. For example, plants generate ethylene, which acts as a hormone. In a large aggregate there can be concentration gradients of such metabolic products as well as nutrients (e.g., oxygen and hormones). These gradients lead to a variety of microenvironments. Cells in different positions in the aggregates may have greatly different biochemical and morphological structures. Although plant cell cultures are pure cultures (i.e., only one species), they may behave as mixed cultures.

Plant cells can be very large with cell diameters typically in the range of 10–100 μm. They grow slowly (doubling times typically ranging from 20–100 hr). Growth is usually nonphotosynthetic with sucrose or glucose supplied exogenously as the carbon and energy sources.

Plant cells contain a higher percentage of water than bacterial cells (95% vs. 80%). This higher water content is due, in part, to the presence of the central vacuole, which can occupy as much as 95% of the intracellular volume in extreme cases. Plant cells sequester many compounds inside the vacuole; very few compounds are secreted. Often, secondary metabolites of commercial interest are found in the vacuole. Many sequestered compounds would be cytotoxic if not removed from the cytoplasm. Their intracellular storage will have important implications in our later discussions on bioreactor strategies.

Cells in callus or suspension culture are generally considered "totipotent," which means that they retain the genetic information and capacity to regenerate the whole plant when the callus or suspension is given the correct environmental signals. This capacity for redifferentiation is critical to artificial seed applications.

The ability of cells to differentiate can be important for the formation of secondary metabolites (non-essential for primary cellular function). Empirically, it has been observed that many products made by differentiated cells in organ culture are not made in callus or suspension culture. This link between biochemical differentiation and structural organization is not absolute; some secondary metabolites can be made at significant levels in suspension cultures even when the product is found only in specialized organs in the parent plant (e.g., vanilla). However, the attainment of some degree of structural organization has often been necessary to obtain good secondary product formation.

Suspension cultures are attractive because of their rapid growth rate, ease of manipulation, their compatibility to traditional cell culture techniques, and their "simplicity" due to a limited number of cell types and less dependence on cell-to-cell interactions. However, these dedifferentiated tissues are inherently unstable genetically. Due to the lack of adequate preservation methods, cells must be maintained by constant subculture. This adds an additional selection toward faster-growing variants. In many cases desired product yields are low and/or retained intracellularly. Primarily for these two reasons many investigators have turned to the use of more structurally differentiated tissues.

Organized Tissues

Whereas fine suspensions are examples of undifferentiated cultures, a range of extents of differentiation can be encountered in plant cultures — from primitive levels of organization to highly differentiated organ cultures.

Simple aggregates in suspension culture are in some sense a primitive form of more recognizable organized tissues. A higher level of organization is a nodule culture. The best characterized nodule cultures are probably those that have been developed in woody species such as poplar, although they are observed in herbaceous plants as well. Nodules are dense cell masses that form roughly spherical cohesive units. Often prominent tissue differentiation (e.g., vascularization) and cellular organization are present (see McCown et al., 1988 for a good review on nodule culture). These nodules can be propagated totally submerged in liquid medium. Nodules can be several millimeters or more in diameter. Growth is both through expansion of existing nodules and generation of new nodules from preexisting ones. Nodules combine many of the advantages of suspension cultures with improved genetic stability, high capacity for regeneration (in at least some cases), and potentially greater capacity for product formation. Nodule cultures have not yet been established for many plant species, and their capacity for secondary metabolite formation has been little tested. [Luckner and Diettrich (1985) describe one example where nodules made a secondary product not made by the corresponding suspension cultures.]

Very high level organization is evident in organ cultures, namely, root and shoot cultures. Root cultures grow as fibrous mats. Historically, growth rates of root cultures have been considered to be lower than the corresponding suspension cultures. However, recent results have demonstrated faster growth in root cultures in a significant number of cases. For many species (some dicots but not monocots) infection with the bacterium *Agrobacterium rhizogenes* results in "hairy roots." Rapid formation and elongation of lateral roots can lead to significant increases in growth rate. Roots often will secrete large amounts of secondary metabolites, and this capacity may make such systems attractive when reuse of the plant biomass is possible (e.g., semicontinuous reactors).

Shoot cultures involve the aerial parts of the plant (stem and leaves). Shoot cultures probably maintain the greatest degree of genetic stability of any plant part and are important components of micropropagation systems. They are also good sources for secondary metabolites. Shoots are grown under illumination, whereas roots and nodules all can be grown in the dark if supplied exogenous carbon sources. Illumination, however, may be important for induction of certain enzymes required for some products.

Organ cultures offer some important advantages over suspensions: (1) biosynthetic capacity returns upon organogenesis; (2) the product spectrum (e.g., for flavors) is more characteristic of the parent plant; (3) product secretion is enhanced in some organs; (4) genetic stability is greatly improved in comparison to suspension or callus cultures; and (5) organ cultures represent a form of self-immobilization compatible with efficient, continuous flow-through systems. The primary disadvantages of organ cultures are (1) little experience in optimizing growth rates; (2) the absence of proven scalable bioreactors; and (3) less direct control of the cellular microenvironment to the plant genome.

Having discussed some general aspects of plant cell culture, we now consider some general implications for bioreactor design.

BIOREACTOR CONSIDERATIONS

Levels of Interaction

The choice of bioreactor design can be used to achieve different levels of cell-to-cell communication. In suspension cultures cells are well dispersed, and if aggregate size is small, most cells are exposed to similar levels of nutrients and metabolic by-products. If cells are artifically forced together by entrapment (e.g., pseudo-tissue), then interchange of metabolic by-products and concentration gradients in nutrients become important. If cells grow in place (e.g., tissue), plasmodesmata can form allowing more direct cell-to-cell interaction. The choice of reactor design can profoundly affect the level of cell-to-cell association and the ultimate biosynthetic response of the culture.

Bioreactors for Suspension Cultures

Plant cells are sensitive to turbulent shear. This sensitivity is moderate, and some cell lines that are quite shear-resistant have been developed. Stirred tanks *designed and operated* as for bacterial fermentations are *not* good choices for plant cells. Plant cells have reduced oxygen requirements (about 10% of *E. coli*) compared to bacteria. Consequently, one may infer that the need for vigorous stirring for oxygen transfer is lessened. However, plant cells can require high critical oxygen levels and grow to high cell densities so that physical dispersion of gases and nutrients can be problematic. Given the constraints of shear sensitivity, the viscosity of a fully developed fermentation, and the needs for oxygen, air-lift designs have been advocated for lower density cultures (i.e., <20g dry wt/L) and slow mechanical agitation with specially designed impellers at higher densities (e.g., marine, sail, or helical impellers).

Because scale-up in geometrically similar vessels results in altered mixing if the same maximum shear is maintained, scale-up of plant cell cultures can be problematic. Even without actual shear damage, changes in mixing will result in changes in the degree of aggregation. A change in aggregate size can result in a change in the cellular microenvironment level of intraaggregate communication, and ultimately biosynthetic capacity.

Although we have focused on oxygen in the gas phase, levels of CO_2 and other volatiles released by plants (e.g., ethylene) are important physiologically. The rate of CO_2 and volatile stripping by gas sparging can be important.

The use of "elicitors" has been widely applied to suspension cultures with significant success in enhancing secondary metabolite formation (DiCosmo and Misawa, 1985). When under attack by pathogens or other stresses, plants will often induce special pathways—many associated with secondary products. Many elicitors are derived from breakdown products of fungal or plant cell walls, but other seemingly unrelated compounds can act as elicitors.

Another approach to improving productivity has been the use of adsorbents or extractants (e.g., Payne et al., 1988). In situ extraction can be effective by removing a target compound and protecting it from degradation. Also, in situ removal of product may relieve feedback inhibition of a critical pathway or reabsorption of product into the cell allowing more product to be made.

For suspended cell systems the operating approach that has been most successful is a two-stage batch fermentation. The medium that supports rapid growth often suppresses product formation. Medium designed to optimize product expression often is deleterious to growth. In the first stage growth is maximized, while in the second stage product formation is maximized.

Many of the problems associated with the culture and scale-up of plant cell cultures can be avoided by immobilization.

Bioreactors with Immobilized Cells

Methods used for cell immobilization are reviewed by Payne et al. (1987). Basically, cells can be entrapped between membranes or screens, entrapped within a gel matrix, or allowed to form a biofilm on or within a porous matrix. Immobilized plant cells often show great differences in physiology when compared to suspension cultures of the same species. These differences probably result from the presence of mass transfer gradients, the formation of plasmodesmata in some cases, and/or the nature of the immobilizing matrix (e.g., type of gel or membrane used). One example of how mass transfer effects can be important has been given by Hallsby and Shuler (1986). They reported that with tobacco cells entrapped between membranes operation with nutrient flow parallel to the cells resulted in large concentration gradients and formation of nodules. When the same medium, pH, and temperature were employed in an identical reactor with transversal flow, concentration gradients were greatly reduced and no nodules were formed.

In addition to mass transfer effects and the presence or absence of plasmodesmata, interactions of cells with the immobilizing matrix may be important. Beaumont and Knorr (1987) have suggested that chitosan celery cultures show apparent elicitation although the exact mechanism is not clear.

The potential advantages of immobilization are: (1) constant high cell density; (2) continous operation facilitated; (3) cell reuse leads to increase efficiency; (4) based on (1), (2), and (3), immobolized cell systems offer significant advantages in terms of volumetric productivities; (5) cells are protected from shear; (6) media can be easily changed; (7) the slow growth and strain instability of plant cells is less important for immobilized cells; (8) continous removal of inhibitory metabolites may enhance overall cellular metabolism; and (9) the biological relationships between aggregation, morphological differentiation, and secondary metabolite production can be rationally exploited.

In contrast to these advantages there are a number of potential limitations. The prime limitation is that the product must be excreted by the cell. Since most of the products of interest are typically stored intracellularly, this constraint is severe. However, techniques are now emerging that will greatly enhance excretion of normally intracellular products (Parr et al., 1987; Payne et al., 1988). Additionally, the product must be synthesized effectively by slowly growing or nongrowing cells, cell viability must be maintained, and there is little experience with large-scale aseptic immobilized cell culture.

Organ cultures are a form of self-immobilization and potentially have the advantages cited above.

Bioreactors for Organ Cultures

The development of scalable, efficient bioreactors for organ culture is just beginning. Nodule and embryo culture would be the easiest to implement. Normal fermenters could be adapted to this purpose, although high shear should be avoided. Nodules can be easily retained in a well-mixed vessel and resemble in many ways the use of gel-entrapped cells.

Roots have been cultivated in a wide variety of laboratory devices on a small scale. In Japan, ginseing roots have been cultured in 20,000L stirred tanks, although details of the cultivation technique are not widely available. Roots form cohesive, intertwined mats. The transfer of gas into the center of such mats is difficult, and most mature mats probably experience significant oxygen limitations in the center. Gas bubbles can be entrapped in such mats causing mats to float and resulting in inefficient contact with the liquid phase. We have successfully cultured onion roots for an extended period (about 2 months) in a reactor (see Shuler, 1986) where the roots were entrapped between two stainless steel screens.

Shoot culture may be the most difficult particularly if light is required. Most shoot cultures have been cultured under only partial submersion or in a nutrient mist. Current dogma has seen that shoot cultures under submerged culture conditions or even in the presence of excess humidity would show vitrification. Vitrification is a condition in which cells fill with water leading to metabolic and morphological abnormalities. Levin et al. (1988) report that they have developed a proprietary procedure that has allowed submerged cultivation of shoot cultures from 14 different plant species. We have tested a number of bioreactor designs for shoot culture of tobacco. Growth under total submergence is as rapid as in traditional culture vessels *if* adequate mixing of the liquid is provided (unpublished results of V. Bringi).

Growth under submerged conditions has a number of advantages (Levin et al., 1988): (1) greater efficiency— one 10L bioreactor does the same work as 100 individual culture vessels, greatly reducing labor input and space requirements; (2) the growth environment is more easily monitored and controlled; (3) the use of a bioreactor reduces handling and exposure to contamination; and (4) the process can be automated easily.

These remarks should orient the reader to some of the general concerns in bioreactor development for plant cell cultures. Now consider some examples specifically in the culture of food flavors and pigments; these examples are not meant to be an exhaustive review.

FOOD FLAVORS

Dziezak (1986) has summarized many of the most direct efforts at production of food flavors, and Mulder-Krieger et al. (1988) recently published a comprehensive listing of attempts to produce essential oils and flavors in plant cell and tissue cultures. Production of the essential oils as secondary metabolites has been extraordinarily difficult in callus and suspension cultures. These products are complex mixtures, and often even when production occurs the product spectrum differs from that for the whole plant.

The techniques that have been most effective in increasing product yields or quality have been (1) use of organized tissues; (2) in situ product extraction; (3) development of specific production media; and (4) optimization of light, temperature, and gas phase composition. These are the strategies we have previously discussed as being general approaches to improve productivity. Although these techniques provide tools that should be manipulated, no general recommendations can be made on the dependency of secondary metabolites on these parameters. For example, exposure to light of a certain wavelength will enhance formation of some products and decrease others. Although organized tissue is almost always better than initial callus or suspension cultures, careful selection of subpopulations of these undifferentiated cells can result in high-yield variants that are more productive than the corresponding organ culture. Optimization of product formation is at present a difficult, time-consuming task, requiring a large number of experiments.

Plant cell tissue culture has been attempted to produce a wide variety of fruit flavors. Dziezak (1986) presents a table listing many of these (apple, apricot, banana, blueberry, cherry, cocoa, coconut, grape, lime, orange, peach, pineapple, raspberry, and strawberry) and the original references. In addition to these flavors work has been reported on peppermint and spearmint (e.g., Lin and Staba, 1961; Kireeva et al., 1978), chocolate aroma (Townsley, 1974a), coffee (Townsley, 1974b), capasaicin, which is the hot flavor component of chili pepper (Holden et al., 1987), celery (Watts et al., 1985) and vanilla (Romagnoli and Knorr, 1988). Although little information is in print, there are indications that projects for the production of vanilla are closer to commercialization than for other flavors. Sweeteners such as stevioside, thaumatin, miraculein, and monellin are potential plant cell tissue culture products (Sahai and Knuth, 1985). Production of the glucoside stevioside from *Stevia rebaudiana* cell cultures has been reported in two Japanese patents (Kibun Co., Ltd., Japan Patent 76-19169, 1976; Chugai Pharmaceutical Co., Ltd., Japan Patent, 80-19009, 1980). However, products such as thaumatin, which is a protein, are also targets for genetic engineering and probably could be made more inexpensively in yeast. *Aspergillus*, or bacteria.

However, an extensive amount of work on flavor production appears to have been done with onion and garlic. The typical flavors of onion and garlic develop when relatively stable flavor precursors (S-alkyl-L-cysteine sulphoxides) are brought into contact with the enzyme allinase (e.g., contact can be brought about by tissue damage). The major cysteine sulfoxide derivatives are:

R = Methyl	(all species of *Allium*)
Propyl	(many species of *Allium*)
Propenyl	(onion or *A. cepa*)
Allyl	(garlic or *A. sativum*)

Much of the early work on flavor development in *Allium* sp. has given negative results. *Allium* sp. form callus, but good dispersed suspension cultures from *Allium* sp. are difficult to obtain. Fridborg (1971) found that callus that formed roots produce the aroma of onion, while undifferentiated callus did not. The root-producing callus of Fridborg (1971) was a relatively newly established line. Davey et al. (1974) reported that both undifferentiated tissue and callus induced to form roots failed to form significant amounts of the characteristic sulfur-containing flavor components of onion. In this case the calluses had been established in culture for more than 3 years. Freeman et al. (1974) showed that callus in culture for 10 months would form flavor compounds *if* differentiated roots were formed. Selby and Collin (1976) found only very small amounts of the flavor precursors in

Table 3.1 Shifts in Flavor Spectrum in Response to Chemical Enhancers
Added to Root Cultures of *Allium cepa*[a]

Supplement to medium	Concentration of flavor precursor[b] (mg/g fresh wt)		
	Methyl	Propyl	Propenyl
Methionine			
0 mM	0.094	0.053	0.50
10 mM	0.34	0.022	0.96
Cysteine			
0 mM	0.094	0.053	0.50
10 mM	1.70	0.55	0.87

[a]Unpublished results from manuscript in preparation by Prince and Shuler.
[b]Methyl, propyl, and propenyl refer to the side group, R, in the cysteine sylfoxide derivatives.

undifferentiated tissue and efforts to increase production by clonal selection failed. Likewise, previous attempts to increase the productivity of undifferentiated onion cultures by the addition of biosynthetic precursors such as cysteine, valine, sulfate, serine, and methacrylic acid have failed (Selby et al., 1980). Turnbull et al. (1981) have observed rapid uptake and transformation of labeled biosynthetic precursors by excised onion shoot tips.

In our laboratory we have found that root cultures of many *Allium* sp. can be established. Acceptable growth rates (<4 day doubling times) can be achieved. Preliminary experiments indicate release of substantial amounts of the flavor precursors in some cases. The levels of flavor precursors in root culture is similar to that in the whole plant on a dry weight basis.

Perhaps the most interesting observation has been that flavor precursor formation can be enhanced and the flavor spectrum altered by feeding chemical precursors that can be transformed into the flavor precursors. Table 3.1 summarizes the results of one particular experiment. These results suggest the possibility that a flavor dependent on several species might be "designed" by feeding different amounts of various chemical precursors.

In addition to direct production of flavors, plant cell tissue cultures may play an important role in the micropropagation of plants important to the production of essential oils, fragrances, and flavors. Stones (1988) has summarized recent progress in this regard.

FOOD COLORS

Of the 11 groups of pigments indigenous to foods all but the heme pigments can be derived from plants (Francis, 1985). However, direct production of pigments in plant tissue culture have only been attempted to any significant extent for the anthocyanins, betalains, and quinones (e.g., shikonin).

Spears (1988) reviewed natural sources for food colors, including animal, algal and fungal, and microbial sources as well as plant sources. He outlined the general disadvantages found with the use of natural colorants as:

1. Pigment content is less well-defined. Variations in pigment exist among different extraction methods and supplier (plant cell tissue should result in greater uniformity).
2. Higher usage levels (and unit cost) are required due to lower tinctorial strength in natural colors than in artificial dyes.
3. Color shade is affected by food product pH (e.g., anthocyanin and curcumin); heat, light, and chemical stability of color are also affected by pH.
4. Some natural colors have characteristic odors and flavors (e.g., paprika, beet root, saffron, but a plant tissue culture process could result in pigment formation with reduced flavor or aroma components).
5. The majority of natural colorings, except anthocyanin and beet root red, are oil soluble and need chemical modification or carriers for use in foods.
6. Natural colors are six to eight times more expensive than artificial dyes (plant cell tissue culture might lower costs, although this is perhaps unlikely with current technology).

However, the advantages of natural colors, primarily in terms of consumer acceptance, outweigh these disadvantages. Spears (1988) reported that over the past 15 years, 350 out of 426 food coloring patent applications related to natural color sources indicating the continued strong interest in developing natural color sources. Plant cell tissue culture is one of the competing techniques for producing such compounds. A good overview of pigment producton by plant cell tissue culture is given by Ilker (1987).

For betalains, the recent review by Bohm and Rink (1988) is particularly instructive. Betalains consist of two major subgroups: betacyanins, which are purple, and the betaxanthins, which are yellow-to-orange. Two recent examples of attempts to produce betalains in tissue culture are given by Sakuta et al. (1986) and Berlin et al. (1986). Berlin et al. (1986) found that cultures of *Chenopodium rubrum* L. could accumulate up to 1% of the dry weight as betacyanins under optimized growth conditions with tyrosine as a precursor. Intracellular degradation of betacyanins was a problem. Sakuta et al. (1986) found a close correlation between aggregate size and betacyanin formation in *Phytolacca americana* L. and highest production in actively dividing cells (in contrast to many secondary metabolites where production occurs most rapidly in slowly growing or stationary phase cells).

Constabel and Nassif-Makki (1971) have noted that betaxanthins can accumulate in small cell clusters that grow as protuberances. Bohm and Rink (1988) also indicate production of the betaxanthins based on unpublished reports from their laboratory. The betaxanthins are of perhaps greater economic interst since no crop plant exists as a competitor to plant cell cultures.

Large amounts of betacyanins are readily available from extraction from red beets. However, the pigment from red beet extraction contains the characteristic unpleasant smell and taste, and consequently plant cell cultures might produce a pigment without this defect.

More work in tissue culture has been done with the anthocyanins. Seitz and Hinderer (1988) have recently reviewed this area.

Anthocyanins are produced in vitro by callus and suspension cultures of many different species, including grape (*Vitis* hybrid) (Yamakawa et al., 1983), Madagascar periwinkle (*Catharanthus roseus*) (Hall and Yeoman, 1986), *Haplopappus gracilis* (Fritsch and Grisebach, 1975), petunia (*Petunia hybrida*) (Colijn et al., 1981), evergreen shrub (*Bupleurum falctum*) (Hiraoka et al., 1986), China aster (*Callistephus chinensis*) (Rau and Forkman, 1986), carrot (*Daucus carota*) (Dougall and Vogelien, 1987), sweet potato (*Ipomoea batatas*) (Nozue et al., 1987), and cornflower (*Centaurea cyanus*) (Kakegawa et al., 1987).

A consistent level of production of the secondary metabolite of interest is necessary. Without selective pressure, however, most secondary products cannot be produced reliably at high levels for extended times in tissue cultures (Collin, 1987). The expression of the synthesis of many secondary products in tissue culture becomes repressed over time (with successive subcultures) [Deus-Neumann and Zenk, 1984]. Although the cells may retain the genes required for the synthesis of the product of interest, proper nutritional and environmental conditions for expression need to be found.

In stock suspension cultures of *Catharanthus roseus*, Hall and Yeoman (1987) found only 10% of the cells actively accumulating anthocyanin, although it was subsequently determined that all cells in the culture were able to synthesize it. Substantial differences in anthocyanin level (up to 30-fold) among the anthocyanin-producing cells were noted.

The mechanisms for differences or reduction of yields are still unknown. Dougall and Vogelien (1987) found that anthocyanin degradation was slow relative to the rate of accumulation and that there was no significant difference in the degradation rates of slow or high accumulating subchlones of *Daucus carota* (carrot). The ability to accumulate anthocyanin appeared to be interconvertable between high and low producing clones. The loss of ability to accumulate anthocyanin from high ability clones is probably primarily influenced by the culture medium.

Two-stage processes with different nutrient regimes have been proposed for the production of many secondary products. Yamakawa et al. (1983) working with Bailey Alicant A grape cultures have examined the effects of hormones (2,4-D and kinetin), nitrogen sources (nitrate and ammonium), sucrose, phosphate, and nitrogen levels, nutrient/hormone ratio, light irradiation, and aeration. They found that high sucrose and low phosphate, and 2,4-D were repressive. Their highest productivity was 0.83 g/L/15 days. Maintenance of proper aeration and illumination were also important. Knobloch et al. (1982) found that very basic nutrients (phosphate- and nitrogen-containing mineral salts) interfere with light-induced formation of anthocyanins in illuminated *Catharanthus roseus* suspension cultures. High sucrose concentrations led to greater than a threefold increase in anthocyanin over 2 weeks. When all phosphate and inorganic nitrogen were removed from a MS medium base, growth was reduced 7.5-fold, but specific anthocyanin productivity was ninefold higher.

In our laboratory we have been working with anthocyanin production from grape callus and suspension cultures derived from Bailey Alicant A *Vitis* hybrid. Although suspension cultures initially derived from callus are highly pigmented, they lose the full capability to sustain high levels of anthocyanin production through repeated subcultures. However, we have formulated a production medium which restores some of the original capability to produce anthocyanin even after extended subculture. The production medium will not maintain growth through several subcultures and cannot replace the maintenance medium. Yamakawa et al. (1983) have also determined media composition for high anthocyanin production from this same variety of grape cells. Primary differences between their high anthocyanin-producing media and our production media are nitrate concentration (30 mM vs. 2.5 mM, respectively), ammonium concentration (30 mM vs. 2.0 mM, respectively), and the hormones used (0.01 mg/L 2,4-D + 0.6 mg/L kinetin vs. 0.5 mg/L 2,4-D alone, respectively). Both media use high sucrose levels and have comparable phosphate and sulfate levels. The production medium of Yamakawa et al. (1983) did not work well with our grape cultures.

Table 3.2 shows the effect of our production medium on anthocyanin production. Note that the number of selections while in callus phase and the number of subcultures in suspension both affect the anthocyanin productivity. Also, a lot of anthocyanin is generated in the second week of culture in production medium. Even with highly selected callus (36 selections), anthocyanin production drops sharply in newly derived suspension cultures. The maintenance medium is Gamborg's B5 (GB5) medium. The major differences between this medium and our production media are given in Table 3.3.

As with flavors, the role of tissue culture is not restricted to direct metabolite production, but also is important in micropropagation. Some examples of selection for suitable cultivars include: (1) Concord variety of Labrusca grape for

Table 3.2 Anthocyanin Levels in Grape Suspension Cultures as a Function of Medium Number of Callus Selections and Subcultures[a]

Number of selections in callus	Number of subcultures in suspension	Anthocyanin (mg/L)		
		Maintenance Medium	Production medium	
			1 Wk	2 Wk
0	41	8		
	50	5		
	54	5		
6	20	25		
	25	11		
	60	7	73	
	70	4	42	
	80	4	16	
21	21	86		781
	23	110	287	
	25	161	320	953
	30	132	320	
36	1	591		
	2	511		
	3	339		
	4	252		
	5	209		

[a]Hirasuna, unpublished results.

anthocyanin; (2) Detroit dark red for beet root pigments (betalains); (3) Paprika selected for less flavor; (4) Gardenia fruits for crocin.

Other plant pigments that are good future targets are: (1) Cape Jasmine (*Gardenia jasminoides*)—mainly for crocin, but also contains other flavonoids and iridoid pigments; (2) Pokeberries (*Phytolacca americana*) —phytolaccanin (identical to betanin); (3) Heavenly blue (morning glory flowers) (*Ipomoea tricolor*) light-stable anthocyanin, rich blue color; (4) Amaranthaceae leaves—red-violet pigments similar to beet root pigments.

PROCESS INTEGRATION

Many factors can alter metabolite production and localization in cell cultures. In some cases these factors can interact synergistically. Asada and Shuler (1989) have presented an example of this approach with respect to the production of ajmalicine

Table 3.3 Media for Grape (Balley Alicant A) Cell Suspensions

	Maintenance medium	Production medium
Major Inorganic Components		
KNO_3	2.500 g/L	0.250 g/L
$CaCl_2 \cdot 2H_2O$	0.150	0.150
$MgSO_4 \cdot 7H_2O$	0.250	0.250
$(NH_4)_2 \cdot SO_4$	0.134	0.134
$NaH_2PO_4 \cdot H_2O$	0.150	0.150
Carbon Source		
Sucrose	20.0 g/L	80.0 g/L

For both media:
Minor inorganics: Salts containing I, B, Mn, Zn, Mo, Cu, Co
Chelated iron: Ferric iron chelated with EDTA
Organics: myo-inositol (100 mg/L), thiamin-HCl (10 mg/L),
   nicotinic acid (1mg/L), pyridoxine-HCl (1 mg/L)
Hormone: 2,4-D 0.5 mg/L
Initial pH = 5.8

(a drug used in treating circulatory problems) from *Catharanthus roseus*. Results of that work are summarized in Fig. 3.1. Although ajmalicine is normally an intracellular compound, substantial excretion from viable cells was obtained by combining the use of a production medium, in situ extraction onto a neutral resin (Amberlite XAD-7), use of an elicitor (a homogenate of the mold, *Phytophthora cactorum*), and calcium alginate immobilization. The use of in situ recovery with reaction integrates the production, separation, and purification components of traditional fermentations.

The use of all of these factors simultaneously has not been attempted for flavor or pigment production. However, a similar strategy may be useful in producing food flavors or colors.

ECONOMICS

Plant cell tissue culture is an inherently expensive technique. Only products of substantial price and volume, coupled with concerns of supply and quality, can be justified as targets for commercial development. Probably the primary determinant of economic viability is the volumeteric productivity of the reactor component of

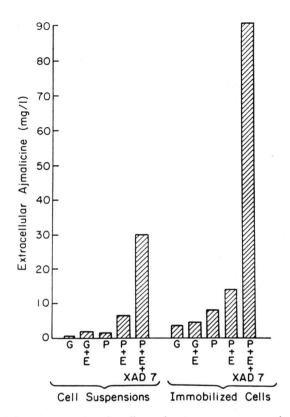

Figure 3.1 This figure summarizes the effects of various treatments on the formation and release of ajmalicine from *C. roseus*. G stands for growth medium, P for production medium. E indicates elicitor addition (2%) of an autoclaved culture of *Phytophthora cactorum* for a 2-day exposure time and the neutral resin, Amberline. XAD-7 indicates the addition of resin. Samples were analyzed 23 days after inoculation. (*From Asada and Shuler, 1989, with permission.*)

the process. Cooney (personal communication, 1988) has found that for a wide range of fermentations the product of volumetric product and wholesale price is 12c/L-d ± 60%.

Table 3.4 lists the volumeteric productivities obtained in some well-studied plant cell culture systems. Based on a volumeteric productivity of 0.9 g/L-d a product would need a wholesale price of $130/kg to achieve 12¢/L-d of revenue, while a productivity of 0.05 g/L-d requires a price of $2400/kg. This analysis, although very crude, places an approximate lower bound on the value a target

Table 3.4 Volumeteric Productivities for Some Products Made in Plant Cell Culture

Product	Cell line	Productivity (gl/L-d)	Investigators
Anthocyanin	Grape (Bailey Alicant A, *Vitis* hybrid)	0.06	Yamakawa et al. (1983)
Berberine[a]	*Thalictrum minus*	0.05	Kobayashi et al. (1988)
Diosgenin	*Dioscorea* spp.	0.75	Based on Table in Fowler and Scrapp (1988)
Podoverine	*Podophyllum versipelle*	0.15	Ulbrich et al. (1988)
Rosmarinic Acid	*Coleus blumei*	0.91	Ulbrich et al. (1988)
Shikonin	*Lithospemum erythrorhizon*	0.15	Tabata and Fujita (1985)

[a]Excreted into the medium where it crystallizes.

product should have to make a plant cell tissue culture product a potentially viable target for plant cell culture. Clearly separation costs and product purity are important considerations in addition to volumeteric productivity. Almost invariably attempts to produce a new compound in plant tissue culture will result in low initial productivities. Usually that productivity can be increased one to two orders of magnitude using a combination of clonal selection, medium optimization, and bioreactor optimization. A program of optimization requires considerable effort. Only products having a large dollar market ($/kg × kg used/year) will justify such a developmental effort.

ACKNOWLEDGMENTS

The work of these authors has been supported, in part, by NSF Grants ECE-8503183 and EET8801492 and by a grant from the Cornell's Biotechnology Program, which is sponsored by the New York State Science and Technology Foundation, the U.S. Army Research Office, and a consortium of industries.

REFERENCES

Asada, M., and Shuler, M. L. 1989. Stimulation of ajmalicine production and excretion from *Catharanthus roseus*: Effects of adsorption *in situ*, elicitors and alginate immobilization. *Appl. Microbiol. Biotechnol.* **30**: 475.

Beaumont, M. D., and Knorr, D. 1987. Effects of immobilizing agents and procedures on viability of cultured celerty *(Apium graveolens)* cells. *Biotechnol. Lett.* **9**: 377.

Berlin, J., Sieg, S., Strack, D., Bokern, M., and Harms, H. 1986. Production of betalains by suspension cultures of *Chenopodium rubrum* L. *Plant Cell Tissue Organ Culture* **5**: 163.

Bohm, H., and Rink, E. 1988. Betalains, Ch. 26. In *Cell Culture and Somatic Cell Genetics of Plants,* Vol. 5. (Ed.) Constabel F. and Vasil, I. K., p. 446. Academic Press, New York.

Colijn, C. M., Johnsson, M. V., Schram, A. W., and Kool, A. J. 1981. Synthesis of malvidin and petunidin in pigmented tissue cultures of *Petunia hybrida. Protoplasma* **107**: 63.

Collin, H. A. 1987. Determinants of yield of secondary products in plant tissue culture. *Adv. Bot. Res.* **13**: 145–187.

Constabel, F., and Nassif-Makki, H. 1971. Betalainbildung in beta-Calluskulturen. *Ber. Dtsch. Bot. Ges.* **84**: 629. As summarized in Bohm and Rink (1988).

Davey, M. R., MacKenzie, I. A., Freeman, G. G., and Short, K. C. 1974. Studies on some aspects of the growth, fine structure and flavour production of onion tissue grown *in vitro. Plant Sci. Lett.* **3**: 13.

Deus-Neumann, B., and Zenk, M. H. 1984. Instability of indole alkaloid production in *Catharanthus roseus* cell suspension cultures. *Plant Med.* **50**: 427.

DiCosmo, F., and Misawa, M. 1985. Eliciting secondary metabolism in plant cell cultures. *Trends Biotechnol.* **3**: 318.

Dougall, D. K., and Vogelien, D. L. 1987. The stability of accumulated anthocyanin in suspension cultures of the parental line and high and low accumulating subclones of wild carrot. *Plant Cell Tissue Org. Culture* **8**: 113.

Dziezak, J. D. 1986. Biotechnology and flavor development plant tissue cultures. *Food Technol.* **40**: 122.

Fowler, M. W., and Scragg, A. H. 1988. Natural products from higher plants and plant cell culture. In *Plant Cell Biotechnology.* (Ed.) Pais, M. S. S., Mavituna, F., and Novais, J. M., p. 165. NATO ASI Series Vol. H18, Springer-Verlag, Berlin.

Francis, F. J. 1985. Pigments and other colorants, Ch. 8. In *Food Chemistry,* 2nd ed. (Ed.) Fennema, O.R., p. 545. Marcel Dekker, New York.

Freeman, G. G., Whenham, R. J., MacKenzie, I. A., and Davey, M. R. 1974. Flavour components in tissue culture of onion (*Allium cepa*). *Plant Sci. Lett.*3:121.

Friborg, G. 1971. Growth and organogenesis in tissue cultures of *Allium cepa* var. proliferum. *Physiol. Plant* **25**: 436.

Fritsch, H., and Grisebach, H. 1975. Biosynthesis of cyanidin in cell cultues of *Haplopappus gracilis. Phytochemistry* **14**: 2437.

Giles, K. L., and Morgan, W. M. 1987. Industrial scale plant micropropagation. *Trends Biotechnol.* **5**: 35.

Hall, R. D., and Yeoman, M. M. 1986. Factors determining anthocyanin yield in cell cultures of *Catharanthus roseus* (L.) G. Don. *New Phytol.* **103**: 33.

Hall, R. D., and Yeoman, M. M. 1987. Intercellular and intracellular heterogeneity in secondary metabolite accumulation in cultures of *Catharanthus roseus*following cell line selection. *J. Exp. Bot.* **38**: 1391.

Hallsby, G. A., and Shuler, M. L. 1986. Altering fluid flow patterns changes pattern of cellular associations in immobilized tobacco tissue cultures. *Biotechnol. Bioeng. Symp. Series* **17**: 731.

Hiraoka, N., Kodama, T., and Tomita, Y. 1986. Selection of *Bupleurum falcatum* callus line producing anthocyanins in darkness. *J. Nat. Prod.* **49**:470–474.

Holden, M. A., Hall, R. D. Lindsey K., and Yeoman, M. M. 1987. Capsaicin biosynthesis in cell cultures of *Capsicum frutescens.* In *Plant and Animal Cells: Process Possibilities.* (Ed.) Webb, C., and Mavituna, F., p. 45. Ellis Horwood Ltd., Chichester, U.K.

Ilker, R. 1987. *In vitro* pigment production: An alternative to color synthesis. *Food Technol.* **41**(4): 70.

Kakegawa, K., Kaneko, Y., Hattori, E., Koike, K., and Takeda, K. 1987. Cell cultures of *Centaurea cyanus* produce malonated anthocyanin in UV light. *Phytochemistry* **26**: 2261.

Kireeva, S. A., Melnikov, U. N. Reznikov, S. A., and Mishcheryakova, N. I. 1978. Essential oil accumulation in peppermint callus culture. *Soviet Plant Physiol.***25**:438.

Knobloch, K. H., Bast, G., and Berlin, J. 1982. Medium and light-induced formation of serpentine and anthocyanins in cell suspension cultures of *Catharanthus roseus. Phytochemistry* **21**: 591.

Kobayashi, Y., Fukui, H., and Tabata, M. 1988. Berberine production by batch and semi-continuous cultures of immobilized *Thalictrum* cells in an improved bioreactor. *Plant Cell Rep.* **7**: 249.

Levin, R., Gaba, V., Tai, B., Hirsch, S., DeMola, D., and Vasil, I. K. 1988. Automated plant tissue culture for mass propagation. *Bio/Technol.* **6**: 1035.

Lin, M. L., and Staba, E. J. 1961. Peppermint and spearmint tissue cultures. I. Callus formation and submerged culture. *Lloydia* **24**: 139.

Luckner, M., and Dietrich, B. 1985. Formation of cardenoilides in cell and organ cultures of *Digitalis lanata.* In *Primary and Secondary Metabolism of Plant Cell Cultures.* (Ed.) Neumann, D., et al., p. 154. Springer-Verlag, Berlin.

McCown, B. H., Zeldin, E. L., Pinkalla, H. A., and Dedolph, R. R. 1988. Nodule culture: A developmental pathway with high potential for regeneration, automated micropropagation, and plant metabolite production from woody plants. In *Genetic Manipulation of Woody Plants*. (Ed.) Hanover, J. W., and Keathley, D. E., p. 149. Plenum Publishing Co., New York.

Mulder-Krieger, Th., Verpoorte, R., Baerheim, A., Svendsen, A., and Scheffer, J. J. C. 1988. Production of essential oils and flavours in plant cell and tissue cultures. A review. *Plant Cell Tissue Organ Cult.* **13:** 85.

Nozue, M., Kawai, J., and Yoshitama, K. 1987. Selection of a high anthocyanin-producing cell line of sweet potato cell cultures and identification of pigments. *J. Plant Phys.* **129:** 81.

Parr, A. J., Robins, R. J., and Rhodes, M. J. C. 1987. Release of secondary metabolites by plant cell cultures. In *Plant and Animal Cells: Process Possibilities*. (Ed.) Webb, C., and Mavituna, F., p. 229. Ellis Horwood Ltd., Chichester, U.K.

Payne, G. F., Shuler, M. L., and Brodelius, P. 1987. Large scale plant cell culture. In *Large Scale Cell Culture Technology*. (Ed.) Lydersen, B. K., p. 193. Hanser Publ., New York.

Payne, G. F., Payne, N. N., Shuler, M. L., and Asada, M. 1988. *In situ* adsorption for enhanced alkaloid production by *Catharanthus roseus. Biotechnol. Lett.***10:**187.

Rau, D., and Forkmann, G. 1986. Anthocyanin synthesis in tissue cultures of *Callistephus chinensis* (China aster). *Plant Cell Rep.* **5:** 435.

Redenbaugh, K., Viss, P., Slade, D., and Fujii, J. A. 1987. Scale-up: Artificial seeds. In *Plant Tissue and Cell Culture*. (Ed.) Green, C. E., Sommers, D., Hackett, W. F., and Biesboer, D. D., p. 473. Alan R. Liss, New York.

Romagnoli, L. G., and Knorr, D. 1988. Effects of ferulic acid treatment on growth, flavor development of cultured *Vanilla planifolia* cells. *Food Biotechnol.* **2:** 93.

Sahai, O., and Knuth, M. 1985. Commercializing plant tissue culture processes: Economics, problems and prospects. *Biotechnol. Prog.* **1:** 1.

Sakuta, M., Takagi, T., and Komamine, A. 1986. Growth related accumulation of betacyanin in suspension cultures of *Phytolacca americana* L. *J. Plant Physiol.***125:** 337.

Seitz, H. U., and Hinderer, W. 1988. Anthocyanins, Ch. 3. In *Cell Culture and Somatic Cell Genetics of Plants 5*. (Ed.) Constabel, F. and Vasil, I. K., p. 49. Academic Press, San Diego.

Selby, C., and Collin, H. A. 1976. Clonal variation in growth and flavour production in tissue cultures of *Allium cepa* P. *Annals Bot.* **40:** 911.

Selby, C., Turnbull, A., and Collin, H. A. 1980. Comparison of the onion plant (*Allium cepa*) and onion tissue culture. II. Stimulation of flavour precursor synthesis in onion tissue cultures. *New Phytol.* **84:** 307.

Shuler, M. L. 1986. Plant tissue culture using membranes. In *Membrane Separations in Biotechnology.* (Ed.) McGregor, W. C., p. 303. Marcel Dekker, New York.

Spears, K. 1988. Developments in food colourings: The natural alternatives. *Trends Biotechnol.* **6:** 283.

Stones, T. W. 1988. The current and future role of plant tissue culture in the natural essential oils, fragrances, and flavour materials industry. In *Flavors and Fragrances: A World Perspective.* (Ed.) Lawrence, B. M., Mookherjee, B. D., and Willis, B. J., p. 115. Elsevier Science Publ., Amsterdam.

Tabata, M., and Fujita, Y. 1985. Production of shikonin by plant cell cultures. In *Biotechnology in Plant Science.* (Ed.) Zatlin, M., Day, P., and Hollaender, A., p. 207. Academic Press, Orlando.

Townsley, P. M. 1974a. Chocolate aroma from plant cells. *Can. Inst. Food Sci. Technol.* **7:** 76.

Townsley, P. M. 1974b. Production of coffee from plant cell suspension cultures. *Can. Inst. Food Sci. Technol.* **7:** 79.

Turnbull, A., Galpin, I. J., Smith, J. L., and Collin, H. A. 1981. Comparison of the onion plant (*Allium cepa*) and onion tissue culture. IV. Effect of shoot and root morphogenesis on flavour precursor synthesis in onion tissue culture. *New Phytol.* **87:** 257.

Ulbrich, B., Osthoff, H., and Wiesner, W. 1988. Aspects of screening plant cell cultures for new pharmacologically active compounds. In *Plant Cell Biotechnology.* (Ed.) Pais, M. S. S., Mavituna, F., and Novais, J. M., p. 461. NATO ASI Series Vol, H18, Springer-Verlag, Berlin.

Walker, K. A., Wendeln, M. L., and Jaworski, E. G. 1979. Organogenesis in callus tissue of *Medicago sativa.* The temporal separation of induction processes from differentiation processes. *Plant Sci. Lett.* **16:** 23.

Watts, M. J., Galpin, I. J., and Collin, H. A. 1985. The effect of greening on flavour production in celery tissue culture. *New Phytol.* **100:** 45.

Yamakawa, T., Kato, S., Ishida, K., Kodama, T., and Minoda, Y. 1983. Production of anthocyanins by *Vitis* cells in suspension culture. *Agric. Biol. Chem.* **47:** 2185.

4

Membrane Bioreactors: Enzyme Processes

Mohamed A. Mehaia

King Saud University
Buriedah, Saudi Arabia

Munir Cheryan

University of Illinois
Urbana, Illinois

INTRODUCTION

Enzymatic reaction processes are commonly used in the production of foods, pharmaceuticals, and many other biological products. Most such enzymatic processes, however, are conducted in traditional batch-type reactors, where the biocatalysts (enzyme or cells) are used in their "free" or soluble form, and the product generally is recovered from the reactor vessel after the reaction is completed. Batch processes have several disadvantages, including their inherent inefficiency compared to continuous processes, batch-to-batch variation leading to inconsistent products, high labor costs, and the need to inactivate or remove the biocatalyst after the reaction, thus increasing bioprocessing costs. Increasingly, however, continuous removal of products is particularly important when the production process is inhibited by high concentration of the product.

During the past years, many attempts have been made to develop, design, and use new types of bioreactor systems in order to improve bioreactor performance and to enhance productivity of enzyme processes. These include immobilization of the biocatalysts (Chibata, 1980a, b; Rosevear, 1984; Keyes and Albert, 1985; Keys and Saraswathi; 1985; Cheryan and Mehaia, 1986) and separator combination system (Takamatsu and Ryu, 1988) among others (Cheryan and Mehaia, 1986). A recent innovation has been the development of membrane bioreactors, which provide a viable alternative to all other conventional and currently researched "high-rate" bioconversion processes.

Membrane bioreactors are advanced applications of membrane technology with the most promising future. The possibility of conducting a desired biological reaction and product separation simultaneously in one device has attracted much attention. Extensive research has led to development of various membrane bioreactor configurations, some of them reaching commercial scale of operation. An enzymatic process using a membrane bioreactor for the production of L-amino acids was developed by Wandrey and Flaschel (1979) and has been employed by the German company Degussa with an annual capacity of 200 tons (Wandrey and Wichmann, 1987). Recently, some reviews of membrane bioreactors, for enzyme reactions and/or fermentation, have appeared (Flaschel et al., 1983; Michaels and Matson, 1985; Schmidt et al., 1986; Drioli, 1986; Matson and Quinn, 1986; Cheryan and Mehaia, 1986; Gekas, 1986; Chang, 1987; Prenosil and Hediger, 1988; Belfort, 1989). In this chapter, we therefore intend to concentrate only on membrane bioreactors for enzyme processes and update this topic.

IMMOBILIZED ENZYMES

Immobilized enzymes were one of the most exciting aspects of biotechnology during the 1970s. The first investigation using fixed enzymes was performed in 1916 by Nelson and Griffin with invertase adsorbed on charcoal; the adsorbed enzyme showed the same activity as native enzyme. However, the first attempt to immobilize an enzyme to improve its properties for a particular application was not made until 1953 by Grubhofer and Schleith (1954). Immobilized enzymes can offer certain advantages over soluble enzymes in areas such as the study of enzymes, analytical biochemistry, preparative pharmacology, and industrial processing including food processing (Table 4.1). For example, in industrial processing, immobilized enzymes are reusable, generally more stable, and more suited to continuous processing design than soluble enzyme. There are, however, several problems with immobilized enzymes, such as losses in activity, steric hindrance, enzyme-substrate orientation, diffusional restriction problems, and the expense of the immobilization step. The cost of the support material itself may be limiting in some cases (Cheryan and Mehaia, 1986). However, the commercialization of immobilized enzymes is slow, and there are only eight known industrial processes that have been commercialized (Table 4.2) since the last reviews of this topic (Olson and Richardson, 1974; Olson and Cooney, 1974; Messing, 1975;

Table 4.1 Immobilized Enzymes in Food Processing: Advantages and Disadvantages

---

Advantages of immobilized enzyme:
  1. Enzymes can be reused.
  2. Stability of enzymes is improved.
  3. Greater pH and thermal stability.
  4. Better control of reaction.
  5. Less apparent product inhibition.
  6. Reactions require less space.
  7. A catalyst can be tailor-made for specific use.
  8. Higher purity and yield of products.
  9. Reaction easily terminated by separating substrate from enzyme.
  10. No contamination due to added enzyme.
  11. Continuous operation becomes practical.
  12. Greater flexibility in reactor design.

Disadvantages of immobilized enzyme:
  1. Lower specific activity.
  2. Inactivation with continued operation.
  3. Need for sanitizing reactors.
  4. Cost of support and immobilization procedure.

Weetall, 1977; Chibata, 1978; Finocchairo et al., 1980; Chibata, 1980a, b; Pitcher, 1980; Keyes and Albert, 1985; Keyes and Saraswathi, 1985; Swaisgood, 1985; Wandrey and Wichmann, 1987).

Immobilization Methods

Methods of immobilization which have been well tried and proven can be roughly divided according to whether the enzyme is chemically bound or physically retained as shown in Fig. 4.1. While the classification in Fig. 4.1 applies to both single enzymes and whole cells, the choice of experimental methods depends very much on the type of biocatalyst (enzyme or microbial cells) to be immobilized (Hartmeier, 1985). One of the most promising techniques for immobilization of biocatalysts is the recently developed "membrane bioreactors," using synthetic semipermeable membranes of the appropriate chemical nature and physical configuration; this will be discussed in detail in this chapter. For more detailed information on enzyme immobilization, other references should be consulted (Cheetham, 1985; Kennedy, 1985a, b).

Table 4.2 Commercial Immobilized Enzyme Reactors

Enzyme	Product	Immobilizing method	Reactor type	Operating mode	Company	Starting date
Aminoacylase[a]	L-Amino acids	Adsorbed	Packed bed	Continuous	Tanabe Seiyaku	1969
Aminoacylase[d]	L-Amino acids	Membrane	MRR[e]	Continuous	Degussa	1981
Aspartase[b]	Aspartate	Entrapped	Packed bed	Continuous	Tanabe Seiyaku	1973
Fumarase[b]	Fumarate	Entrapped	Packed bed	Continuous	Tanabe Seiyaku	1974
Glucose isomerase	HFCS[c]	Adsorbed	Stirred tank	Batch	Clinton Corn	1972
		Covalent	Packed bed	Continuous	Novo	1974
					Novo	1975
Lactase	Lactose-free milk	Entrapped	Stirred tank	Batch	Snamprogetti	1977
Penicillin acylase	6-Amino-penicillinic acid	Adsorbed	Stirred tank	Batch	Squibb	1966
		Covalent	Stirred tank	Batch	Astra	1973
		Covalent	Stirred tank	Batch	Beecham	1974
		Entrapped	Packed bed	Continuous	Snamprogetti	1975
Steroid dehydrogenase[b]	Prednisolone	Heat-treated			Squibb	1964

[a]Lilly (1978).
[b]Immobilized cells.
[c]High-fructose corn syrup.
[d]Wandrey and Flaschel (1979), Wandrey and Wichmann (1987).
[e]Membrane recycle reactor.

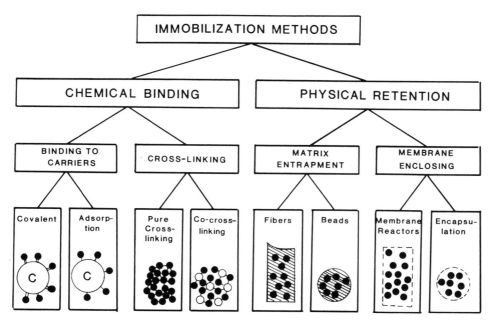

Figure 4.1 Major classes of immobilized biocatalysts: (O) carrier; (●) biocatalyst (enzyme molecule or microbial cell).

Types of Reactors

A number of reactor configurations, for containment of soluble or immobilized enzyme during processing of a liquid, are available (Fig. 4.2). They include batch reactors, continuous stirred tank reactors (CSTR), fluidized-bed reactors, and fixed-bed or plug-flow reactors (PFR). Most other reactors are combinations or modifications of these types. The simplest design is a batch reactor; however, this type is not the most efficient and is more difficult to control than a continuous process (Swaisgood, 1985). Many modifications of the batch reactor have been designed to simplify recovery and reuse of the enzyme (Weetall and Pitcher, 1986). The other reactors shown in Fig. 4.2 are all variants of continuous systems, the characteristics of which have been reviewed in detail by Weetall and Pitcher (1986), Cooney (1983), Wang et al. (1979), Lilly and Dunnill (1976), Vieth et al. (1976), Pitcher and Messing (1975), and Vieth and Venkatasubramanian (1974). Swaisgood (1985) has discussed a number of factors that should be considered in the choice of a reactor type, such as effect of back-mixing, diffusion limitations, and substrate or product inhibition effects.

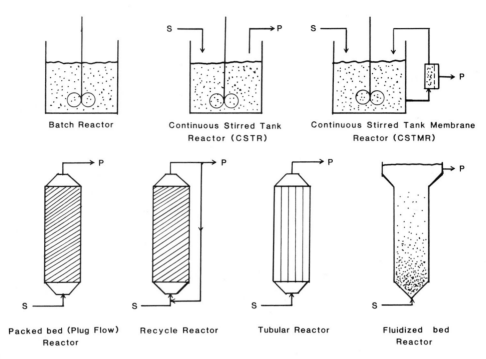

Figure 4.2 Examples of reactors used with immobilized biocatalysts (enzyme or microbial cell).

Some examples of continuous-flow systems are shown in Fig. 4.2. The most widely studied for immobilized enzymes — and used in large-scale commercial operation — is the plug-flow reactor. The plug-flow reactor is kinetically more efficient, on the basis of product formed per unit time per unit of enzyme activity, than a well-mixed reactor system. Table 4.2 lists commercial immobilized enzyme reactors. Many types of tubular reactors have been suggested in which the enzyme will be immobilized on the inner walls of the tubes (e.g., hollow fiber reactor). Free enzymes or enzymes bound to a soluble polymer can be used in a continuous-stirred tank reactor to which an ultrafiltration unit has been coupled (see next section).

The choice of reactor systems for a particular process depends on several factors, such as reactor utilization and cost, enzyme reuse and operation requirements (Wang et al., 1979; Martensson, 1982). Table 4.3 summarizes the general characteristics of the main reactor types.

Table 4.3 Characteristics of Various Reactor Types

Reactors	Continuous operation	Plug flow	Ease of use
Stirred tank (batch)			+
CSTR			
(single stage)	+		+
(multistage)	+	almost	+
Packed bed	+	+	+
Fluidized bed	+	+	+
Tubular	+	+	+
Membrane reactor			
(CSTMR)	+		+
(Hollow fiber)	+	+	+

DESIGN OF ENZYME MEMBRANE REACTORS

The concept of the enzyme membrane reactor is based on the potential of synthetic semipermeable membranes to retain the biocatalyst (enzymes or microbial cells) but not the products formed during the reaction. This implies that the molecular weight of the products should be considerably lower than that of the biocatalyst. By selecting a membrane module with an appropriate cut-off, it is possible to separate low molecular weight components such as sugars and salts, and retain high molecular weight components such as enzymes, starch, cellulose, and proteins. The advantages of such a system are that the biocatalysts are retained in the reactor for further reaction, while the products are continuously removed from the system. Enzyme-membrane reactors can be classified according to the state of mixing or configuration as follows:

Continuous Stirred Tank Membrane Reactors (CSTMR)

Three types of CSTMR are in use today: the dead-end stirred cells (DESC), the membrane recycle reactors (MRR), and cascade membrane reactors (CMR).

Dead-end stirred cell (DESC) Most of the literature on enzyme-membrane reactors deals with DESC (Cheryan, 1986). Fig. 4.3 shows a schematic layout of a typical dead-end stirred cell system. In operation, the cell is charged with the enzyme solution and the substrate solution is continuously fed in under pressure. The permeate (containing the products) is continuously withdrawn through the

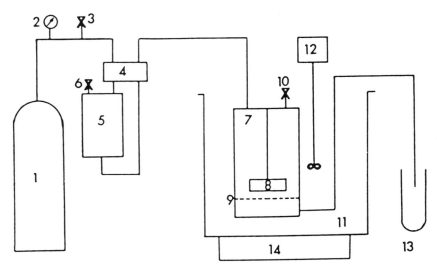

Figure 4.3 Schematic layout of a typical dead-end stirred cell system: (1) gas cylinder to provide transmembrane pressure, (2) pressure gauge, (3) main gas valve, (4) three-way directory valve, (5) substrate reservoir tank, (6) pressure release safety valve, (7) UF cell where reaction takes place, (8) magnetic stirrer for agitation, (9) membrane, (10) safety valve, (11) water bath to maintain temperature, (12) heater/stirrer for water bath, (13) product collection vessel, and (14) magnetic stirrer.

membrane. For improving the reaction rate and minimizing concentration polarization, agitation is required. The membrane should be chosen to retain the enzymes and pass the product molecules. The dead-end stirred cell reactors are less efficient than the other types of continuous stirred tank reactors. The concept of a particular operation and the mechanism of enzyme action have been studied (Drioli et al., 1975; Greco et al., 1979a, b; Roozen and Pilnik, 1979; Henley et al., 1980; Scardi et al., 1980; Hong et al., 1981; Greco et al., 1981a, b; Alfani et al., 1982a, b; Ohlson et al., 1984a, b; Darnoko et al., 1989).

Membrane Recycle Reactor (MRR)   The concept of a continuously operated membrane recycle reactor, shown in Fig. 4.4, was first demonstrated by Porter and Michaels (1972), wherein a solution of a enzyme confined within a continuous stirred-tank reactor (CSTR) is continuously circulated through an ultrafiltration module, from which enzyme-free ultrafiltrate is continuously removed while fresh substrate for the enzymatic conversion is continuously fed to the reactor. This results in continuous generation and isolation of the enzymatically transformed product, without loss of enzyme. However, the membrane should be chosen to

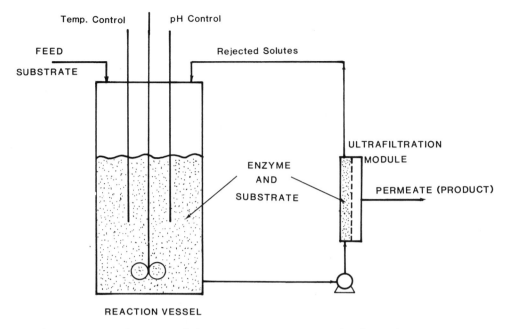

Figure 4.4 Membrane recycle bioreactor (continuous stirred tank membrane reactor, CSTMR).

retain the enzyme while minimizing retention of the product molecules. Since most unmodified enzymes are of the order of 10,000–100,000 in molecular weight, ultrafiltration membrane with these MW cut-off can be used. Table 4.4 is a listing of some commercially available ultrafiltration membranes and their nominal rejection characteristics. The major advantage of enzyme membrane reactors is that the enzyme is theoretically retained within the system and can be reused a number of times; this increases the productivity of the system. However, there is no limitation to the size of the substrate if it can be completely converted into products of low molecular weight. Therefore, the CSTR membrane reactor is the only configuration possible for the hydrolysis of macromolecules such as proteins, starch, or cellulose. One of the main advantages of CSTR membrane reactors is that the catalyst can be homogeneously distributed, thereby avoiding transport limitations. Therefore, this type of reactor should be potentially suitable for the depolymerization of natural macromolecules, reactions requiring pH control and especially suitable for operations with coenzyme-dependent enzyme systems (Flashel et al., 1983). In fact, the concentration of product in the reactor outlet will be the same as the concentration in the reaction vessel at any given time. The

Table 4.4 Some Commercially Available Ultrafiltration Membranes

Designation	Nominal molecular weight cutoff	Apparent pore diameter (nm)	Water flux[a] (liters/m²/hr)	Rejection (%)				
				D-Alanine	Sucrose	Myoglobin	IgM[b]
Amicon (flat sheet)[c]							
UM 05	500	2.1	17	15	70	>95	>98
UM 2	1,000	2.4	34	0	50	>95	>98
UM 10	10,000	3.0	102	0	25	95	>98
PM 10	10,000	3.8	935	0	0	80	>98
PM 30	30,000	4.7	850	0	0	35	>98
PM 50	50,000	6.6	1500	0	0	30	>98
PM 100	100,000	12.0	1200	0	0	0	>98
XM 50	50,000	6.5	425	0	0	20	>98
XM 100A	100,000	11.0	1105	0	0	0	>98
XM 300	300,000	48.0	2215	0	0	0	>98
A/G Technology (hollow fibers)[d]							
UFP -5	5,000	—	18	—	—	—	—
UFP-10	10,000	—	30	—	—	99.5	—
UFP-30	30,000	—	84	—	—	—	—
UFP-100	100,000	—	156	—	—	—	99.4
UFP-500	500,000	—	780	—	—	—	—

[a]At 368 kPa.
[b]Immunoglobulin M.
[c]From Amicon Corporation.
[d]From A/G Technology Corporation. Flux measured at 70 kPa, 25°C.

kinetics and performance of CSTR membrane have been studied (Wandrey and Flaschel, 1979; Cheryan and Deeslie, 1980, 1983; Deeslie and Cheryan, 1981a, b; 1982; Flaschel et al., 1983; Park et al., 1985; Davis, 1986; Trujillo, 1987; Bressollier et al., 1988; Prenosil and Hediger, 1988; Belfort, 1989; Mannheim, 1989). A CSTR-membrane reactor has some advantages compared to dead-end stirred cell reactors, for example the reaction vessel is separated from the ultrafiltration unit, and it is more flexible for adjusting operating conditions to attain and maintain a high level of conversion in the reactor system.

Cascade Membrane Reactors In CSTR-membrane reactors, the rate of conversion is very low. This is due to low concentration of substrate in the system, and the rate of reaction is proportional to the substrate concentration. Consequently, this system must be significantly larger than a plug-flow reactor used for the same conversion. To overcome this limitation, one can set up several CSTR-membrane recycle series in a type of cascade fashion. There are two possible configurations shown in Fig. 4.5 and 4.6. One utilizes a membrane separation unit in each stage (Fig. 4.5), and the other uses a separation unit only after the last stage (Fig. 4.6). Using membrane in each stage the substrate solution is fed to the first stage while a partially converted solution leaves it across the ultrafiltration module as permeate. The solution is pumped into the next stage and so on. This type of reactor is generally more expensive, but it may be useful if one is trying to conduct multienzyme reactions, such as the production of high-fructose corn syrup from starch (Cheryan and Mehaia, 1986). The other reactor, using a separation unit only after the last stage, is simple, and the enzyme is retained in the separation unit, concentrated, and recycled to the first stage. The advantage of this design is that standard reactors can be used in conjunction with only one separation unit. Wandrey and Flaschel (1979) have described the application of a

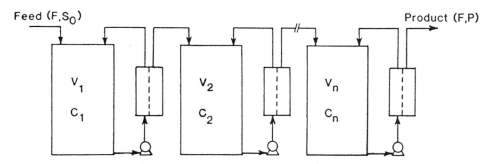

Figure 4.5 Schematic of cascade membrane bioreactors in series: cascade system with a membrane module in each stage. F, feed flow rate; V, volume; C, biocatalyst concentration; P, product concentration; S_0, inlet substrate concentration.

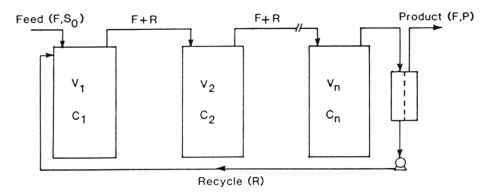

Figure 4.6 Schematic of cascade membrane bioreactors in series: cascade system with a membrane module in the last stage. R, recycle flow rate.

three-stage cascade system with one separation unit for the production of L-methionine using acylase. Cheryan and Deeslie (1984) have developed a semicontinuous cascade system (Fig. 4.7), using membranes with different pore sizes in each stage, and instead of the permeate flowing down the cascade as in Fig. 4.5, the retentate was pumped through a series of membranes with increasing pore sizes, and the permeate containing the desired fraction was withdrawn from the system at each system.

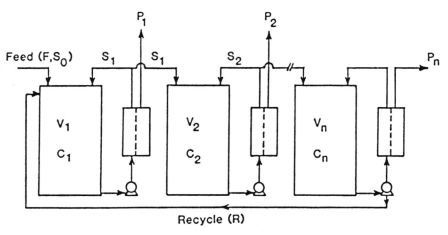

Figure 4.7 Schematic of semicontinuous cascade system for producing protein hydrolysate fractions of different molecular sizes. (*From Cheryan and Deeslie, 1984.*)

Plug-Flow (Hollow Fiber) Reactors (HFER)

Hollow fiber membrane devices have been used as enzyme reactors during the past 15 years because they have some advantages, such as large surface area–volume ratio, immobilizing soluble enzyme in intact form, isolation of enzymes from the substrate stream, continuous removal of inhibitory products higher volumetric productivities, and the membrane is self-supporting and can be backflushed easily.

The idea of a hollow fiber enzyme reactor was first suggested by Rony (1971), who later demonstrated its usefulness (Rony, 1972). Since then many researchers have examined the application of this enzyme reactor (Chambers et al., 1976; Cheryan, 1986; Cheryan and Mehaia, 1986; Gekas, 1986). There are two basic configurations of hollow fiber devices available. One is the beaker type (Fig. 4.8) and the other is the tubular type (Fig. 4.9). Enzymes can be loaded or trapped in either the shell side or the tube side. Several theoretical models for enzyme catalysis in hollow fiber enzyme reactors have appeared (Waterland et al., 1974; Lewis and Middleman, 1974; Georgakis et al., 1975; Mashelkas and Ramchandran, 1975; Kim and Cooney, 1976; Kawakami et al., 1980; Katoaka et al., 1980; Kohlwey and Cheryan, 1981; Miyawaki et al., 1982a, b, c; Kim and Chang, 1983a, b; Davis and Watson, 1981; Frennesson et al., 1985; Davis, 1986; Trujillo, 1987; Bressollier et al., 1988; Ishikawa et al., 1989a, b). If the enzyme is in the tube side and the feed is pumped through the shell side, the residence time distribution more closely approximates a completely mixed model than the plug flow model (Katoaka et al., 1980). Simple mathematical models and experiments have

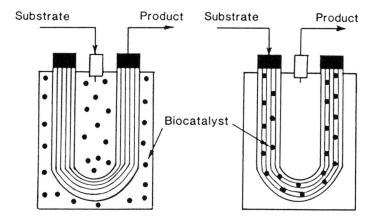

Figure 4.8 Schematic of a breaker-type hollow fiber devices. The biocatalyst can be trapped either in the tube side or in the shell side.

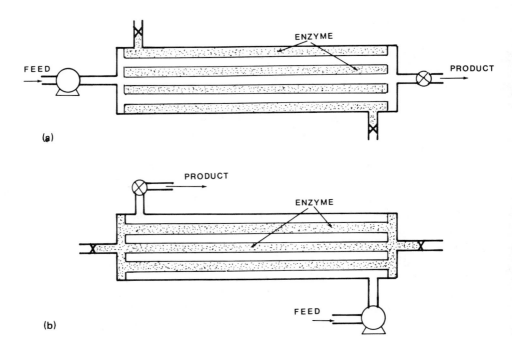

Figure 4.9 Schematic of tubular-type hollow fiber bioreactor. The biocatalyst (enzyme) can be trapped either in the (a) shell side or in the (b) tube side.

indicated that it is advantageous for most applications to have the enzyme in the shell side and the substrate flowing through the lumen/tube side (Kawakami et al., 1980). The selection of the appropriate hollow fiber membrane should take into account a number of factors. These factors include permeability of the substrates and products, within the hollow fiber membrane, the nominal molecular weight cut-off, membrane anisotropy, and membrane compatibility with enzyme and substrate solutions.

Microporous Membrane Reactors (MMR)

Synthetic ultrafiltration membranes have been used for immobilization of enzymes by using them as a means of retaining the enzyme, thus separating it from the low molecular weight reaction products. The enzyme is in soluble form and combines the advantages of free enzymes with the possibility of reuse which characterizes the insoluble (bound) enzymes (Matson, 1979; Matson and Quinn, 1986; Lopez et al., 1988; Matson and Lopez, 1988; Belfort, 1989).

Another approach is to immobilize or to load the enzymes on to the microporous membrane (Hoq et al., 1984; 1985a, b, 1986; Matson and Quinn, 1986) or onto the macroporous side of anistropic hollow fiber membranes (Engasser et al., 1980a, b) and to follow various modes of feeding the substrate and carrying out the reaction. Fig. 4.10 shows a schematic of a microporous membrane reactor. This method is considered as physical immobilization as far as simple containment of the enzyme within porous structure of the membrane is concerned. Cross-linking of the contained enzyme through glutaraldehyde (Breslau and Kilcullen, 1978; Miyairi, 1979; Silmon, 1982; Jancsik et al., 1982; Gekas, 1986; Nakajima et al., 1988) or even covalent coupling can also be used (Gregor and Rauf, 1975; Gregor, 1975, 1977; Gekas, 1986; Lopez-Leiva and Gekas, 1986; Bardeletti and Coulet, 1987), such as a microporous plastic sheet with immobilized enzyme developed by Amerace Corporation (Goldberg, 1978, 1979; Amerace Corporation, 1980). In fact, this implies special preparation and treatment of the membrane, but on the other hand, it can make possible operation of the system under pressure-driven conditions in the ultrafiltration mode, which otherwise is impossible. In the case of low molecular weight compounds direct contact with the bound enzymes is possible without the diffusion limitations that occur in the other modes of operation. The mathematical model and analysis of the immobilization process of enzyme immobilization in a porous membrane have been reported by Hossain and Do (1985). Hausser and Goldberg (1983) have concluded that the use of Amerace's continous immobilized dual enzyme reactor system, containing two enzymes (α-amylase/glucoamylase), for the production of maltose-containing corn syrups has been shown to be a viable process.

Matson and Quinn (1986) and Matson and Lopez (1988) have developed a single membrane-entrapped enzyme bioreactor (i.e., without covalently linking the enzyme to the membrane) between aqueous and organic flow streams. This system is appropriate for enzymatic conversions that involve sparingly soluble substrate or product or cases where the product exerts a feedback inhibition on the enzyme action. Lopez et al. (1988) also prefer physical extrampent of the enzyme within a porous asymmetric or skinned membrane so that the degraded enzyme can be replaced easily when desired. Excellent resolution of over 90% enantiomeric excess in both phases was obtained for the N-benzyl tyrosine ethyl ester conversion in such a hollow fiber membrane system using chymosin, 1-octanol, and phosphate buffer at pH 7.8.

Laminated Enzyme Membrane Reactor (LEMR)

The concept of a laminated biomembrane enzyme system was first demonstrated by Matson (1979) and recently reviewed by Belfort (1989). The enzymatic

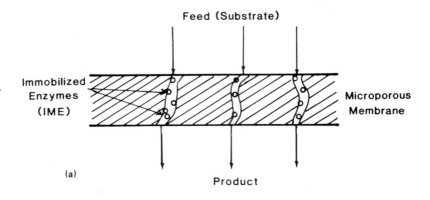

(a)

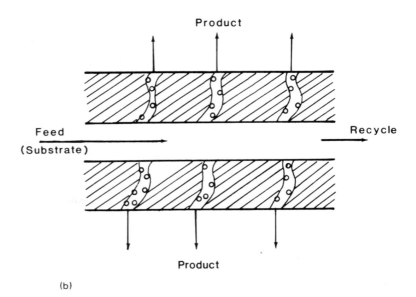

(b)

Figure 4.10 Schematic of a microporous membrane bioreactor. (a) Microporous film with immobilized enzymes used in a dead-end system. (b) Microporous film with immobilized enzymes used in a crossflow configuration.

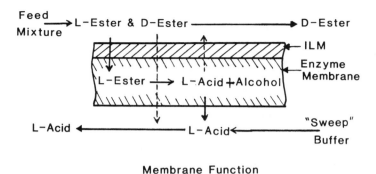

Membrane Function

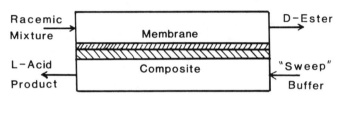

Reactor Operation

Figure 4.11   Schematic of laminated enzyme membrane reactor. Enzymatic resolution of a racemic amino acid derivative. (*After Matson and Quinn, 1986.*)

resolution of a racemic amino acid derivative, N-acyl-tyrosine ethyl ester, was chosen in order to explain the concept (Fig. 4.11). A separating membrane such as a liquid membrane (a porous Teflon membrane containing an organic solvent) is laminated with an enzyme membrane (containing the immobilized enzyme chymotrypsin). Both esters dissolve easily into the immobilized liquid membrane (ILM) and diffuse to the enzyme region where only the L-ester is converted to its acid form, L-acid. Since the L-acid is charged at the solution pH, it cannot dissolve into the ILM and diffuse back into the feed stream and is forced to migrate into the "sweep" buffer. By controlling the feed and "sweep" buffer flow rates, a substantial concentration of the product is possible. Lopez et al. (1988) have investigated the utility of LEMR in the enzymatic deacylation of benzyl penicillin to produce 6-amino penicillinic acid, and they obtained encouraging results. However, LEMR is relatively complicated and does not, at least with the systems described, utilize convective transport (Belfort, 1989).

Cofactor Immobilization and Regeneration in Membrane Reactors

Cofactors, such as nicotinamide adenine dinucleotide phosphate (NADP), coenzyme A, or adenosine triphosphate (ATP) are interesting for industrial processes since they catalyse oxido-reductive, group transfer, energy transfer, and synthetic reactions. However, cofactors are rather expensive substances, and for industrial use they have to be regenerated and reused.

Methods of binding cofactors have been known in principle for more than 15 years. Recently, Klua and Wandrey (1987), Kulbe et al., (1987a, b, c), Kulbe and Chimel (1988), Howaldt et al. (1988), Schmidt et al. (1986), Wandrey and Wichmann (1985, 1987), and Flaschel et al. (1983) have reviewed the immobilization methods for coenzymes. Three approaches have been used with membrane reactors (Cheryan and Mehaia, 1986): (a) NAD has been bound to a water-soluble polymer and then entrapped in a membrane reactor with cojugated enzymes (Yamazaki et al., 1976; Morikawa et al., 1978; Wichmann et al., 1981; Schmidt et al., 1986; Kulbe et al., 1987a, b, c; Wandrey and Wichmann, 1987; Bossow and Wandrey, 1987; Berke et al., 1988 (Fig. 4.12), (b) native NAD and/or conjugated enzyme has been entrapped in a membrane reaction in which the membrane MW cut-off was lower than that of NAD (Katoaka et al., 1980; Chambers et al., 1981; Miyawaki et al., 1982a, b, c; Nakamura, 1988), and (c) the conjugated enzymes have been immobilized but cofactor has not (Waterland et al., 1974; Fink and Rodwell, 1975; Miyawaki et al., 1982; Ishikawa et al., 1989a, b) (Fig. 4.13). This last approach appears to be simpler than the immobilized cofactor method, although a good system of recovery of the cofactor is needed to reduce overall cost. Ishikawa et al. (1989a, b) have studied the theoretical analysis and experimental investigation of a glucose-6-phosphate production and simultaneous ATP regeneration by conjugated enzymes in an ultrafiltration hollow fiber reactor using the last approach.

Recently, Kulbe and Chimel (1988) and Howaldt et al. (1988) have used a different approach for coenzyme-dependent carbohydrate conversion using membrane bioreactors. In this case, a sulfonated polysulfone membrane "ionic ultrafiltration membrane" was used to retain the negatively charged native NAD(H) or NADP(H) by repulsion because the membrane was also negatively charged. The properties, requirements, advantages, disadvantages, and possible improvements of the ionic ultrafiltration membrane concept for retainment of native cofactors are discussed (Kulbe et al., 1988).

APPLICATIONS OF ENZYME MEMBRANE REACTORS IN FOOD PROCESSING

Enzymes have been utilized by human beings since ancient times, well before their nature was understood. Enzymic catalysis plays a fundamental role in many in-

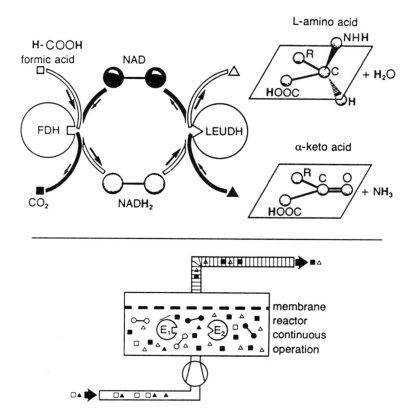

Figure 4.12 Enzyme membrane reactor concept for continuous enzymatic reactions with cofactor regeneration. In this concept the cofactor is immobilized to a macro-molecule too large to permeate through the membrane. L-leucine dehydrogenase (LEUDH) is used, for example, as the enzyme for L-leucine production by the reductive amination of the corresponding α-keto acid. The coenzyme NAD bound to polyethy-lene glycol is reduced under enzymatic catalysis by formate hydrogenase (FDH) with formic acid. (*From Wandery and Wichmann, 1987.*)

dustrial processes, particularly in food processing and pharmaceuticals production. The efforts over the past two decades to develop processes for enzyme immobilization have been made with several goals in mind: to achieve the advantages inherent in heterogeneous enzymic catalysis, a more reproducible continuous process, smaller plant sizes due to higher productivity, and a cleaner product stream, hopefully free of enzyme activity. However, as mentioned earlier, very few immobilized enzyme processes have achieved industrial scale. It is possible that the alternate technology of membrane reactors may overcome many

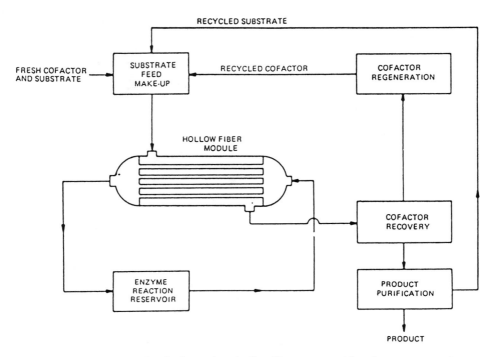

Figure 4.13 Generalized scheme for a hollow fiber reactor with cofactor regeneration. The enzymes are trapped in the tubes, while substrate and cofactor are fed in from the shell side, (*From Fink and Rodwell, 1975.*)

of the limitations of conventional immobilization of enzymes. The major advantage of enzyme membrane reactors is that the enzyme is retained within the system and can be reused several times, thus increasing productivity and capacity of the system, measured as amount of product per unit amount of enzyme per unit time. Further advantages of the membrane reactor are that the product stream is free from enzymes and other particulate materials, which should reduce downstream processing costs.

Protein Hydrolysis

The enzymatic hydrolysis of proteins seems to be of important industrial interest since the supplementation of beverages with soluble protein is of particular interest; in addition, enzymatic hydrolysis of proteins is a convenient means of improving certain functional properties of protein without diminishing nutritional values (Cheryan and Deeslie, 1983). Enzymatic hydrolysis has several advantages over chemical hydrolysis (Cheryan and Deeslie, 1980), but traditional batch methods have some disadvantages, such as the relatively high cost of enzymes and their inherent inefficiency compared to continuous processes, resulting in low yields and productivity; in addition, too much hydrolysis tends to make proteins bitter. In principle, enzyme membrane reactors should solve some of these problems (Cheryan, 1986; Cheryan and Mehaia, 1986). Table 4.5 summarizes the applications of enzyme membrane bioreactors in protein hydrolysis.

Animal proteins Animal proteins tend to be less difficult to hydrolyze than plant proteins. This may be because plant proteins apparently possess "hydrophobic cores" in their structures that are difficult to break down. The gradual buildup of unhydrolyzed protein in the enzyme reactors will eventually lead to problems. Careful selection of the enzymes and good membrane design with adequate control of concentration polarization is critically important with such proteins.

The first application of enzyme membrane reactors was reported by Blatt (1971), who demonstrated the rapid and continuous ultrafiltration of products, obtained from whey hydrolysis by α-chymotrypsin, in a dead-end stirred cell reactor (Fig. 4.14). Gel chromatography was used to demonstrate an increase in lower molecular weight components with time of hydrolysis. Boudrant and Cheftel (1976) have studied the hydrolysis of casein with Alcalase in a batch and dead-end stirred cell reactors. To reduce the loss of enzymatic activity during operation, the enzyme was chemically stabilized using glutaraldehyde. The stabilized Alcalase could be reused for 7 days at 40°C and achieve 75% substrate conversion. The reaction does not follow exactly first-order kinetics with respect to the substrate concentration, and it is inhibited by the reaction products. Gel chromatography showed the molecular weight range of the peptides to be less than 2,000.

Mannheim and Cheryan (1989) designed a continuous stirred tank membrane reactor for hydrolysis of casein using a hollow fiber ultrafiltration module. Four performance variables — reactor volume, permeate flux, enzyme, and substrate concentration — were studied. More than 90% conversion could be achieved after one hour at 50°C and pH 8 (Table 4.6).

Visser et al. (1989) have studied the degradation of bovine β-casein by plasmin in a membrane recycle reactor for the continuous production and isolation of peptide fractions. Sixteen peptide fragments were identified, which together accounted for virtually all the potential cleavage sites in β-casein. In the system studied, the N-terminal half of β-casein appeared to be more sensitive to plasmic hydrolysis than the rest of the molecule.

The use of a proteolytic enzyme for the hydrolysis of fish protein concentrate in a continuous stirred tank membrane reactor was described by Cheftel et al. (1971) and Cheftel (1972). The purpose was to produce soluble proteins (low

Table 4.5 Enzymatic Hydrolysis of Protein Using Membrane Bioreactors

Enzyme	Substrate	Reactor type	Reference
1. Animal proteins:			
Alcalase	casein	MRR	a
Alcalase	casein	DESC	b
Alcalase	Bovine plasma protein (hemoglobin)	MRR	c
Chymotrypsin	whey protein	DESC	d
Chymotrypsin	casein	MMR	e
Pancreatine	whey proteins	MRR	f
Plasmin	β-casein	MRR	g
Proteases	fish protein concentrate	MRR	h
Trypsin	fish protein	DESC	i
2. Plant proteins:			
Acid protease	soy protein	MRR	j
Acid protease, pepsin, molsin	soy protein	DESC	k
Acid protease, pepsin, esperase	*Vicia faba protein*	DESC	l
Alcalase, pronase	soy protein	MRR	m
Esperase	potato protein	DESC	l
*Penicillium duponi*enzyme	soy protein	MMR	n
Proteases	cottonseed protein	BHFER	o
Trypsin	alfalfa protein	DESC	p
3. Self-cleaning membrane:			
Chymotrypsin	benzyl-L-tyrosine ethyl ester	MMR	q
Papain	cheese whey	MMR	r
Proteases	cheese whey, hemoglobin	MMR	s
Proteases	albumin, hemoglobin	MMR	t
Proteases	nonfat dry milk	MMR	u
Proteases	raw sewage	MMR	v

[a]Mannheim (1989), Mannheim and Cheryan (1989).
[b]Boudrant and Cheftel (1976).
[c]Bressollier et al. (1988).
[d]Blatt (1971).
[e]O'Neil et al. (1971).
[f]Roger and Maubois (1981); Maubois and Brule (1982), Maubois (1988).
[g]Visser et al. (1989).
[h]Cheftel et al. (1971); Cheftel (1972).
[i]Bhumiratana et al. (1977).
[j]Iacobucci et al. (1974).
[k]Roozen and Pilnik (1973, 1979).
[l]Roozen and Pilnik (1979).
[m]Deeslie and Cheryan (1981a, b, 1982, 1988); Cheryan and Deeslie (1983, 1984).
[n]Adu-Amankwa et al. (1981).
[o]Cunningham et al. (1978).
[p]Payne et al. (1978).
[q]Gregor and Rauf (1975).
[r]Velicangil and Howell (1977).
[s]Velicangil and Howell (1981).
[t]Howell and Velicangil (1982).
[u]Wang et al. (1980).
[v]Jenq et al. (1980).
MRR; Membrane recycle reactor; DESC: dead-end storred cell; MMR: Membrane microporous reactor; BHFER: beaker hollow fiber enzyme reactor.

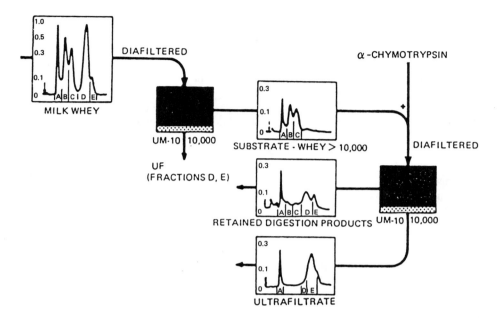

Figure 4.14. Substrate preparation and continuous extraction of proteolytic products, as formed, from cheese whey. Whey was first ultrafiltered through a UM-10 membrane. The retentate was reacted with α-chymotrypsin in a stirred cell with a UM-10 membrane. Chromatographic profiles indicated the breakdown of proteins in the retentate and the appearance of breakdown products in the permeate. (*From Blatt, 1971.*)

molecular weight fish proteins) to make a more suitable protein supplement for use in certain food products. Bhumiratana et al. (1977) have studied the hydrolysis of fish proteins by trypsin in a batch, semi-batch, and dead-end stirred cell reactor. They obtained about 75% conversion, but they experienced rapid membrane fouling. This is the main problem with using dead-end UF cells as membrane reactors with highly insoluble protein substrate.

Bressollier et al. (1988) studied the effect of four operational variables — reaction volume, ultrafiltration flux, enzyme, and substrate concentration — on capacity and conversion rate of a hollow fiber CSTR for Alcalase hydrolysis of bovine plasma (hemoglobin) proteins. They obtained about 72.5% conversion and 77 hr^{-1} capacity at steady state, but after 16 hr of operation the conversion was only 15% (Table 4.6).

Maubois and Brule (1982), Roger and Maubois (1981), and Maubois (1988) used pancreatin to continuously hydrolyze whey proteins and casein in a

Table 4.6 Operating Conditions of Membrane Bioreactors used for
Continuous Protein Hydrolysis

Variables	Soy protein	Soy protein	Casein	Casein	Bovine plasma protein
Data source	a	b	c	d	e
Enzyme	Pronase	Fungal acid protease	Alcalase	Alcalase	Alcalase
Membrane module	CSTR-UF	Plate-recycle	CSTR-UF	DESC	CSTR-UF
pH of reaction	8.0	3.7	8.0	8.8	7.5
Temperature (°C)	50	60	50	40	45
Protein concentrations (% w/w)	0.5-3.0	1.5	5	5	2.1
Continuous run periods (hr)	90	70	14	150	16
Overall yields (%)	85-94	85-95	90	90	15
Residence time (min)	17-305	58-60	71	375	240
Productivity (g hydrolyzate/g enzyme):					
Batch	10–15	9.4	25	—	—
Membrane system	72–108	18–28	320	—	—

[a]Cheryan and Deeslie (1983, 1984); Deeslie and Cheryan (1981a, b, 1982, 1988).
[b]Iacobucci et al. (1974)
[c]Mannheim (1989); Mannheim and Cheryan (1989)
[d]Boudrant andCheftel (1976).
[e]Bressollier et al. (1988).

CSTMR using a 5000 MWCO membrane reactor. The hydrolysates obtained by this process had a free amino acid content less than 10% and most of the peptides had between 2 and 10 residues. This process has since been commercialized.

Plant proteins Examples of the fairly successful applications of the membrane recycle reactor (MRR) for soy protein hydrolysis are the works of Cheryan and Deeslie (1983, 1984); Deeslie (1980); Deeslie and Cheryan (1981a, b, 1982, 1988), and Iacobucci and coworkers (1974). Fig. 4.15 is a schematic flow diagram of the Cheryan-Deeslie process for continuous proteins hydrolysis. Table 4.6 summarizes the reaction conditions and some results from the Cheryan-Deeslie process and the Iacobucci-coworkers process.

Cheryan and Deeslie (1983, 1984) reported that an adequate pretreatment of the protein substrate, such as preheating (to improve reaction rates) and prefiltration

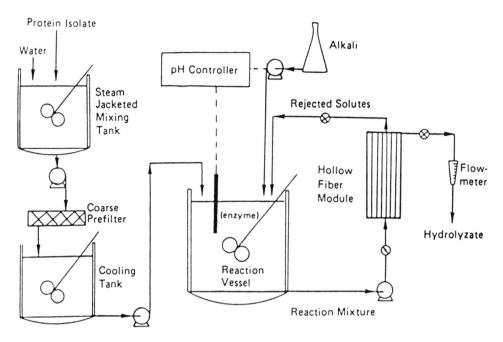

Figure 4.15 Schematic of a CSTR membrane enzyme reactor system for the continuous hydrolysis of proteins. (*From Deeslie and Cheryan, 1981.*)

(to remove coarse insolubles, such as fiber), is recommended to improve overall performance of the membrane recycle reactor. They observed that the steady state of the system can be achieved in less than one hour, this depending on the enzyme/substrate ratio and the residence time. The important variables that need to be optimized are the enzyme concentration, the substrate concentration, volume of the reaction zone, and the flux. Factors affecting the long-term stability of the system were also studied (Deeslie and Cheryan, 1982). Cheryan and Deeslie (1983) concluded their process to be better by 5- to 10-fold than traditional batch processes. They managed to obtain 91% conversion and identified at least four peptide fractions of MW 2500 or less (Deeslie and Cheryan, 1988). Iacobucci et al. (1974) also reported high productivity, but they based their calculations on the enzyme replaced, that is, the amount of enzyme added to the system to maintain a preset level of activity.

Roozen and Pilnik (1973, 1979) have hydrolyzed potato, *Vicia faba*, and soy proteins using various proteases in a dead-end stirred cell. An increase in peptide concentration in the permeate was observed during hydrolysis, but there was a rapid buildup of solids in the retentate so that permeation stopped after about 6 hr. This may be due to poor control of polarization and fouling. They concluded that the usefulness of ultrafiltration for obtaining bland protein hydrolysates seems to be limited to short-term processes with cellulose acetate membrane.

Adu-Amankwa et al. (1981) have hydrolyzed a soy protein with immobilized *Penicillium duponti* enzyme in a recycle reactor. The enzyme was immobilized on reconstituted collagen by macromolecular complexation, impregnation, and covalent cross-linking techniques. They obtained about 57% conversion of soluble protein in 10 hr. The average productivity of the reactor based on a stable enzyme activity was five times better than a batch reactor.

Cunningham et al. (1978) hydrolyzed a cottonseed protein using pepsin or molsin in a hollow fiber beaker. The protein and enzymes were mixed and reacted in the shell side, and the product was pulled off continuously through the lumen (tube side). However, the physical configuration of this particular type of hollow fibers is not suitable for large-scale production (Cheryan and Mehaia, 1986).

Payne et al. (1978) attempted to hydrolyze alfalfa leaf protein concentrate with trypsin in a dead-end stirred cell, but they had a fouling problem. They obtained about 55% conversion of insoluble starting material after 4 hr of operation.

Self-cleaning membrane   In order to reduce the severe flux losses encountered during ultrafiltration of protein solutions, proteases were immobilized on ultrafiltration membranes to hydrolyze the deposited solute molecules (Velicangil and Howell, 1977; Wang et al., 1980; Jenq et al., 1980; Howell and Velicangil, 1982). Venlicangil and Howell (1977) immobilized papain on an Amicon membrane. Using the immobilized papain membrane to concentrate cheddar cheese whey, they obtained a 20% improvement in flux. They also developed a method to attach proteases onto UF membrane by producing a primary adsorbed layer of enzyme, which then retards the rate of gel formation on the ultrafilter (Howell and Velicangil, 1982). This resulted in 25–78% improvement in flux when processing 0.5% albumin or hemoglobin. A more remarkable improvement in the flux, when processing a nonfat dry milk solution, was achieved by Wang et al (1980).

Ultrafiltration of raw sewage was performed using multiple enzymes (proteases) immobilized by physical sorption on noncellulosic UF membranes (Jenq et al., 1980). An increase of 12% in the permeate flux rate was observed due to the action of the immobilized proteases.

Carbohydrate Hydrolysis

Tables 4.7 and 4.8 summarize the application of enzyme membrane bioreactors in carbohydrate hydrolysis.

Starch hydrolysis   Starch hydrolysates are usually used in the production of glucose, high-fructose sweeteners, and brewing syrups and as fermentation

Table 4.7 Enzymatic Hydrolysis of Polysaccharides in Membrane Bioreactors

Enzyme	Substrate	Membrane type	Reference
1. **Starch hydrolysis:**			
α-Amylase	starch	DESC	a
α-Amylase and glucoamylase	starch-corn	MMR	b
β-Amylase	sweet potato	DESC, TC	b
β-Amylase	starch	MRR	d
β-Amylase	starch	Tubular	e
β-Amylase and pullulanase	starch	MRR	f
Amylases	starch	DESC	g
Amylases	starch	Tubular	h
Amylases	starch	MRR	i
Amyloglucosidase	starch	HFER	j
Glucoamylase	corn starch	MRR	k
Glucoamylase	cassava starch	DESC	l
Glucoamylase	starch	DESC	m
Glucoamylase	starch	Tubular	n
Glucoamylase	starch	MRR	o
2. **Cellulose hydrolysis:**			
Cellulase	cellulose	DESC	p
Cellulase	filter paper	DESC	q
Cellulase	Mun. sludge	DESC	r
Cellulase	solka floc	DESC	s
Cellulase	solka floc	MRR	t
Cellulase	solka floc, sugarcane bagasse, sorghum stubble, peanut shells	DESC, MRR	u
Cellulases of *Sporotrichum cellulophilum*	cellulose	DESC	v
Cellulase and cellobiase	pretreated sallow	DESC	w
Cellulase and β-glucosidase	cellulose, olive husk	DESC	x
Cellulase (celluclast 2L)	wheat straw	PBR-UF	y

TC: Thin channel cell; HFER: hollow fiber enzyme reactor; PBR-UF: packed-bed reactor with UF membrane.

[a]Butterworth et al. (1970).
[b]Hausser and Goldberg (1983).

(*Table continued*)

Table 4.7 (*continued*)

[c]Azhar and Hamdy (1981).
[d]Engasser et al. (1980a).
[e]Closset et al. (1974)
[f]Iwasaki et al. (1974).
[g]Marshall and Whelan (1971).
[h]Tachauer et al. (1974).
[i]Larsson and Mattiasson (1984).
[j]Engasser et al. (1980b)
[k]Sims and Cheryan (1989).
[l]Darnoko et al. (1989).
[m]Fisher (1973).
[n]Madgavkar et al. (1977).
[o]Stavenger (1971)
[p]Hahn-Hagerdal et al. (1980), Alfani et al. (1982a, b), Ohlson et al. (1984a).
[q]Hartmeier (1980).
[r]Cinq-Mars and Howell (1977).
[s]Ghose and Kostick (1970).
[t]Tjerneld et al. (1985).
[u]Henley et al. (1980).
[v]Kinoshita et al. (1986).
[w]Ohlson et al. (1984, b); Frennesson et al. (1985).
[x]Alfani et al. (1983).
[y]Gonzalez et al. (1989).

Table 4.8 Enzymatic Hydrolysis and Conversion of Oligosaccharides in Membrane Bioreactors

Enzyme	Substrate	Membrane type	Reference
1. **Oligosaccharides hydrolysis:**			
α-Galactosidase	raffinose	HFER, MMR	a
β-Glucosidase	cellobiose	DESC	b
β-Glucosidase	cellobiose	DEUSMER	c
Amyloglucosidase	maltose	HFER	d
Amyloglucosidase	maltose	DESC	e
Amyloglucosidase	maltose	DEUSMR	f
Amylogcosidase	maltose	MMR	g
Glucoamylase	maltose	MMR	h
β-Galactosidase	lactose	HFER	i
β-Galactosidase	lactose	CAMFER	j
β-Galactosidase	lactose	HFER, MRR	k
β-Galactosidase	lactose	MRR	l
β-Galactosidase	lactose	DEUSMR	m
β-Galactosidase	lactose, ONPG	MRR	n
β-Galactosidase	lactose, ONPG	MSR	o
β-Galactosidase	lactose, milk	MMR, FE, HFER	p
β-Galactosidase	ONPG	DESC	q
β-Galactosidase	ONPG	HFER	r
β-Galactosidase	lactose, whey permeate, skim milk	HFER	s

β-Galactosidase	whey permeate	MRR	t
β-Galactosidase	whey permeate	DESC	u
β-Galactosidase	whey, whey permeate	MMR	v
Invertase	sucrose	HFER	w
Invertase	sucrose	TC, DESC	x
Invertase	sucrose	DEUSMR	y
Invertase	sucrose	MRR, DESC	z
Invertase	sucrose	GEMR	aa
Invertase	sucrose	MMR	bb

2. High-fructose syrup production:

Glucose isomerase	glucose	HFER, MMR	cc
Glucose isomerase	glucose	FE, MMR	dd
Glucose isomerase of *S. griseus*	glucose	DHFBR	ee

[a]Korus and Olson (1977b, 1978); Smiley et al. (1976); Silman et al. (1980).

[b]Hong et al. (1981); Gianfreda and Greco (1981a).

[c]Gianfreda and Greco (1981b).

[d]Engasser et al. (1980a).

[e]Swanson et al. (1973).

[f]Greco et al. (1979a, b).

[g]Furusaki and Asai (1983).

[h]Bardeletti and Coulet (1987).

[i]Korus and Olson (1978); Jones and White (1984).

[j]Prenosil and Hediger (1988); Peterson et al. (1989a, b).

[k]Huffman-Reichenbach and Harper (1982); Korus and Olson (1978).

[l]Kowalewska et al. (1978); Roger et al. (1976, 1977); Norman et al. (1978).

[m]Greco et al. (1981a, b).

[n]Park et al. (1985); Kim and Change (1983a, b).

[o]Vorsilak et al. (1975).

[p]Korus and Olson (1978); Pastore et al. (1974, 1976).

[q]Lee and Hong (1978).

[r]Kohlwey and Cheryan (1981); Lewis and Middleman (1974); Breslau (1981); Waterland et al. (1975).

[s]Presnosil et al. (1985).

[t]Huffman-Reichenbach and Harper (1985).

[u]Mertens and Juyghebaert (1987).

[v]Jancsik et al. (1982); Lopez-Leiva and Gekas (1986); Amerace Corporation (1980); Goldberg (1978, 1979).

[w]Katoaka et al. (1980); Alfani et al. (1983), Silman et al. (1980); Korus and Olson (1978).

[x]Hartmeier (1980), Bowski and Ryu (1974); Bowski et al. (1972).

[y]Greco et al. (1979a, b), Scardi et al. (1980).

[aa]Imai et al. (1983).

[bb]Nakajima et al. (1988).

[cc]Korus and Olson (1977a, 1978).

[dd]Pansolli et al. (1976).

[ee]Chung et al. (1987).

TC: Thin channel cell; HFER: hollow fiber enzyme reactor; CAMFER: capillary membrane fixed enzyme reactor; DEUSMR: dead-end unstirred membrane reactor; GEMR: gelled enzyme membrane reactor; FE: fiber-entrapped; DHFBR: dual hollow fiber bioreactor; MSR: membrane sandwich reactor; ONPG: O-nitrophenyl-β-D-galactopyranoside.

substrates. Recently, several reports on the use of enzyme membrane reactors for starch hydrolysis have been published (Table 4.7). Most of them have used dead-end stirred cells (Butterworth et al., 1970; Marshall and Whelan, 1971; Fisher, 1973; Azhar and Hamdy, 1981; Darnoko et al., 1989) and some of them have used membrane recycle reactor (Engasser et al., 1980a, b; Larrson and Mattiasson, 1984; Iwasaki et al., 1989; Sims and Cheryan, 1989), and the others have used hollow fiber enzyme reactors (Engasser et al., 1980a, b) or tubular reactors (Closset et al., 1974; Tachauer et al., 1974; Madgavkar et al., 1977) or microporous membrane reactors (Hausser and Goldberg, 1983; Peterson, 1989a, b).

Butterworth et al., (1970) studied the conversion of starch by α-amylase and glucoamylase. They obtained a quasi–steady-state operation in which the total carbohydrate of the permeate equalled that of the feed after about 70 hr of operation. However, the flux declined steadily during operation, owing to the buildup of solids and their compaction on the membrane. Nevertheless, the productivity was superior to a comparable batch operation. An empirical correlation for flux decline has been given by Azhar and Hamdy (1981) for the degradation of starch by β-amylase. In this case, the reactor was coupled to a thin-channel system. They reported that a thin-channel system showed a superior performance compared with a stirred cell system for immobilizing β-amylase during hydrolysis of sweet potato slurry. Marshall and Whelan (1971) suggested that a mixture of α-amylase and glucoamylase would give better conversions to glucose, but no data were presented in their article.

Closset et al. (1974) attempted to develop a "tubular membrane reactor." The membrane was an ultrafiltration membrane cast onto a rigid porous support tube. The reaction and separation were to occur during passage of the starch-enzyme mixture through the tube. They had some problems with concentration polarization and starch aging during operation. Tachauer et al. (1974) used a mixture of α- and β-amylases in a tubular membrane, and they obtained better results. Still better results were obtained with glucoamylase or amyloglucosidase (Madgavkar et al., 1977), which prevented the formation of limit dextrins and their buildup as a gel layer on the membrane. These studies clearly point out the importance of careful selection of the enzyme.

Egnasser et al. (1980a) used a hollow fiber enzyme reactor for maltose and soluble starch hydrolysis using amyloglucosidase. They reported that the performance of the reactor may be affected by diffusional limitations and glucose inhibition. Larsson and Mattiasson (1984) recently used a membrane recycle reactor for continuous conversion of starch by α-amylase and glucoamylase, using a combination of an aqueous two-phase system and an ultrafiltration unit. They reported that there was no tendency for fouling that could be seen during a 100-hr period of operation. Hausser and Goldberg (1983) used microporous plastic sheet for immobilized α-amylase and glucoamylase in continuous production of high conversion maltose-containing corn syrups. They reported that the use of Amerace's continuous immobilized dual enzyme reactor system for the production of maltose-containing corn syrups has been shown to be a viable process.

A continuous UF reactor for hydrolysis of cassava starch gave interesting results. Darnoko et al. (1989) studied the hydrolysis of extruded cassava starch with glucoamylase in a dead-end stirred cell reactor. Starch conversions were 64% at 22°C and 97% at 55°C. Productivity of UF reactor was 10–11 times better than a batch reactor over a 24-hr operating period.

The hydrolysis of corn starch in a CSTMR has been studied by Sims and Cheryan (1989) with a view to alleviating the current bottleneck in the corn-refining industry. The liquefaction step presently takes about 2 hr, while the saccharification step may take as much as 48 hr. Iwasaki et al. (1989) used a CSTR-UF membrane reactor for starch hydrolysis using β-amylase and pull-ulanase. They reported that the CSTR-UF system was found to have high maltose productivity with relatively short residence time, being easily controlled by transmembrane pressure.

Cellulose hydrolysis   The cellulose in biomass can serve as a valuable energy resource for fermentation. Cellulose has first to be hydrolyzed to glucose by enzymatic or chemical means. Enzymatic methods, using the mixture of enzymes collectively known at cellulase, have several benefits over chemical methods. One of the first studies on the breakdown of cellulose to lower molecular weight sugars in a dead-end stirred cell reactor was by Ghose and Kostick (1970). The system was run semicontinuously for a period of 10 days, resulting in an overall 71% conversion of cellulose to glucose. Hahn-Hagerdal et al. (1980) used a dead-end stirred cell for studying the conversion of cellulose to ethanol with immobilized enzymes (e.g., cellulases, β-glucosidase) and microorganisms (e.g., *S. cerevisiae*). They concluded that bioconversion has to be optimized with respect to productivity as well as product concentration. When taking membrane technology into account, the optimization may concentrate more on productivity since the resulting dilute product stream can be upgraded in a subsequent membrane operation. Another study of coimmobilized yeast with cellulolytic enzymes in dead-end stirred cell reactor for conversion of filter paper to ethanol was reported by Hartmeier (1980). The degradation of filter paper is done in an ultrafiltration device with soluble cellulase followed by treatment in a packed bed with the yeast/cellulase coimmolbilized leading to ethanol as end product.

Henley et al. (1980) showed that a membrane reactor gave better conversion (87–92%) than a simple CSTR (67–67%) for the hydrolysis of Solka-Floc and other cellulosic materials by cellulase. By replacing the dead-end cell with a hollow fiber cartridge, performance is further improved. They suggested to minimize damage to the hollow fibers and to avoid plugging problems, the raw material (cellulose) would have to be ground or milled to a particale size of 50–100 μm or less. This has been recommended for any membrane application (Cheryan, 1986).

Tjerneld et al. (1985) have studied enzyme recycle in cellulose hydrolysis by combined use of aqueous two-phase systems and ultrafiltration. Cellulose (Solka Floc) was semicontinuously enzymatically hydrolyzed over a period of 1200 hr in an aqueous two-phase system based on crude dextran and polyethylene glycol. The sugars were recovered from the top phase after phase separation using an ultrafiltration membrane unit. Enzyme and phase forming polymers were recycled. Cellulose was added intermittently to keep a constant concentration of about 80 g/L in the phase system. The degree of conversion was 94% of substrate. The enzymes were added at the beginning of the hydrolysis, and no further additions of enzymes were made.

Ohlson et al. (1984a, b) and Frennesson et al. (1985) have studied the hydrolysis of sodium hydroxide–pretreated sallow in a dead-end stirred cell reactor using cellulase and cellobiase. The degree of conversion was improved from 40% in a batch hydrolysis to 95% in an ultrafiltration membrane reactor within 20 hr, and the initial hydrolysis rate was increased up to seven times. The amount of reducing sugars produced was 25.7 g/g enzyme compared to 4.7 g/g enzyme in a batch hydrolysis.

Kinoshita et al. (1986) have studied the continuous hydrolysis of cellulose by cellulases of *Sporotrichum cellulphilum* in an ultrafilter membrane reactor (DESC type). Glucose accounted for 55–60% of the product. The reactor was five times more efficient per unit weight of enzyme used than the batch reaction. In the long-time hydrolysis at 30–60°C, the production of cumulative total sugar was highest at 50°C. Other examples of cellulose hydrolysis in membrane bioreactors is shown in Table 4.7.

Recently, Gonzalez et al. (1989) have studied the hydrolysis of pretreated wheat straw by cellulose in an upflow packed-bed reactor. The reactor outlet was connected to an ultrafiltration unit (hollow fiber), which allowed the continuous removal of the sugars produced and the recycling of the enzyme solution. The enzyme load, residence time, and recycling effects on reaction performance have been studied. They obtained about 90% yield of the potential sugars. The productivity of the column reactor with enzyme recycling was increased twofold compared to the batch process. They reported that in order to increase sugar concentration in the product, enzymatic hydrolysis of successive substrate loads should be used. Sugar concentration in the end solution was limited by product inhibition, reaching a final value of around 42 g/L.

Oligosaccharide hydrolysis Considerable research has been done, using enzyme membrane reactor for sugar hydrolysis, such as raffinose, cellobiose, maltose, lactose, and sucrose (Table 4.8).

Raffinose, the sugar found in beet, has been hydrolyzed in a hollow fiber enzyme reactor using α-galactosidase and invertase (Smiley et al., 1976; Korus and Olson,1977b, 1978; Silman et al., 1980). Korus and Olson (1977b, 1978) reported that using hollow fiber enzyme reactor nearly complete substrate conversion was possible with no detectable enzyme leakage through the membrane. Enzyme stabilities in polysulfone hollow fibers which have been preconditioned with bovine albumin approach the stabilities of the free enzymes (Korus and Olson, 1977b).

Maltose, known as malt sugar, has been hydrolyzed by amyloglucosidase in membrane bioreactors (Swanson et al., 1977; Greco et al., 1979a; Furusaki and Asai, 1983; Engasser et al., 1980a; Gianfreda and Greco, 1981b). Engasser et al. (1980b) reported that the overall productivity of a hollow fiber bioreactor depends on the reactor hydrodynamics, its external and internal transport capacities and the kinetics of the enzyme reactions. Greco et al. (1979a) have studied the activity and stability of maltase in an unstirred ultrafiltration membrane reactor. Furusaki and Asai (1983) have reported a new immobilization technique of amyloglucosidase for maltose hydrolysis. It involves immobilizing enzyme on a porous polytetra-fluoroethylene membrane with a nonporous polyurethane coat by the use of an electrostatic force, i.e., the Coulomb force. Recently, Bardeletti and Coulet (1987) have reported a sequential bienzyme system asymmetrically immobilized on a permeable membrane. Glucoamylase was bound to the one side of the surface of a porous membrane, and glucose oxidase was bound to the other side. Maltose was hydrolyzed into glucose by the first enzyme and the glucose converted to gluconic acid and the hydrogen peroxide by the second enzyme. They concluded that, compared with an asymmetric monoenzyme system, a strong amplification (80-fold) of the concentration effect occurs.

Lactose, the sugar found in milk and whey, and its corresponding hydrolase, lactase (β-galactosidase), have been the subject of an extensive research in the past decade. Partly, this is because of the development of the enzyme immobilization technique which gives new and interesting possibilities for the utilization of many food compounds (Gekas and Lopez-Levia, 1985).

Maximum hydrolysis is usually achieved in a batch system where β-galactosidase is in free solution. However, batch methods are expensive since the enzyme cannot be recovered (Huffman-Reichenbach and Harper, 1985). Ultra-filtration membranes can be employed to separate the hydrolyzed permeate from the enzyme-rich retentate in a batch or continuous enzyme membrane reactors. Much of the lactose hydrolysis with enzyme membrane reactors has recirculated β-galactosidase and lactose in the ultrafiltration mode (Roger et al., 1976, 1977; Norman et al., 1978; Kowalewska et al., 1978; Huffman-Reichenbach and Harper, 1985; Kim and Chang, 1983a, b; Park et al., 1985). An alternative technique using hollow fiber membranes is to entrap β-galactosidase in the shell side of a hollow fiber and then operate the reactor in the recycle mode (Lewis and Middleman, 1974; Waterland et al., 1975; Korus and Olson, 1978; Kohlwey and Cheryan, 1981;

Breslau, 1981; Huffman-Reichenbach and Harper, 1982, 1985; Prenosil andHeeiger, 1985). Huffman-Reichenbach and Harper (1982, 1985) have tested β-galactosidase retention and activity in hollow fiber enzyme reactors under different operational conditions (Fig. 4.16). They reported that the method of loading β-galactosidase in a hollow fiber membrane unit influenced both enzyme retention and loss of activity during operation. Operated in the normal ultrafiltration mode with enzyme in the recirculating retentate, there was about 5% leakage across the membrane and approximately 50% loss of enzymatic activity after 2 hr. Loading enzyme in the sponge layer of the membrane markedly increased leakage across the membrane, but there was minimal loss of activity (<10%) of the enzyme retained. Average leakage was 30, 40, and 7% when the membrane reactors were run in the recycle, backflush, and static modes, respectively. Leakage of enzymes across the membrane was not related to a change in molecular weight (100,000 daltons) of the enzyme nor to defects in the membrane. Pretreatment of hollow fibers with a protein (e.g., bovine serum albumin, BSA) was necessary to minimize enzyme inactivation (Kohlwey and Cheryan, 1981; Huffman-Reichenbach and Harper, 1985). In general, it is important to consider the mode of reactor operation and the type of solute, solvent, and membrane, since enzyme retention characteristics depend in particular on the diffusion rate and convective flux, fluid shear stresses, axial flow rate, and enzyme characteristics (concentration, shape, size, deformability, etc.) (Kleinstreuer and Poweigha, 1984).

Pastore et al. (1974, 1976) reported that β-galactosidase entrapped into cellulose acetate fibers is used in a successful commercial operation for hydrolyzing lactose in milk. The processed milk is claimed to retain its original organoleptic properties and exhibits only a slightly sweeter taste.

Lopez-Leiva and Gekas (1986) have reported a new module concept for the hydrolysis of lactose in whole whey. The module uses a commercially available microporous sheet, with immobilized β-galactosidase, in cross-flow configuration. In this case the microporous sheet itself acts as a microfilter retarding the detrimental effect of fat upon the immobilized enzyme. They also showed that a better effect is obtained if a UF-membrane is used on top of the microporous sheet. In this case an almost constant transmembrane flux was obtained during a period of 20 hr.

Recently, Peterson et al. (1989a) studied the hydrolysis of lactose by immobilized β-galactosidase in a continuous flow capillary bed reactor, manufactured by the Amerace Corporation, operating at 30°C (Fig. 4.17). Lactose conversions ranging from 24% to greater than 99% were achieved at reactor space times ranging from 0.06 to 6.3 min. These conversion data were successfully modeled in terms of a plug flow reactor model and a form of Michaelis-Menten kinetics, which included competitive inhibition by both the alpha and beta forms of galactose. Peterson et al. (1989b) also studied the effect of temperature on the reactor and concluded that the thermal deactivation of immobilized β-galactosi-

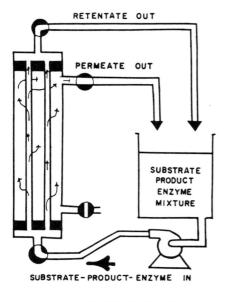

(a) ULTRAFILTRATION MODE

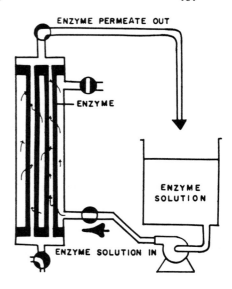

(b) BACKFLUSH MODE

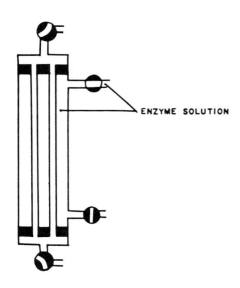

(c) STATIC MODE

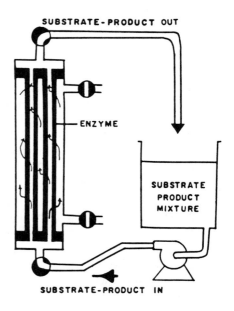

(d) RECYCLE MODE

Figure 4.16 Schematic of hollow fiber membrane reactors showing four modes of operation. (*From Huffman-Reichenbach and Harper, 1982.*)

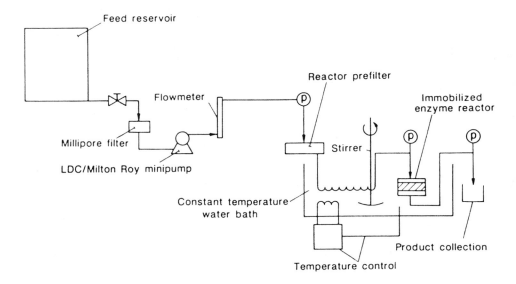

Figure 4.17 Schematic diagram of continuous-flow capillary bed reactor. (*From Peterson et al., 1989.*)

dase behaved as a first-order decay process. Based on estimated thermal deactivation rate constants, at an operating temperature of 40°C, 10% of the enzyme activity would be lost in one year.

The hydrolysis of sucrose by invertase in different types of membrane bioreactors have been studied (Table 4.8). Bowski et al. (1972) and Bowski and Kyer (1974) studied the mathematical modeling of sucrose hydrolysis by invertase in a continuous flow reactor utilizing a thin-channel ultrafiltration system. They also studied the effects of ultrafiltration on the activity of invertase and observed that enzyme activity decreases with continuous and repeated use, possibly because of the alterations in the tertiary structure of enzyme but not necessarily because of the effects of shear, enzyme adsorption, and concentration polarization. Silman et al. (1980) and Korus and Olson (1977a, b; 1978) have studied the hydrolysis of raffinose in a hollow fiber enzyme reactor using α-galactosidase and invertase. Silman et al. (1980) reported that the hollow fiber reactor is well suited for the conduct of multienzyme conversions. Nakajima et al. (1988) reported a new module concept for the hydrolysis of sucrose. Invertase was immobilized physically within the inner surface of an asymmetric ceramic membrane by filtering enzyme solution from the side of the porous-support layer.Substrate flow from the skin-layer side was favorable for preventing plugging and also for highconversion. Enzyme activity yields of up to 46% were obtained.

High-Fructose Syrup Production The use of immobilized glucose isomerase to prepare high-fructose syrup has received a great deal of attention over the past several years. An immobilization technique that offers several advantages for food systems is the use of hollow fibers. Korus and Olson (1977a b) have studied the immobilization of glucose isomerase in several types of asymmetric hollow fibers (Table 4.8). They concluded that the stability of glucose isomerase in a hollow fiber reactor approached that of the free enzyme, although polysulfone fibers had to be preconditioned with an inert protein. Hoehn et al. (1983) have produced high-fructose syrup by hydrolysing inulin contained in Jerusalem artichoke tubers. The products of the hydrolysis, glucose and fructose, were separated from the high molecular weight substance by ultrafiltration followed by evaporation. They were mixed with a 42% (w/v) fructose corn syrup to form the commercially desirable 55% (w/v) fructose syrup. Recently, Chung et al. (1987) reported a new process using a dual hollow fiber bioreactor (DHFBR) system for whole cell enzyme immobilization of *Streptomyces griseus* with glucose isomerase activity. The productivity of fructose by DHFBR was increased 12-fold over that of the batch system.

Lipid Conversions

Modifications of starch and protein by enzymatic means are well established as industrial processes. Unfortunately, the use of enzymes to modify lipids is still in a developing state. Several reports have published on the hydrolysis of fat by lipase in free (Ishida et al., 1981; Kobayashi et al., 1983; Linfield et al., 1984), immobilized (Lieberman and Ollis, 1975; Kilara et al., 1977; Lavayre and Bratti, 1982; Kimura et al., 1983), or cell-bound states (Bell et al., 1981). All of these studies were carried out in conventional emulsion systems. The conventional emulsion systems have certain drawbacks with respect to maintaining fine emulsion, product separation, continuous operation, control of water/glycerol concentration, and restricted access to the immobilized lipase (Hoq et al., 1984). Hoq et al., (1984, 1985a, b, 1986) have developed a microporous hydrophobic membrane bioreactor with which continuous production of glycerides by lipase (Hoq et al., 1984) and continuous hydrolysis of olive oil by lipase (Hoq et al., 1985a, b, 1986; Yamane et al., 1987) were successfully carried out. In the case of continuous production of glycerides (Table 4.9), a plate-type dialyzer was used as the membrane bioreactor at 40°C. The glycerol solution, containing lipase and water, was supplied continuously to one side of a sheet of microporous polypropylene membrane (strongly hydrophobic), and the effluent was recycled while undiluted liquid fatty acid (oleic or linoleic) was fed continuously to the

Table 4.9 Enzymatic Conversion of Lipids in Membrane Bioreactors

Enzyme	Substrate	Reactor type	Reference
1. **Oil hydrolysis:**			
Lipase	olive oil	MMR	a
Lipase	olive oil	MMR, GEMR	b
Lipase	soybean oil	HHFMR	c
2. **Synthesis of glycerides:**			
Lipase	glycerol and fatty acids	MMR	d

[a]Hoq et al. (1985a, b, 1986); Yamane et al. (1987).
[b]Drioli (1986).
[c]Pronk et al. (1988).
[d]Hoq et al. (1984), vander Padt and van't Riet (1989).
HHFMR: Hydrophilic hollow fiber membrane reactor.

opposite side of the membrane and came in contact with a glycerol-water-lipase solution to cause the reaction. The product, glycerides, was obtained at the outlet in a pure state. Highest conversion was 90% when the water content of the glycerol solution was 3–4%. In the case of continuous hydrolysis of olive oil by lipase in a microporous hydrophobic membrane bioreactor, olive oil and buffer solution, fed continuously through two compartments partitioned by membrane, caused reaction at the interface of lipase-adsorbed membrane and buffer solution. Fatty acids were obtained in a single phase without being mixed with components of other phases. At all mean residence times, countercurrent flow mode was superior to cocurrent one. The hydrolysis seemed to be limited by diffusion of fat or fatty acids through the micropores of the membrane at higher interfacial enzyme concentrations. Satisfactory performance of the membrane bioreactor was obtained in a long-term continuous operation which lasted for 24 days by feeding buffer-glycerol (18%) solution over the adsorbed lipase. The operational half-life of the adsorbed enzyme was 15 days at 40°C (Hoq et al., 1985). Pronk et al. (1988) have developed a hydrophilic hollow fiber membrane for soybean oil hydrolysis using lipase. They reported that the immobilized lipase exhibited a high stability; the half-life time was 43 days at 30°C. Furthermore, it proved that kinetic studies can be carried out with this system, operated in a batch or continuous mode.

L-Amino Acid Production

Amino acids have been produced by single or multienzyme reaction systems with a simultaneous coenzyme regeneration, using membrane bioreactors (Table 4.10)

Table 4.10 Amino Acid Production in Enzyme Membrane Bioreactors

Enzyme	Substrate	Membrane type	Reference
1. L-Alanine:			
Acylase	N-acetyl-DL-alanine	MRR	a
L-Alanine dehydrogenase	pyruvic acid	MRR	a
Aspartase β-decaroxylase	fumaric acid	MRR	b
Lactate dehydrogenase/ alanine dehydrogenase	D-L-lactate pyruvate, ammonium sufate	MRR	c
2. L-Aspartic Acid:			
Aspartase	ammonium formate,	MRR	a,b,d
Aspartase	ammonium formate	DESC	e
3. L-Leucine:			
Leucine dehydrogenase/ formate dehydrogenase	α-ketoiso caproate, ammonium formate	MRR	a,f
Hydroxy isocaproic acid dehydrogenase/leucine dehydrogenase	hydroxyisocaproate, ketoisocaproate, ammonium sulfate	MRR	c,g
4. L-Tertiary leucine:			
Formate dehydrogenase/ leucine dehydrogenase	dimethylpyruvate, ammonium formate	MRR	c,h
5. L-Methionine:			
Acylase	N-acetyl-DL-methionine	MRR	a,i
Hydroxy isocaproic acid dehydrogenase/leucine dehydrogenase	hydroxymethionine, ketomethionine, ammonium sulfate	MRR	c
6. L-Phenylalanine:			
Acylase	N-acetyl-DL-phenylalanine	MRR	j
Chymotrypsin	D,L-phenylalanine methyl ester	MRR	k
Phenylalnine dehydrogenase/ formate dehydrogenase	phenylpyruvate, ammonium formate	MRR	c
7. L-Tryptophan:			
Acylase	N-acetyl-DL-tryptophane	MRR	a
8. L-Tyrosine:			
Chymotrypsin	N-acetyl-L-tyrosine	MRR	k,l
9. L-Valine:			
Acylase	N-acetyl-DL-valine	MRR	a

[a]Leuchtenberger et al. (1984a, b), Wandrey and Wichmann (1987).
[b]Jandel et al. (1982); Wandrey and Wichmann (1987).
[c]Schmidt et al. (1986).
[d]Koyama et al. (1987).
[e]Lee and Hong (1988).
[f]Ohshima et al. (1985).
[g]Bossow and Wandrey (1987).
[h]Wandrey and Wichmann (1987).
[i]Wandrey and Flaschel (1979).
[j]Leuchtenberger et al. (1984).
[k]Jones and Beck (1976); Matson and Quinn (1986).
[l]Lopez et al. (1988)

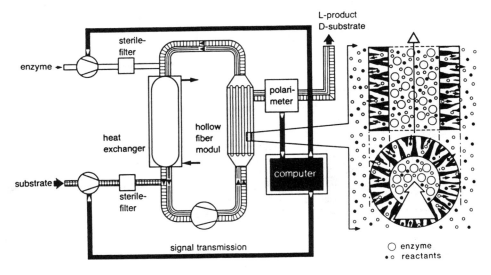

Figure 4.18 Flow diagram of an enzyme membrane reactor controlled by a process computer and used for L-amino acid production. (*From Wandrey and Wichmann, 1987.*)

(Figs. 4.12, 4.13, and 4.18) The first industrial application of an oligoenzyme system with cofactor regeneration in membrane reactors was the production of L-amino acids from cheap keto acids, a process developed by Wandrey's group and Kula's group at KFA, Julich, FRG and Braunschweig-Stockheim, respectively, together with Degussa AG (Wandrey and Flaschel, 1979; Flaschel and Wandrey, 1983; Hummel et al., 1984; Wandrey, 1984; Wandrey and Wichmann, 1987). To date about 200 tons per year of L-amino acid have been produced by racemate resolution in membrane bioreactors (Leuchtenberger et al., 1984a, b; Wandrey and Wichmann, 1987).

Recently, Lee and Hong (1988) have studied the production of aspartic acid by aspartase reaction on ammonium fumarate using a membrane reactor (DESC type) coupled with electrophoresis. They concluded that the aspartate concentration in the permeate stream was 20% higher than that in the reaction solution and the steady-state conversion was 60%. Instead of aspartate, aspartic acid can be recovered directly from the permeate stream by controlling the circulation of buffer electrolyte in the anode compartment. They also suggested that the electrophoresis coupled membrane reactor has a potential application for the enzyme reaction systems that produce electrically charged products and if the reaction is governed by product inhibition.

Other Enzyme Processes

The enzyme membrane bioreactor systems have been used for several other applications, as indicated in Table 4.11. These include production of acetaldehyde, 6-amino-penicillinic acid, 3-fluoro-L-alanine, D-galacturonic acid, gluconic acid, glucose-6-phosphate, malic acid, mannitol, and sorbitol and hydrolysis of *p*-nitrophenylphosphate and urea, among others.

3-Fluoro-L-alanine (antibacterial substance) was selectively and continuously produced from 3-fluoropyruvate and ammonium formate in an enzyme membrane reactor by the multienzyme reaction system of alanine dehydrogenase and formate dehydrogenase with a simultaneous coenzyme regeneration (Ohshima et al., 1989). The average conversion and space-time yield were 73% and 75 g/L day, respectively.

Kulbe et al. (1987a) have studied the production of D-galacturonic acid (used for enzymatic synthesis of L-ascorbic acid) from pectin with pectin-methylesterase and polygalacturonase in two-stage tubular enzyme membrane process by removing the product via electrodialysis. After 10 hours of operation, the concentration of D-galacturonic acid had increased threefold in the concentrate cell of the electrodialyser.

Kulbe and coworkers (1987a, b, c, 1988) have applied membrane bioreactors for carbohydrate conversion processes. Carbohydrate can be converted simultaneously to the higher valued products, gluconic acid and mannitol, by the coupled enzyme system of glucose dehydrogenase (GDH) and mannitol dehydrogenase (MDH). Process techniques for the mannitol-gluconic acid process have been developed using flat membrane and CSTR and hollow fiber membrane reactors (Kulbe and Chmiel, 1988; Kulbe et al., 1987a, b, c; Howaldt et al., 1988). They reported that at high conversion, higher space-time yields were attainable in the HFER than in the CSTR. This may be due to mannitol-inhibited reaction. Overall, for the GDH/MDH system, there was good agreement between simulation and experiment in a CSTR and in a tube reactor with recycle. Recently, Kulbe and coworkers have developed an ionic ultrafiltration membrane system for retainment of native cofactors, and they have applied this concept for conversion of glucose-fructose mixtures to mannitol and gluconic acid. They obtained about 90% conversion with 99% coenzyme rejection.

CONCLUSIONS

In this chapter we have covered several different membrane bioreactor concepts and their applications in food processing. In general, membrane bioreactors have been shown (1) to improve the efficiency of enzyme-catalyzed bioconversions by simultaneously combining the two unit operation of enzymatic conversion product

Table 4.11 Other Enzyme Reactions in Membrane Bioreactors

Enzyme	Substrate	Membrane type	Reference
1. Acetaldehyde production:			
Alcohol dehydrogenase	ethanol	HFER	a
2. Aminohydrolysis reaction:			
Allosteric enzyme	deoxycytidine-5-monophosphate	GEMR	b
3. 6-Aminopenicillanic acid production:			
Penicillin acylase	benzyl penicillin	DEUSMR	c
Penicillin acylase	G-penicillin-K	GEMR	d
Penicillin acylase	benzyl penicillin	LEMR	e
4. Deacetylation of 7-amino cephalosporanic acid (7-ACA)			
Cephalosporin acetylesterase	7-ACA	DESC	f
5. 3-Fluoro-L-alanine production:			
Alanine dehydrogenase/formate dehydrogenase	3-fluoropyruvate, ammonium formate	DESC	g
6. β-Galactose oxidation:			
β-Galactose dehydrogenase	β-glactose	DEUSMR	h
7. D-Galacturonic acid production:			
Pectin-methyl-esterase/polygalacturonase	pectin	MRR, Tubular	i
8. Gluconic acid production:			
Glucose dehydrogenase/mannitol dehydrogenase	glucose/fructose	MRR	j
Glucose dehydrogenase/mannitol dehydrogenase	glucose/fructose	MRR, HFER	k
Glucose oxidase	glucose	MMR	l
9. Glucose-6-phosphate production:			
Glucokinase/acetate kinase	glucose/ATP/acetyl phosphate	HFER	m
Glucokinase/acetate kinase	glucose/ATP acetyl phosphate	MRR DESC	n
10. Glucose oxidation:			
Glucose oxidase	glucose	MRR	o
Glucose oxidase	glucose	paralleled plates	p
Glucose oxidase/catalase	glucose	DHFBR GEMR	r
11. γ-Glutamyl-cystein production:			
γ-Glutamyl-cysteine synthetase	Glutamine/cysteine/ATP	DESC	n
12. Malic acid:			
a. *Production:* Fumarase	fumaric acid	MRR	s
b. *Decarboxylation:* Malic enzyme	malic acid	ECMR	t

Table 4.11   *(continued)*

13. **Mannitol and gluconic acid production:**

Invertase, glucose dehydrogenase/mannitol dehydrogenase	starch, molasses, sucrose	MRR	u
Glucose dehydrogenase/mannitol	glucose-fructose	MRR	u,v
Glucose dehydrogenase/mannitol dehydrogenase	glucose-fructose	MRR, HFER	k

14. **p-Nitrophenylphosphate hydrolysis (PNPP)**

Alkaline phosphatase	PNPP	HFER	w
Acid phosphatase	PNPP	DEUSMER, MMR	h,s,x

16. **Sorbitol and gluconic acid production:**

Glucose dehydrogenase/aldose reductase	glucose	MRR	z

17. **Urea hydrolysis:**

Urease	urea	BHFER	aa
Urease	urea	DEUSMER	bb
Urease	urea	MMR	cc
Urease	urea	TC	dd

---

[a]Chambers et al. (1981).
[b]Iorio et al. (1981).
[c]Veronese et al (1984), Greco et al. (1983).
[d]Jancsik et al. (1982).
[e]Lopez et al. (1988).
[f]Abbott et al. (1976)
[g]Ohshima et al. (1989).
[h]Gianfreda and Greco (1981a, b); Greco et al. (1981a, b).
[i]Kulbe and Chmiel (1988), Kulbe et al. (1987a), Heinzler et al. (1987).
[j]Kulbe et al. (1987b), Kulbe and Chmiel (1988), Kulbe et al. (1987c).
[k]Kulbe and Chmiel (1988), Howaldt et al. (1988).
[l]Bardeletti and Coulet (1987).
[m]Ishikawa et al. (1989a, b)
[n]Berke et al. (1988).
[o]Bardeletti et al. (1985) (Bardeletti and Coulet, (1987), Ishimori et al. (1981).
[p]Hashimoto et al. (1983).
[q]Chang et al. (1987).
[r]Kozhukharova et al. (1988).
[s]Leuchtenberger et al. (1984a, b).
[t]Iorio et al. (1984, 1985), Hayakawa et al. (1985).
[u]Kulbe et al. (1987b), Kulbe and Chmiel (1988).
[v]Kulbe et al. (1987c).
[w]Davis (1974)
[x]Greco and Gianfreda (1981), Greco et al. (1979a, b; 1981a).
[z]Kulbe and Chmiel (1988). [aa]Kawakami et al. (1980).
[bb]Greco et al. (1979a, b); Scardi et al. (1980).
[cc]Staude et al. (1982).
[dd]Gacesa et al. (1983).
LEMR: Laminated enzyme membrane reactor; ECMR: enzyme capillary membrane reactor.

separation, (2) to increase product yields in bioconversions that are inhibited by product, (3) to be easily scaled up and in general to apply the benefit of progress in the field of membrane operations. Furthermore, such bioreactors can be used to maximize product yields, to facilitate multiphase processing, and to permit reuse of expensive cofactors. However, there are some potential limitations. In a membrane recycle bioreactor operating essentially as a CSTR, the outlet stream will have the same concentration as the reaction mixture, which means the system will be ideally operating with a high product concentration at all times. Thus, the CSTR-type recycle reactor is more suited for substrate-inhibited reactions than product-inhibited reactions, when the conversion is high. The key factor in successful long-term operation of membrane bioreactors is the performance of the membrane module. It is important not only to have the appropriate physical configuration of the module, but also the correct chemical nature of the membrane to minimize the interactions between the membrane and components in the reaction mixture. Sterilization of the membrane is also critical to its application in continuous bioreactors. At the very least, it should be compatible with common chemical sterilants. These problems, however, are being seriously considered by some progressive membrane manufacturers.

Among the applications that have been studied are conversion of substrates important for the food and pharmaceutical industry. Production of L-amino acids using membrane bioreactors has been in commercial use since 1981. Membrane bioreactors also have potential applications in the production of agricultural chemicals, flavor and fragrances, specialty organic chemicals, and pharmaceuticals.

ACKNOWLEDGMENTS

Some of the work described in this chapter was supported by funds provided by the Illinois Agricultural Experimental Station, the University of Illinois and the State of Illinois Value Added Program, administered through the Department of Food Science at the University of Illinois. The secretarial assistance of Ms. Becky Jackson is greatly appreciated.

REFERENCES

Abbott, B. J., Cerimele, B., and Fukuda, D. S. 1976. Immobilization of a cephalosporin acetylesterase by containment within an ultrafiltration device. *Biotech. Bioeng.* **18**: 1033–1042.

Adu-Amankwa, B., Constantinides, A., and Vieth, W. R. 1981. Enzymatic modification of vegetable protein: Immobilization of *Penicillium duponti* enzyme on reconstituted collagen and the use of the immobilized-enzyme complex for solubilizing vegetable protein in a recycle reactor. *Biotech. Bioeng.* **23:** 2609–2627.

Alfani, F., Cantarella, M., Erto, L., and Scordi, V. 1983. Enzymatic saccharification of cellulose and cellulosic materials. In *Energy from Biomass.* A. Strub, P. Chartier, and G. Schleser (Ed.), p. 1000. Applied Science Publishers, New York.

Alfani, F., Albanesi, D., Canarella, M., and Scardi, V. 1982a. Effect of temperature and shear on the activity of acid phosphate in a membrane reactor. *Enzyme Microb. Technol.* **4:** 181–184.

Alfani, A., Albanesi, D., Cantarella, M., Scardi, V., and Vetromile, A. 1982b. Kinetics of enzymatic saccharification of cellulose in a flat-membrane reactor. *Biomass* **2:** 245.

Amerace Corporation. 1980. APH pilot reactor for hydrolysis of whey permeate. New Product Bull. 02283.

Azhar, A., and Hamdy, M. K. 1981. Alcohol fermentation of sweet potato. Membrane reactor in enzymatic hydrolysis. *Biotech. Bioeng.* **23:** 1297–1307.

Bardeletti, G., and Coulet, P. R. 1987. Sequential bienzyme system asymmetrically immobilized on a permeable membrane: Final product compartmentalization. *Enzyme Microb. Technol.* **9:** 652–657.

Bardeletti, G., Maisterrena, B., and Coulet, P. R. 1985. Boundary layer effect on product flux-splitting in an immobilized enzyme membrane system. *J. Membrane Sci.* **24:** 285–296.

Belfort, G. 1989. Membranes and bioreactors: A technical challenge in biotechnology. *Biotechnol. Bioeng.* **33:** 1047–1066.

Bell, G., Todd, J. R., Blain, J.A., Patterson, J. D. E., and Shaw, C. E. L. 1981. Hydrolysis of triglyceride by solid phase lipolytic enzymes of *Rhizopus arrhizus* in continuous reactor system. *Biotech. Bioeng.* **23:** 1703–1719.

Berke, W., Schiz, H.-J. Wandrey, C., Morr, M. Denda, G. and Kula, M.-R. 1988. Continuous regeneration of ATP in enzyme membrane reactor for enzymatic syntheses. *Biotech. Bioeng.* **32:** 130–139.

Bhumiratana, C. G., Hill, C.G., and Amundson, C. H. 1977. Enzymatic solubilization of fish protein concentrate in membrane reactors. *J. Food Sci.* **42:** 1016–1021.

Blatt, W. F. 1971. Membrane partition chromatography: A tool for fractionation of protein mixtures. *J. Agric. Food Chem.* **19:** 589.

Bossow, B., and Wandrey, C. 1987. Continuous enzymatically catalyzed production of L-leucine from the corresponding racemic hydroxy acid. *Annals N.Y. Acad. Sci.* **506:** 325–336.

Bowski, L., and Ryu, D. Y. 1974. Determination of invertase activity during ultrafiltration. *Biotech. Bioeng.* **16:** 697–699.

Bowski, L., Shah, P.M., Ryu, D. Y., and Vieth, W. R. 1972. Process simulation of sucrose hydrolysis of invertase in a continuous flow stirred tank/ultrafiltration reaction system. *Biotech. Bioeng. Symp.* **3:** 229–239.

Boudrant, J., and Cheftel, C. 1973. Hydrolyse continue du saccharose par de I-invertase immobisee sous forms soluble. *Biochimie* **55:** 413–420.

Boudrant, J., and Cheftel,C. 1976. Continuous proteolysis with a stabilized protease. II. Continuous experiments. *Biotech. Bioeng.* **18:** 1735–1749.

Breslau, B. R. 1981. U.S. Patent 4,266,026.

Breslau, B. R., and Kilcullen, B. M. 1978. Hollow fiber enzymatic reactions: An engineering approach. In *Enzyme Engineering,* Vol. 3, E.K. Pye and H. H. Weetall (Ed.)., p. 179. Plenum Press, New York.

Bressollier, P., Petit, J. M., and Julien, R. 1988. Enzyme hydrolysis of plasma proteins in a CSTR ultrafiltration reactor: Performance and modeling. *Biotechnol. Bioeng.* **31:** 650–658.

Butterworth, T. A., Wang, D. I. C., and Sinskey, A. J. 1970. Application of ultrafiltration for enzyme retention during continuous enzymatic reaction. *Biotech. Bioeng.* **12:** 615–631.

Chambers, R. P., Cohen, W., and Baricos, W. 1976. In *Methods in Enzymology.* K. Mosbash (Ed.), p. 291. Academic Press, New York.

Chambers, R. P., Ford, J. R., Allender, J. W., Baricos, W. H., and Cohen, W. 1981. In *Enzyme Engineering.* E. K. Pye and L. B. Wingard (Ed.), v. 2, p. 195. Plenum, New York.

Chang, H. N. 1987. Membrane bioreactors: Engineering aspects. *Biotechnol. Adv.* **5:** 129–145.

Chang, H. N., Kyung, Y-S, and Chung, B. H. 1987. Glucose oxidation in a dual hollow fiber bioreactors with a silicone tube oxygenator. *Biotech. Bioeng.* **29:** 552–557.

Cheetham, P. S. J. 1985. Principles of industrial enzymology. Basis of utilization of soluble and immobilized enzymes in industrial processes. In *Handbook of Enzyme Biotechnology.* A Wiseman (Ed.), p. 54. Halsted Press: A Division of John Wiley and Sons, New York.

Cheftel, C. 1972. Continuous enzymic solubilization of FPC. Studies with recycling enzyme. *Annales de Technologie Agricole* **21**(3): 423.

Cheftel, C., Ahren, M., Wang, D. I. C., and Tannenbaum, S. R. 1971. Enzymatic solubilization of FPC: Batch studies applicable to continuous enzyme recycling process. *J. Agr. Food Chem.* **19:** 155.

Cheryan, M. 1983. Application of membrane processing in the soy protein industry. In *Soybean Research in China and the United States.* B. J. Irwin, J. B. Sinclair, and W. Jin-Ling (Ed.), p. 102. Int. Agr. Pub., Urbana, IL.

Cheryan, M. 1986. *Ultrafiltration Handbook,* Ch. 8. Technomic, Lancaster, PA.

Cheryan, M., and Deeslie, W. D. 1980. Production of protein hydrolyzates in ultrafiltration-enzyme reactors. In *Ultrafiltration Membranes and Applications.* A. R. Cooper (Ed.), p. 591. Plenum Press, New York.

Cheryan, M., and Deeslie, W. D. 1983. Soy protein hydrolysis in membrane reactors. *J. Am. Oil Chem. Soc.* **60:** 1112–1115.

Cheryan, M., and Deeslie, W. D. 1984. Protein hydrolysis. U.S. Patent 4,443,540.

Cheryan, M., and Mehaia, M. A. 1986. Membrane bioreactors. In *Membrane Separation in Biotechnology.* W. C. McGregor (Ed.), Ch. 10. Marcel Dekker, New York.

Chibata, I. 1978. *Immobilized Enzymes.* Halsted Press, Division of John Wiley and Sons, NY.

Chibata, I. 1980a. *Immobilized Enzymes: Research and Development.* John Wiley and Sons, New York.

Chibata, I. 1980b. Development of enzyme engineering. Application of immobilized cell systems. *In Food Process Engineering.* v. 2, p. 1, P. Linko and J. Larinkari (Ed). Applied Science, London.

Chung, B. H., Chang, H. N., and Kho, Y. H. 1987. Dual hollow fiber membrane bioreactor for whole cell enzyme immobilization of *Streptomyces griseus* with glucose isomerase activity. *J. Ferment. Technol.* **65:** 575–581.

Cinq-Mars, G. V., and Howell J. 1977. Enzymatic treatment of primary nunicipal sludge with *Trichoderma viride* cellulase. *Biotech. Bioeng.* **18:** 377–385.

Closset, G. P., Cobb, J. T., and Shah, Y. T. 1974. Study of performance of a tubular membrane reactor for an enzyme catalyzed reaction. *Biotech. Bioeng.* **16:** 345–360.

Cooney, C. I. 1983. Bioreactors: Design and operation. *Science* **219:** 728.

Cunningham, S. D., Cater, C. M. and Mattil, K. J. 1978. Cottonseed protein modification in an ultrafiltration cell. *J. Food Sci.* **43:** 1477.

Darnoko, D., Cheryan, M., and Artz, W. E. 1989. Saccharification of cassava starch in an ultrafiltration reactor. *Enzyme Microb. Technol.* **11:** 154–159.

Davis, J. C. 1974. Kinetics studies in a continuous steady state hollow fiber membrane enzyme reactor. *Biotechnol. Bioeng.* **16:** 1113–1122.

Davis, M. E. 1986. Mathematical modeling of annular reactors. *Chem. Eng. J.* **33:** 133–142.

Davis, M. E., and Watson, L. T. 1985. Analysis of a diffusion-limited hollow fiber reactor for the measurement of effective substrate diffusivities. *Biotechnol. Bioeng.* **27:** 182–186.

Deeslie, W. D. 1980. Enzymatic modification of soy proteins in an ultrafiltration-enzyme reactor system. Ph.D. thesis, University of Illinois, Urbana, IL.

Deeslie, W. D., and Cheryan, M. 1981. Continuous enzymatic modification of proteins in an ultrafiltration reactor, *J. Food Sci.* **46:** 1035–1042.

Deeslie, W. D., and Cheryan, M. 1981b. A CSTR-hollow fiber system for continuous hydrolysis of proteins. I. Performance and kinetics. *Biotechnol. Bioeng.* **23:** 2257–2271.

Deeslie, W. D., and Cheryan, M. 1982. A CSTR-hollow fiber system for continuous hydrolysis of proteins. Factors affecting long-term stability of the reactor. *Biotechnol. Bioeng.* **24:** 69–82.

Deeslie, W. D., and Cheryan, M. 1988. Functional properties of soy hydrolysates from a continuous ultrafiltration reactor. *J. Agric. Food Chem.* **36:** 26–31.

Drioli, E. 1986. Membrane process in the separation, purification, and concentration of bioactive compounds from fermentation broths. *ACS Symposium Series* **314:** 52–66.

Drioli, E., Gianfreda, L., Palescandolo, R., and Scardi, V. 1975. Activity of acid phosphate as a gel layer on an ultrafiltration cellulose acetate membrane. *Biotechnol. Bioeng.* **17:** 1365–1367.

Engasser, J. M., Caumon, J., and Marc, A. 1980a. Hollow fiber enzyme reactors for maltose and starch hydrolysis. *Chem. Eng. Sci.* **35:** 99–105.

Engasser, J. M., Caumon, J., and Marc, A. 1980b. Hollow fiber enzyme reactor for sugar and starch processing. In *Food Process Engineering.* v. 2, P. Linko and J. Jarinkari (Ed.), p. 175. Applied Sci., London.

Fink, D. J., and Rodwell, V. W. 1975. Kinetics of a hollow fiber dehydrogenase reactor. *Biotech. Bioeng.* **17:** 1029–1050.

Finocchiaro, T., Olson, N. F., and Richardson, T. 1980. Use of immobilized lactase in milk systems. *Adv. Biochem. Eng.* **15:** 71–88.

Fisher, E. E. 1973. U.S. Patent 3,720,583.

Flaschel, F., Wandrey, C, and Kula, M-R. 1983. Ultrafiltration for the separation of biocatalysts. *Adv. Biochem. Engr. Biotechnol.* **26:** 73–142.

Frennesson, I., Tragardh, G., and Hagerdal, B. H. 1985. An ultrafiltration membrane reactor for obtaining experimental reaction rates at defined concentrations of inhibiting sugars during enzymatic saccharification of alkali-pretreated sallow: Formulation of a simple empirical rate equation. *Biotechnol. Bioeng.* **27:** 1328–1334.

Furusaki, S., and Asai, N. 1983. Enzyme immobilization by the Coulomb force. *Biotech. Bioeng.* **25:** 2209–2219.

Gacesa, P., Eisenthal, R., and England, R. 1983. Immobilization of urease within a thin channel ultrafiltration cell. *Enzyme Microb. Technol.* **5:** 191–195.

Gekas, V. C. 1986. Artificial membranes as carriers for the immobilization of biocatalysts. *Enzyme Microb. Technol.* **8**: 450–460.

Gekas, V., and Lopez-Levia, M. 1985. Hydrolysis of lactose: A literature review. *Process Biochem.* **20**: 2–12.

Georgakis, P. C., Chan, H., and Avis, R. 1975. The design of stirred reactors with hollow fiber catalysts for Michaelis-Menten kinetics. *Biotechnol. Bioeng.* **17**:99–106.

Ghose, T. K., and Kostick, J. T. 1970. A model for continuous enzymatic saccharification of cellulose with simultaneous removal of glucose syrup. *Biotech. Bioeng.* **12**: 921–946.

Gianfreda, L., and Greco, G., Jr. 1981a. The stabilizing effect of soluble macromolecules on enzyme performance. *Biotech. Letters* **3**: 33–38.

Gianfreda, L., and Greco, G. 1981b. Agricultural waste treatment by means of ultrafiltration membrane, enzymatic reactors. In *Energy From Biomass*. W. Palz, P. Chartier, and W. Hall (Ed.), p. 306. Appl. Sci. Publ., London.

Goldberg, B. S. 1978. Immobilized proteins. U.S. Patent 4,103,746.

Goldberg, B. S. 1979. Method for immobilized proteinaceous substances. U.S. Patent 4,169,014.

Gonzalez, G., Caminal, G., de Mas, C., and Lopez-Santin, J. 1989. Enzymatic hydrolysis of wheat straw: Studies on a column reactor with enzyme recycling. *Process Biochem.* **24**(2): 62–70.

Greco, G., Jr., and Gianfreda, L. 1981. Enzyme stabilization by linear chain polymers in ultrafiltration membrane reactors. *Biotech. Bioeng.* **23**: 2199–2210.

Greco, G., Albanesi, D., Cantarella, M., Gianfreda, L., Palescandolo, R., and Scardi, V. 1979a. Enzyme inactivation and stabilization studies in an ultrafiltration reactor. *Eur. J. Microbiol. Biotechnol.* **8**: 249–261.

Greco, G., Alfani, F., Iorio, G., Cantarella, M., Formisano, A., Gianfreda, L. K., and Palescandolo, R. 1979b. Theoretical and experimental analysis of soluble enzyme membrane reactors. *Biotelchnol. Bioeng.* **21**: 1421–1438.

Greco, G. Gianfreda, L., Albanesi, D., and Cantarella, M. 1981a. Thermal inactivation and enzyme kinetics in an ultrafiltration membrane reactor. *J. Appl. Biochem.* **3**: 233–245.

Greco, G., Jr., Livolsi, A. M., Scarfi, M. R., Mansi, F., and Gianfreda, L. 1981b. Ultrafiltration membrane enzymatic reactors, polarized versus stirred reactor performance. *Eur. J. Appl. Microbiol. Biotechnol.* **13**: 251–253.

Greco, G., Veronese, F., Largajolli, R., and Gianfreda, L. 1983. Purified penicillin acylase performance in a stabilized ultrafiltration membrane reactor. *Eur. J. Appl. Microbiol. Biotechnol.* **18**: 333–338.

Gregor, H. 1975. U.S. Patent 4,033,822.

Gregor, H. 1977. U.S. Patent 2,050,920.

Gregor, H., and Rauf, P. 1975. Enzyme-coupled ultrafiltration membrane. *Biotechnol. Bioeng.* **17:** 445–449.

Grubhofer, N., and Schleith, L. 1954. The coupling of proteins to diazotised polyamiostyrene. *Hoppe-Seyler Z. Physiol. Chem.* **297:** 108–112.

Hahn-Hagerdal, B., Lopez-Leiva, M., and Mattiasson, B. 1980. Membrane technology applied to bioconversion of macromolecular substrates and upgrading of products. A model study on ethanol production from cellulose. *Desalination* **35:**365.

Hartmeier, W. 1985. Immobilized biocatalysts. From simple to complex systems. *Trend in Biotechnol.* **3:** 149–153.

Hartmeier, W. 1980. Basic trials on the conversion of cellulosic material to ethanol using yeast coimmobilized with celluloytic enzymes. In *Advances in Biotechnology.* M. Moo-Young (Ed.), v. 3, p. 377. Pergamon Press, New York.

Hashimoto, M., Tsukamoto, T., Morita, S., and Okada, I. 1983. *Chem. Pharm. Bull.* **34:** 1.

Hausser, A. G., and Goldberg, B. S. 1983. An immobilized two-enzyme system (fungal α-amylase/glucoamylase) and its use in the continuous production of high conversion maltose-containing corn syrups. *Biotech. Bioeng.* **25:** 525–539.

Hayakawa, K., Uraba, I. and Okada, H. 1985. Operational stability of a continuous enzyme reactor containing poly(ethylene glycol) bound NAD and thermostable dehydrogenases. *J. Ferment. Technol.* **63:** 245.

Heinzler, A., Gudernatsch, W., Knopki-Fobo, G., and Kulbe, K.D. 1987. D-Galacluronic acid and oligogalacturonic acid by enzymatic pectin degradation, chromatography, and membrane reactor. In *Pro. 4th European Congr. on Biotechnology.* O. M. Neijssel, R. R. van der Meer, and K. Luyben (Ed.), v. 2, p. 609. Elsevier, Amsterdam.

Henley, R. G., Yang, R. Y. K., and Greenfield, P. F. 1980. Enzymatic saccharification of cellulose in membrane reactors. *Enzyme Microb. Technol.* **2:** 206–208.

Hoehn, E., McKay, C. J., and Marray, D. E. 1983. U.S. Patent 4,421,852.

Hoq, M. M., Koike, M., Yamane, T., and Shimizu, S. 1985a. Continuous hydrolysis of olive oil by lipase in microporous hydrophobic hollow fiber bioreactors. *Agric. Biol. Chem.* **49:** 3171–3178.

Hoq, M. D., Yamane, T., Shimizu, S., Funada, T., and Ishida, S. 1985b. Continuous hydrolysis of olive oil by lipase in microporous hydrophobic membrane bioreactor. *J. Am. Oil Chem. Soc.* **62:** 1016–1021.

Hoq, M. M., Yamane, T., Shimizu, S., Funada, T., and Ishida, S. 1984. Continuous synthesis of glycerides by lipase in a microporous membrane bioreactor. *J. Am. Oil Chem. Soc.* **61:** 776–781.

Hoq, M. M., Yammane, T., and Shimizu, S. 1986. Role of oleic acid solubilized in buffer-glycerol solution on adsorbed lipase during continuous hydrolysis of olive oil in a microporous hydrophilic membrane bioreactor. *Enzyme Microb. Technol.* **8**: 236–240.

Hong, J., Tsao, G. T., and Wankat, P. C. 1981. Membrane reactor for enzymatic hydrolysis of cellobiose. *Biotech. Bioeng.* **23**: 1501–1516.

Hossain, M. M., and Do, D. D. 1985. Modeling of enzyme immobilization in porous membrane. *Biotechnol. Bioeng.* **27**: 1126–1135.

Howaldt, M., Gottlob, A., Kulbe, K. D., and Chmiel, H. 1988. Simultaneous conversion of glucose/fructose mixtures in a membrane reactor. *Ann. N.Y. Acad. Sci.* **542**: 400–405.

Howell, J. A., and Velicangil, O. 1982. Theoretical considerations of membrane fouling and its treatment with immobilized enzymes for protein ultrafiltration. *J. Appl. Poly. Sci.* **27**: 21–32.

Huffman-Reichenbach, L-M, and Harper, W. J. 1982. Beta-galactosidase retention by hollow fiber membrane. *J. Dairy Sci.* **65**: 887–898.

Huffman-Reichenbach, L. M., and Harper, W. J. 1985. Lactose hydrolysis in batch and hollow fiber membrane reactors. *New Zealand J. Dairy Sci. Technol.* **20**: 57.

Hummel, W., Schutte, H., and Kula, M.-R. 1984. New enzymes for the synthesis of chiral compounds. *Ann. N.Y. Acad. Sci.,* **434**: 194–205.

Iacobucci, G. A., Myers, M. J., Emi, S., and Myers, D. V. 1974. Large-scale continuous production of soybean hydrolysate in a membrane reactor. *Proc. 4th Int. Congr. Food Sci. Technol.* **5**: 83–95.

Imai, K., Shiomi, T., Sato, K., and Fujishima, A. 1983. Preparation of immobilized invertase using poly (vinyl alcohol) membrane. *Biotechnol. Bioeng.* **25**: 613–617.

Iorio, G., Drioli, and Molinari, R. 1981. Characterization and control of an immobilized allosteric enzyme system. *Ann. N.Y. Acad. Sci.* **369**: 235–243.

Iorio, G., Catapano, G., Drioli, E., Rossi, M., and Rella,R. 1984. Characterization of an enzymatic capillary membrane reactor. *Ann. N.Y. Acad. Sci.,* **434**: 123–126.

Iorio, G., Catapano, G., Drioli, E., Rossi, M., and Rella, R. 1985. Malic enzyme immobilization in continuous capillary membrane reactors. *J. Membrane Sci.* **22**: 317–324.

Ishikawa, H., Tanaka, T., Takase, S., and Hikita, H. 1989a. Theoretical analysis of G6P production and simultaneous ATP regeneration by conjugated enzymes in an ultrafiltration hollow-fiber reactor. *Biotech. Bioeng.* **34**: 357–368.

Ishikawa, H., Takase, S., Tanaka, T., and Hikita, H. 1989b. Experimental investigation of G6P production and simultaneous ATP regeneration by conjugated enzymes in an ultrafiltration hollow-fiber reactor. *Biotech. Bioeng.* **34:**369–379.

Ishimori, Y., Karube, I., and Suzuki, S. 1981. Mechanical control of the activity of glucose oxidase immobilized on porous polyvinylchloride membrane. *Biotechnol. Bioeng.* **23:** 2601–2608.

Iwasaki, K., Nakajima, M., Nabetani, H., and Watanabe, A. 1989. Continuous hydrolysis of soluble starch by free β-amylase and pullulanase with CSTR-UF membrane reactor. *5th Intr. Congr. Eng. and Food.* Cologne, West Germany, May 28–June 3, p. 49.

Jancsik, V., Beleznai, Z., and Keleti, T. 1982. Enzyme immobilization by poly (vinyl alcohol) gel entrapment. *J. Mol. Catal.* **14:** 297–306.

Jandel, A-S, Hustedt, H., and Wandrey, C. 1982. Continuous production of L-alanine from fumarate in a two-stage membrane reactor. *Eur. J. Appl. Microbiol. Biotelchnol.* **15:** 59–63.

Jenq, C. Y., Wang, S. S., and Davidson, B. 1980. Ultrafiltration of raw sewage using an immobilized enzyme membrane. *Enzyme Microb. Technol.* **2:** 145–149.

Jones, J. B., and Beck, J. F. 1976. Asymmetric synthesis and resolution using enzyme. In *Application of Biochemical Systems in Organic Chemistry, Part 1.* J. B. Jones, C. J. Sih, and D. Perlman (Ed.), p. 112. Techniques of Chemistry, VX. Wiley Interscience, New York.

Jones, C. K. S., and White, E.T. 1984. Enzymatic hydrolysis of lactose in a hollow-fiber reactor. *Ann. N.Y. Acad. Sci.* **434:** 119–122.

Katoaka, H., Saigusa, T., Mukataka, S., and Takahashi, J. 1980. Effect of exial dispersion and mass transfer resistance on conversion in a hollow fiber enzyme ractor. *J. Ferment. Technol.* **58:** 431.

Kawakami, K., Hamada, T., and Kusunoki, K. 1980. Performance of a hollow fiber beaker device for continuous enzymic reactions. *Enzyme Microb. Technol.* **2:** 295–298.

Kennedy, J. F. 1985a. Principles of immobilization of enzymes. In *Handbook of Enzyme Biotechnology.* A. Wiseman (Ed.), p. 147. Halsted Press: A Division of John Wiley and Sons, New York.

Kennedy, J. F. 1985b. Data on techniques of enzyme immobilization and bioaffinity procedures. In *Handbook of Enzyme Biotechnology.* A. Wiseman (Ed.), p. 380. Halsted-Press: A Division of John Wiley and Sons, New York.

Keyes, M. H., and Albert, D. 1985. Immobilized enzymes. In *Encyclopedia of Polymer Science and Engineering.* 2nd ed., v. 6, p. 189. Wiley Interscience, NY.

Keyes, M. H., and Saraswathi, S. 1985. Immobilized enzymes. In *Bioactive*

Polymeric Systems. C. G. Gebelein and C. E. Carraher, Jr. (Ed.), p. 249. Plenum Press, New York.

Kilara, A., Shahani, K. M., and Wagner, F. W. 1977. Preparation and properties of immobilized papain and lipase. *Biotech. Bioeng.* **19**: 1703–1714.

Kim, I. H., and Chang, H. N., 1983a. Variable-volume fiber enzyme reactor with plusatile flow. *AIChE J.* **29**: 910–914.

Kim, I. H., and Chang, H. N. 1983b. A theoretical study of a membrane/enzyme reactor with sinusoidal ultrafiltration. *J. Chem. Eng. Japan.* **16**: 67–71.

Kim, S. S., and Cooney, D. 1976. An improved theoretical model for hollow-fiber enzyme reactors. *Chem. Eng. Sci.* **31**: 289–294.

Kimura, Y., Tanaka, A., Sonomoto, K., Nihira, T., and Fukui, S. 1983. Application of immobilized lipase to hydrolysis of triacylglyceride. *Eur. J. Appl. Microbiol. Biotechnol.* **17**: 107–112.

Kinoshita, S., Chua, J. W., Kato, N., Yoshida, T., and Taguchi, H. 1986. Hydrolysis of cellulose by cellulases of *Soprotrichum cellulophilum* in an ultrafilter membrane reactor. *Enzyme Microb. Technol.* **8**: 691–695.

Kleinstreure, C., and Poweigha, T. 1984.Modeling and simulation of bioreactor process dynamics. *Adv. Biochem. Eng./Biotechnol.* **30**: 91–146.

Klua, M-R., and Wandrey, C. 1987. Continuous enzymatic transformation in an enzyme-membrane reactor with simultaneous NADH regeneration. In *Methods in Enzymology,* v. 136, K. Mosbach (Ed.), p. 9. Academic Press, New York.

Kobayashi, T., Mukataka, S., Kataoka, H., and Takahashi, J. 1983. *Abstract of Annual Meeting of the Soc. Ferment. Technol.* Japan, p. 21.

Kohlwey, D. K., and Cheryan, M. 1981. Performance of a β-D-galactosidase hollow fiber reactor. *Enzyme Microb. Technol.* **3**: 64–68.

Korus, R. A., and Olson, A. C. 1977a. Use of glucose isomerase in hollow fiber reactors. *J. Food Sci.* **42**: 258–260.

Korus, R. A., and Olson, A. C. 1977b. The use of α-galactosidase and invertase in hollow fiber reactors. *Biotelch. Bioeng.* **19**: 1–8.

Korus, R. A., and Olson, A. C. 1978. Use of α-galactosidase, β-galactosidase, glucose isomerase and invertase in hollow fiber reactors. In *Enzyme Engineering.*E. K. Pye and H. H. Weetal (Ed.), v. 3, p. 543. Plenum Publ., New York.

Kowalewska, J., Poznanski, S., Bednarske, W., and Wulima, K. 1978. The application of membrane techniques in enzymatic hydrolysis of lactose and repeated use of β-galactosidase. *Nordeuropaeisk Mejeri Tidsskrift.* **1/78**: 20.

Klyama, Y., Shimazaki, K., Kawahara, Y., Kubota, K. and Yoshii, H. 1987. Production of L-aspartic acid by membrane reactor. In *Proc. 4th European Congr. on Biotechnology.* O. M. Neijssel, R. R. van der Meer, and K. Luyben (Ed.), v. 1, p. 119. Elsevier,

Kozhukharova, A., Kirova, N., Popova, Y., Batsalova, K., and Kunchev, K. 1988. Properties of glucose oxidase immobilized in gel of polyvinylalcohol. *Biotech. Bioeng.* **32:** 245–248.

Kulbe, K. D., and Chmiel, H. 1988. *Annals N.Y. Acad. Sci.* **542:** 444–465.

Kulbe, K. D., Heinzler, A., and Knopki, G. 1987a. Enzymatic synthesis of L-ascorbic acid via-D-uronic acids; Membrane-reactor integrated recovery of D-galacturonic acid from pectin hydrolysates. *Ann. N.Y. Acad. Sci.* **506:** 543–551.

Kulbe, K. D., Schwab, U., and Gudernatsch, W. 1987b. Enzyme-cataluyzed production of mannitol and gluconic acid. Product recovery and various procedures. *Ann. N.Y. Acad. Sci.* **506:** 552–568.

Kulbe, K. D., Schwab, U., and Howaldt, M. 1987c. Conjugated NAD(H)-dependent dehydrogenases for the continuous production of mannitol and glucanic acid from glucose-fructose mixtures in a membrane reactor. *Annals N.Y. Sci.* **501:** 216–223.

Larsson, M. and Mattiasson, B. 1984. Continuous conversion of starch to ethanol using a combination of an aqueous two-phase system and an ultrafiltration unit. *Ann. N.Y. Acad. Sci.* **434:** 144–147.

Lavayre, J., and Bratti, J. 1982. Preparation and properties of immobilized lipases. *Biotech. Bioeng.* **24:** 1007–1013.

Lee, C. K., and Hong, J. 1987. Enzyme reaction in a membrane cell coupled with electrophoresis. *Ann. N.Y. Acad. Sci.* **506:** 499–510.

Lee, C. K., and Hong, J. 1988. Membrane reactor coupled with electrophoresis for enzymatic production of aspartic acid. *Biotech. Bioeng.* **32:** 647–654.

Leuchtenberger, W., Karrenbauer, M., and Plocker, U. 1984a. *Sonderheft Biotechnol. Forum Mikrobiol.* **7:** 40.

Leuchtenberger, W., Karrenbauer, M., and Plocker, U. 1984b. Scale-up of an enzyme membrane reactors process for the manufacture of L-enantiomeric compounds. *Ann. N.Y. Acad. Sci.* **434:** 78–86.

Lewis, W., and Middleman, S. 1974. Conversion in hollow fiber/enzyme reactor. *AIChE J.* **20:** 1012–1014.

Lieberman, R. B., and Ollis, D. F. 1975. Hydrolysis of particulate tributyrim in a fluidized lipase reactor. *Biotech. Bioeng.* **17:** 1401–1419.

Lilly, M. D. 1978. *Dechema-Mongr.* **82:** 1693.

Lilly, M. D., and Dunnill, P. 1976. In *Methods in Enzymology.* K. Mosbach (Ed.), v. 44, p. 717. Academic Press, New York.

Linfield, w. M., O'Brien, D. J. Serota, S, and Barauskas, R. A. 1984. Lipid-lipase interactions. 1. Fat splitting with lipase from *Candida rugosa. JAOCS* **61:** 1067–1071.

Lopez-Leiva, M. and Gekas, V. 1986. A crossflow immobilized enzyme reactor for the hydrolysis of whey. *Process Biochem.* **21:** 27–29.

Lopez, J. L., Matson, s. L., Stanley, T. S., and Quinn, J. A. 1988. Liquid-liquid extractive membrane reactors. In *Extractive Bioconversions.* B. Mattiasson and O. Hollst (Ed.), v. 2. Marcel Dekker, New York.

Madgavkar, A. M., Shah, Y. T. and Cobb, J. T. 1977. Hydrolysis of starch in a membrane reactor. *Biotech. Bioeng.* **19:** 1719–1726.

Mannheim, A. 1989. Continuous hydrolysis of milk proteins in a membrane reactor. M. Sc. thesis, University of Illinois, Urbana, IL.

Mannheim, A., and Cheryan, M. 1989. Continuous hydrolysis of milk protein in a membrane reactor. *J. Food Sci.* In press.

Marshall, J. J., and Whelan, W. J. 1971. A new approach to the use of enzymes in starch technology. *Chem. Eng. Ind.* **19:** 701.

Martensson, K. 1982. Choosing an enzyme reactor. Theoretical background and considerations as illustrated in a case study. *Appl. Biochem. Biotech.* **7:** 11–18.

Mashelkar, R. A., and Ramchandran, P. A. 1975. A new model for a hollow fiber enzyme reactor. *J. Appl. Chem. Biotechnol.* **25:** 867–880.

Matson, S. L. 1979. Membrane reactors. Ph.D. thesis, University of Pennsylvania, Philadelphia, PA.

Matson, S. L., and Quinn, J. A. 1986. Membrane reactors in bioprocessing. *Annals N.Y. Academic Sci.* **469:** 152–165.

Matson, S. L., and Lopez, J. L. 1988. Multiphase membrane reactors for enzymatic resolution: Diffusional effects on stereoselectivity. In *Proceedings of Conference on Frontiers in Biotechnology.* Boulder, CO. CRC Press, Cleveland, OH.

Maubois, J. L. 1988. Whey, its biotechnological signification. *8th International Biotechnology Symposium,* Paris, July 17–22, p. 513.

Maubois, J. L., and Brule, G. 1982. Utilisation des techniques a membrane pour la separation, la purification et la fragmentation des proteines laitieres. *Le Lait* **62:** 484–510.

Mertens, B., and Huyghebaert, A. 1987. Lactose hydrolysis in an enzymatic membrane reactor. *Milchwissenschaft* **42:** 640–645.

Messing, R. A. (Ed.) 1975. *Immobilized Enzymes for Industrial Reactors.* Academic Press, New York.

Michaels, A. S., and Matson, S. L. 1985. Membrane in biotechnology: State of the art. Desalination **53:** 231–258.

Miyairi, S. 1979. An enzyme-polymer film prepared with the use of poly vinyl alcohol) bearing photosensitive aromatic azido groups. *Biochim. Biophys. Acta* **571:** 374–377.

Miyawaki, O., Nakamura, K., and Yano, T. 1982a. Theoretical study of continuous NAD recycling by conjugated enzyme immobilized in ultrafiltration hollow fiber. *J. Chem. Eng. Jap.* **15**: 142–147.

Miyawaki, O., Nakamura, K., and Yano, T. 1982b. Experimental investigation of continuous NAD recycling by conjugated enzymes immobilized in ultrafiltration hollow fiber. *J. Chem. Eng. Jpn.* **15**: 224–228.

Miyawaki, O., Nakamura, K., and Yano, T. 1982c. Dynamic ATP recycling for continuous NADP production. *Agric. Biol. Chem.* **46**: 2725–2733.

Morikawa, Y., Karube, I., and Suzuki, S. 1978. NAD recycling in the collagen membrane. *Biochem. Biophys. Acta* **523**: 263–267.

Nakajima, M., Jimbo, N., Nishizawa, K., Nabetani, H., and Watanabe, A. 1988. Conversion of sucrose by immobilized invertase in an asymmetric membrane reactor. *Process Biochem.* **23**: 32–35.

Nakamura, A., Minani, H., Urabe, I., and Okada, H. 1988. Properties of glucose-dehydrogenase-poly (ethylene glycol)-NAD conjugate as an NADH-regeneration unit in enzyme reactors. *J. Fermnt. Technol.* **66**: 267–272.

Nelson, J. M., and Griffin, E. G. 1916/. Adsorption of invertase. *J. Amer. Chem. Soc.* **38**: 1109–1115.

Norman, B. E., Severinsen, S. G., Nielsen, T. and Wagner, J. 1978. Traitement enzymatique de perment de lactoserum avec recuperation de la lactase par ultr-filtration. *Nordeuropaeisk Mejeri Tidsskrift* **5-6/78**: 129.

Ohlson, I., Tragardh, G., and Hagerdal, B. H. 1984a. Enzymatic hydrolysis of sodium-hydroxide-pretreated sallow in an ultrafiltration membrane reactor. *Biotechnol. Bioeng.* **26**: 647–653.

Ohshima, T., Wandrey, C., and Conrad, D. 1989. Continuous production of 3-fluoro-L-alanine with alanine dehydrogenase. *Biotech. Bioeng.* **34**: 394–397.

Ohshima, T., Wandrey, C., Kula, M. R., and Soda, K. 1985. Improvement for L-leucine production in a continuously operated enzyme membrane reactor. *Biotech. Bioeng.* **27**: 1616–1618.

Olson, A. C., and Cooney, C. L. 1974. *Immobilized Enzymes in Food and Microbial Processes.* Plenum Press, New York.

Olson, N. F., and Richardson, T. 1974. Immobilized enzymes in food processing and analysis. *J. Food Sci.* **39**: 653.

O'Neill, S. P., Wykes, J. R., Dunnill, P., and Lilly, M. D. 1971. An ultrafiltration-reactor system using a soluble immobilized enzyme. *Biotelch. Bioeng.* **13**:319–322.

vander Padt, A., and van't Riet, K. 1989. Enzymatic acylglycerol synthesis in a membrane bioreactor. In *Fifth Intr. Congr. on Engr. and Food* ICEF 5, Cologne, Federal Republic of Germany, May 28–June 3, 2.03.

Pansolli, P., Giovenco, S., Dinelli, D., and Morisi, F. 1976. In *Analysis and Control of Immobilized Enzyme Systems.* D. Thomas and J. Kernevez (Ed.)., p. 237. North Holland Press, Amsterdam.

Park, T. H., Kim, I. H., and Chang, H. N. 1985. Recycle hollow fiber enzyme reactor with flow swing. *Biotechnol. Bioeng.* **27**: 1185–1191.

Pastore, M., Morisi, F., and Leali, L. 1976. Reduction of lactose of milk by entrapped *b*-galactosidase. IV. Results of long-term experiments on pilot plant. *Milchwissenchaft* **31**(6): 362–365.

Pastore, M., Morisi, F., and Viglia, A. 1974. Reduction of lactose of milk by entrapped β-galactosidase. II. Conditions for an industrial continuous process. *J. Dairy Sci.* **57**: 269–272.

Payne, R. E., Hill, C. G., and Amundson, C. H. 1978. Enzymatic solubilization of leaf protein concentrate in membrane reactors. *J. Food Sci.* **43**: 385–389.

Peterson, R. S., Hill, C. G., and Amundson, C. H. 1989a. Lactose hydrolysis by immobilized β-galactosidase in capillary bed reactor. *Biotech. Bioeng.* **34**: 438–446.

Peterson, R. S., Hill, C. G., and Amundson, C. H. 1989b. Effects of temperature of the hydrolysis of lactose by immobilized β-galactosidase in a capillary bed reactor. *Biotech. Bioeng.* **34**: 429–437.

Pitcher, W. H., Jr. (Ed.) 1980. *Immobilized Enzymes for Food Processing,* CRC Press, Boca Raton, FL.

Pitcher, W. H., and Messing, R. A. 1975.

Pizzichini, M., and Leonardi, M. 1984. *Acqua Aria* **3**: 263.

Porter, M. C., and Michaels, A. S. 1972. Membrane ultrafiltration. Part 5. A useful adjunct for fermentation and enzyme processing of foods. *Chemtech* **2**: 56–61.

Prenosil, J. E., and Hediger, T. 1985. Scale-up of membrane fixed enzyme reactors: Modelling and experiments. *Desalination* **53**: 265–278.

Prenosil, J. E., and Hediger, T. 1988. Performance of membrane fixed biocatalyst reactors. I. Membrane reactor systems and modelling. *Biotelchnol. Bioeng.* **31**: 913–921.

Pronk, W., Kerkhof, P. J. A. M., van Helden, C., and Van't Riet, K. 1988. The hydrolysis of triglycerides by immobilized lipase in a hydrophilic membrane reactor. *Biotech. Bioeng.* **32**: 512–518.

Roger, L., and Maubois, J. L. 1981. Actualities dans le domaine des technologies a membrane pour la preparation et la separation des proteines laitieres. *Rev. Lait Fr.* **400**: 67–75.

Roger, L., Thapon, J. L., Maubois, J. L., and Brule, G. 1976. Hydrolyse du lactose contenu dans l'ultrafiltrat de lait ou de lactoserum en reacteur enzymatique a membrane. *Le Lait* **56**: 551–575.

Roger, L., Thapon, J. L., Brule, G., and Maubois, J. L. 1977. Continuous lactose hydrolysis in an enzymatic membrane reactor. *Nordeuropaeisk Mejeri-Tidsskrift* **1-2/77**: 38.

Rony, P. R. 1971. Multiphase catalysis. II. Hollow fiber catalysts. *Biotechnol. Bioeng.* **13**: 431–477.

Rony, P. R. 1972. Hollow fiber enzyme reactors. *J. Am. Chem. Soc.* **94**: 8247–8248.

Roozen, J. P., and Pilnik, W. 1973. Ultrafiltration controlled enzymatic degradation of soy protein. *Process Biochem.* **8**: 24–27.

Roozen, J. P., and Pilnik, W. 1979. Enzymatic protein hydrolysis in a membrane reactor related to taste properties. *Enzyme Microb. Technol.* **1**: 122–124.

Rosevear, A. 1984. Immobilized biocatalysts-a critical review. *J. Chem. Tech. Biotechnol.* **34B**: 127–150.

Scardi, V., Canteralla, M., Gianfreds, L., Palescandolo, r., Alfani, F., and Greco, G. 1980. Enzyme immobilized by means of ultrafiltration techniques. *Biochime* **62**: 635–643.

Schmidt, E., Bossow, B., Wichmann, R., and Wandrey, C. 1986. The enzyme membrane reactor. An alternative approach for continuous operation with enzymes. *Kem. Ind.* **35**: 71–77.

Silman, R. W., Black, L. T., McGhee, J. E., and Bagley, E. B. 1980. Hydrolysis of raffinose in a hollow-fiber reactor using an unrefined mixture of α-galactosidase and invertase. *Biotech. Bioeng.* **22**: 533–541.

Silmon, S. 1982. Israel Patent 54,116.

Sims, K., and Cheryan, M. 1989. Unpublished data. Department of Food Science, University of Illinois, Urbana, IL.

Smiley, K. L., Hensley, D. E., and Gasdrof, H. J. 1976. Alpha-galactosidase production and use in a hollow-fiber reactor. *Appl. Environ. Microbial.* **31**: 615.

Staude, B., Jorisch, W., and Ansorge, W. 1982. Reactions with enzymes covalently bonded to heterogeneous ultrafiltration membranes. *J. Membrane Sci.* **11**: 289–296.

Stavenger, P. L., 1971. Putting semipermeable membrane to work. *Chem. Eng. Prog.* **67(3)**: 30–37.

Swaisgood, H. E. 1985. Immobilized of enzymes and some applications in the food industry, Ch. 1. In *Enzymes and Immobilized Cells in Biotech-nology*. A. I. Laskin (Ed.) The Benjamin/Cummings Publishing Company, Menlo, CA.

Swanson, S. J., Emery, A., and Lim, H. C. 1977. Kinetics of maltose hydrolysis by glucoamylase. *Biotech. Bioeng.* **19**: 1715–1718.

Tachauer, E., Cobb, J. T., and Shah, Y. T. 1974. Hydrolysis of starch by a mixture of enzymes in a membrane reactor. *Biotech. Bioeng.* **16**: 545–550.

Takamatsu, S., and Ryu, D. D. Y. 1988. New recirculating bioreactor-separator combination system for continuous bioconversion and separation of products. *J. Enzyme Microb. Technol.* **10**: 593–600.

Tjerneld, F., Persson, I., and Albertsson, D. 1985. Enzyme recycling in cellulose hydrolysis by combined use of aqueous two-phase systems and ultrafiltration. *Biotelch. Bioeng. Symp.* **15**: 419–429.

Trujillo, E. M. 1987. Transient response of encapsulated enzymes in hollow fiber reactors. *Biotechnol. Bioeng. Symp.* **29**: 529–543.

Velicangil, O., and Howell, J. A. 1977. Protease-coupled membranes for ultrafiltration. *Biotech. Bioeng.* **19**: 1891–1894.

Velicangil, O., and Howell, J. A. 1981. Self-cleaning membrane for ultrafiltration. *Biotech. Bioeng.* **23**: 843–854.

Veronese, F. M., Boccu, E., Schiavon, O., Greco, G., Jr., and Gianfreda, L. 1984. Immobilization of purified penicillin acylase in a polarized ultrafiltration membrane reactor. *Ann. N.Y. Acad. Sci.* **434**: 127–130.

Vieth, W. R., and Venkatasubramaian, K. 1974. Enzyme engineering. Part IV. Process engineering for immobilized enzyme systems. *Chemtech* July: 434–438.

Vieth, W. R., Venkatasubramanian, K., Constantinides, A., and Davidson, B. 1976. *Appl. Biochem. Biotech.* **1**: 221.

Visser, S., Noorman, H. J., Slangen, C. J., and Rollema, H. S. 1989. Action of plasmin on bovine β-casein in a membrane reactor. *J. Dairy Res.* **56**: 323–333.

Vorsilak, P., McCoy, B. J., and Merson, R. L. 1975. Enzyme immobilized in a membrane sandawich reactor. *J. Food Sci.* **40**: 431.

Wandrey, G. 1984. In *Biotelch Europe 84*, p. 391. Online Publications Pinner, London.

Wandrey, C., and Wichmann, R. 1985. Coenzyme regeneration. In *Enzymes and Immobilized Cells in Biotechnology,* Ch. 8. A. I. Laskin (Ed.). The Benjamin/Cummings, Pub. Co., London, UK.

Wandrey, C., and Flaschel, E. 1979. Process development and economic aspects in enzyme engineering. Acylalse L. Methionine system. *Adv. Biochem. Engr.* **12**: 147–218.

Wandrey, C., and Wichmann, R. 1987. Productiono of L-amino acids in the membrane reactor. In *Biotec 1 Microbial Genetic Engineering and Enzyme Technology.* C. P. Hollenberg and H. Sahm (Ed.), pp. 85–92. Gustav Fischer, Stuttgart.

Wang, D. I. C., Cooney, C. I., Demain, A. L., Dunnill, P., Humphrey, A. E., and Lilly, M. D. 1979. Enzyme reactors, Ch. 14. In *Fermentation and Enzyme Technology* John Wiley and Sons, New York.

Wang, S. S., Davidson, B., Gillespie, Harris, L. R., and lent, D. S. 1980. Dynamics enhanced protein ultrafiltration using an immobilized protease. *J. Food Sci.* **45:** 700–702.

Waterland, L. R., Michaels, A. S., and Robertson, C. R. 1974. A theoretical model for enzymatic catalysis using asymmetric hollow fiber membrane. *AIChE J.* **20:** 50–59.

Waterland, L. R., Robertson, C. R., and Michaels, A. C. 1975. Enzymatic ctalysis using asymmetric hollow fiber membranes. *Chem. Eng. Commun.* **2:** 37–47.

Weetall, H. H. 1977. In *Biotechnological Applications of Proteins and Enzymes.* Z. Bohak and N. Sharon (Ed.), p. 103. Academic Press, New York.

Weetall, H. H., and Pitcher, W. H., Jr. 1986. Scaling up an immobilized enzyme system. *Science* **232:** 1396–1403.

Wichmann, R., Wandrey, C., Buckmann, A. F., and Kula, M. R. 1981. Continuous enzymatic transformation in an enzyme membrane reactor with simultaneous NAD(H) regeneration. *Biotech. Bioeng.* **23:** 2789–2802.

Yamane, T., Hoq, M. M., and Shimizu, S. 1987. Continuous hydrolysis of olive oil by lipase in microporous hydrophobic membrane bioreactor. *Annals N.Y. Acad. Sci.* **501:** 224–228.

Yamazaki, Y., Naeda, H., and Suzuki, H. 1976. Application of polylysine bound succinyl-NA to a membrane reactor. *Biotech. Bioeng.* **18:** 1761–1775.

5

Food Freeze Concentration

Henry G. Schwartzberg

University of Massachusetts
Amherst, Massachusetts

FREEZE CONCENTRATION

In freeze concentration: (a) cooling is used to selectively freeze solvent from liquid solutions: (b) solute concentrations in residual solution consequently increase; and (c) concentrated solution and solidified solvent are separated when desired concentrations are reached. Freeze concentration is used to produce concentrated solutions or recover purified solvent. We will discuss its use for concentrating aqueous solutions, particularly food solutions. It is also used to separate nonaqueous mixtures, e.g., mixtures of xylenes.

Joslyn (1961) described early attempts to concentrate liquid foods by freezing. Deshpande et al. (1984) describe recently used freeze concentration processes, and Chowdhury et al. (1988) describe recent developments. Underlying principles and design and operation considerations for currently used processes and equipment are discussed in this chapter. Methods for improving these processes will be examined. Thijssen (1974,1975), who contributed much to the development of the process, addressed these issues in earlier reviews, which are well worth consulting.

Freeze concentration has been used to concentrate fruit juices, coffee extract, vinegar, beer, wine, pickling brines, and liquid smoke, to recover potable water from brackish water and sea water, and to concentrate toxic wastes and paper-mill black liquor. Concentration of whole milk, skim milk, whey, tea extract, and food processing wastes is currently being investigated. Major advantages include retention of volatile flavors and aromas, ability to concentrate alcoholic beverages without loss of alcohol, minimization of thermal damage, and provision of cold concentrates. Disadvantages include: potential precipitation of sparingly soluble solutes, high capital cost, and long holdup and startup times. Energy efficiencies and maximum concentrations are somewhat lower than provided by efficient multi-effect and mechanical–vapor recompression evaporators. However, freeze concentration equipment costs have progressively dropped and energy efficiency has progressively increased. Further improvements in efficiency and reductions in cost can be expected.

Typical Systems

A single-stage freeze concentration system is depicted schematically in Fig. 5.1. Feed solution passes through a freezer, where small, branched ice crystals form. Cooling causes crystal growth, and concentration increases. When the desired concentration is reached, concentrate is drawn off. After crystals are large enough to be separated cleanly, ice is separated and discharged.

Bulk subcooling can be used to grow suitably large crystals; however, small crystals are usually produced in scraped-surface freezers (SSF) instead. These small crystals melt and promote growth of larger, rounded crystals. The process (ripening) will be analyzed later. Agitation is used to facilitate growth and rounding-off, but also causes frictional heating and some melting. Concentrate is recirculated to provide slurries that are fluid enough to handle. After startup, the amount of recirculating concentrate remains constant, and the sum of the ice and concentrate discharge rates equals the feed rate. Because of concentrate retention, solute residence times are long.

Wash columns are used to separate solution from discharged ice. The ice advances upward through a stationary layer of wash water, which displaces entrained solution. Displaced solution drains through strainers at the bottom or sides of the column and recycles to the freezer. Clean ice discharges from the top of the column. Centrifugal filters are occasionally used for ice-solution separations instead of wash columns.

Vapor compression refrigeration is used in freeze concentration systems for foods. Refrigerant evaporates at low pressure P_e in jackets of freezers, withdrawing heat from the system. The vapor generated is compressed and condenses at higher

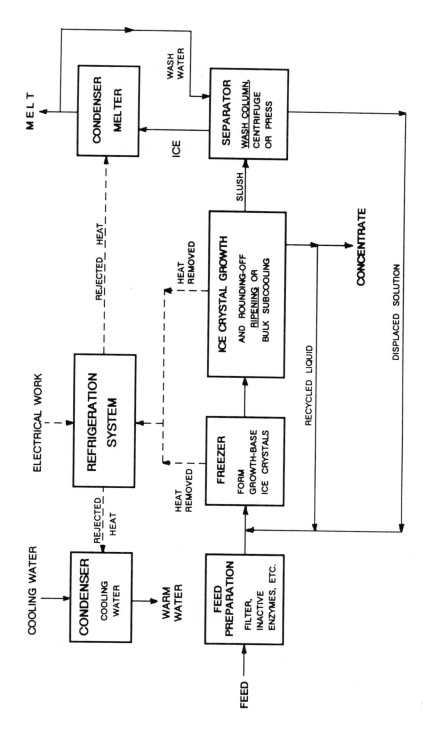

Figure 5.1 Single-stage freeze concentration system—basic arrangement.

129

pressures P_c, partly through use of cooling water. Some vapor is condensed using ice discharged from the system. This reduces energy required for refrigeration.

Feeds must be suitably prepared (Helferich, 1988), e.g., pasteurized in some cases. Enzymes that cause spoilage have to be inactivated. Bubbles interfere with clean separation of ice, and residual oxygen may cause off-flavors, so degassing and deaeration is often necessary. Precipitatable solids and solutes have to be removed.

Single-stage freeze concentration is still occasionally employed, but three to six stages are usually used in modern, large-scale systems (Van Pelt and Jansen, 1988). A three-stage system is depicted in Fig. 5.2. In single-stage systems, ice grows solely at the final product concentration, where growth is very slow. In multistage systems, ice grows mainly at low and intermediate concentrations, where growth is much more rapid. Viscosities are much lower at these concentrations and less

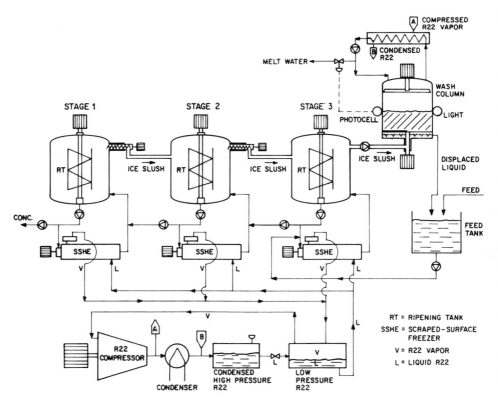

Figure 5.2   Three-stage countercurrent freeze concentration system.

frictional heat is generated. When a single coolant temperature is used (the usual case), larger heat-transfer ΔT and higher rates of heat removal per unit of freezer area are provided by multistage systems. Separation is carried out more efficiently and rapidly when ice moves countercurrent to concentrate flow and discharges from the dilute end of the system. Total residence times are lower in multistage systems than in single-stage systems (Van Pelt and Jansen, 1988).

Costs and Energy Consumption

Freeze concentration plant costs per unit of capacity dropped 92% between 1975 and 1988 (Van Pelt and Jansen, 1988). Plants that can separate 22,000 kg of water per hour are available. Energy requirements have also dropped dramatically. In 1975, 269 kWh of electrical energy were used to remove 1000 kg of water; in 1988, 74 kWh/1000 kg were required (Van Pelt and Jansen, 1988). Rates of 48 kWh/1000 kg have been claimed for some equipment (Johnson, 1986) but have not yet been substantiated for sustained production at rated concentration. The reductions in cost and energy use are largely due to use of large, countercurrent, multistage systems, but progressive optimization has also played a significant role.

Freeze concentration plants built in the mid-1980s cost roughly $440,000 per 1000 kg/hr of water removed. Four-effect, falling-film evaporators cost roughly $25,000 per 1000 kg/hr of water removed.

Roughly three units of thermal energy are needed to produce a unit of electrical energy. Therefore, current electrical energy use for freeze concentration corresponds to thermal energy use of 222 kWh/1000 kg. Based on energy use for heating, pumping, and operating steam jets that provide vacuum and product cooling, and allowing for thermal inefficiency, about 258 kWh of effective thermal energy are used per 1000 kg water evaporated in four-effect juice evaporators (Chen, 1982).

FREEZING OF SOLUTIONS

Freezing Point Depression

Solutes depress freezing points of solutions. Figure 5.3 depicts T, the freezing point of soluble coffee extract, vs. n_s, the solute mass fraction in the extract. Water is selectively removed from solutions when ice forms, so solute concentration increases and solution volume decreases. As concentration increases, freezing point depression $(T - T_o)$ increases. At constant pressure (e.g., atmospheric pressure), freezing points are governed by Eqs. (1) and (2)

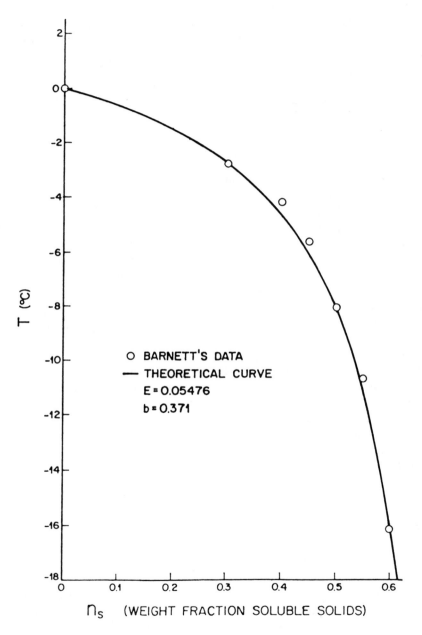

Figure 5.3 Freezing point vs. solute concentration plot for soluble coffee extract.

$$\frac{\partial(\ln a_w)}{\partial T} = \frac{18.02 \ \Delta H}{RT^2} \tag{1}$$

$$\ln (a_w) = \frac{18.02 \ \Delta H_m (T - T_0)}{RT \ T_0} \tag{2}$$

where a_w is the thermodynamic activity of water in the solution, T is in degrees K, $T_0 = 273.16$ K, the freezing point of pure water, and ΔH is the latent heat of fusion of ice at T. At T_0, $\Delta H = \Delta H_0 = 333.5$ kJ/kg. $\Delta H = \Delta H_0 + (C_w - C_l)(T - T_0)$. C_w, the heat capacity of liquid water, = 4.18 kJ/kg.K. C_l, the heat capacity of ice, = 2.09 kJ/kg.K. ΔH_m, the mean latent heat of fusion between T_0 and T, = ΔH_0 + $0.5(C_w - C_l)(T - T_0)$. Since ΔH_m and T decrease in a roughly parallel fashion, Eq. (2) can be replaced by Eq. (3).

$$\ln (a_w) \approx \frac{18.02 \ \Delta H_0 \ (T - T_0)}{RT_0^2} = 0.009687(T - T_0) \tag{3}$$

The maximum error caused by this approximation is usually less than 2%. Usually $a_w > 0.88$. Therefore, Raoult's law applies reasonably well; and $a_w \approx X_w$, the mole fraction of solvent water in the residual unfrozen solution. Food solutes bind water, which consequently is not part of the solvent water

$$a_w \approx X_w = \frac{(n_w - bn_s)}{(n_w - bn_s) + En_s} \tag{4}$$

where n_w and n_s are the respective mass fraction of water and solute in the solution, b is the mass of water bound per unit mass of solute, $E = 18.02/M_s$, and M_s is the effective molecular weight of the solute. Further (Schwartzberg, 1976, 1981),

$$\frac{En_s}{(n_w - bn_s)} \approx \frac{18.02 \ \Delta H_0 \ (T_0 - T)}{RT_0^2} = 0.009687(T_0 - T) \tag{5}$$

$n_w = 1 - n_s$. Therefore, for T in °C,

$$T = \frac{-(103.23)En_s}{(n_w - bn_s)} = \frac{-(103.23)En_s}{[1 - (1 + b) \ n_s]} \tag{6}$$

Figure 5.4 depicts how well experimental freezing point data for sucrose and glucose are fitted by Eqs. (5) and (6). If several solutes are present, $E = 18.02 \ \Sigma(n_i M_i)/\Sigma n_i$, where n_i are the weight fractions and M_i the molecular weights of the solute components. E and b can be treated as empirical constants and determined

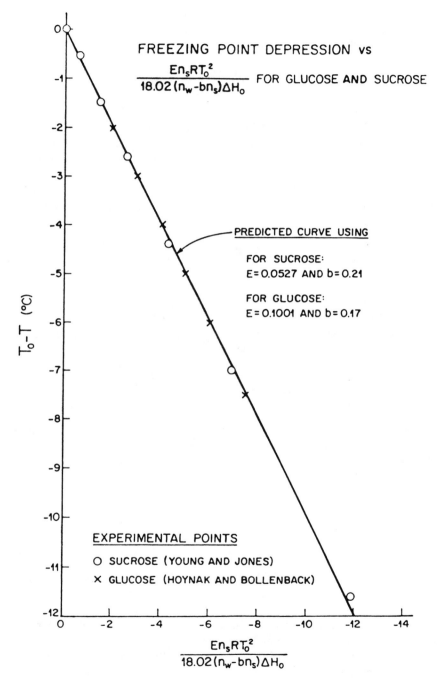

Figure 5.4   Freezing point concentration—fit for glucose and sucrose.

by regression using Eq. (5) or (6) and experimental freezing point data. For example, for the coffee extract data used in Fig. 5.3, $E = 0.0548$ and $b = 0.371$. If E is known, b can be determined from a single, reasonably large value of $(T - T_0)$ and corresponding n_w or n_s, but it is better to use several $(T - T_0)$ vs. n_s pairs and determine b by regression.

Other Freezing Point Depression Relationships

Freezing point depression data can also be correlated by formulas based on use of activity coefficients without explicitly accounting for bound water. These are roughly as accurate as Eqs. (5) and (6), but less convenient for computation. Eq. (7) (Riedel, 1949,1951), correlates $(T_0 - T)$ for fruit and vegetable juices and juice concentrates.

$$(T_0 - T) = 10n_s + 50ns_s^3 \qquad (7)$$

Deviations rarely exceed 0.5°C except for rhubarb juice, which is exceptionally poor in sugars. Mean-square differences between $(T_0 - T)$ from Eq. (6) and Eq. (7) are minimized when $E = 0.0988$ and $b = 0.0524$. The sugars in juices tested by Riedel were primarily monosaccharides; for juices richer in sucrose, e.g., orange juice, $(T_0 - T)$ is smaller. T_i is the freezing point prior to freeze concentration. Table 5.1 lists $(T_0 - T_i)$ vs. n_s for single-strength juices. Monosaccharide and disaccharide mass fractions, n_m and n_d, respectively, are also listed for some juices. Figure 5.5 depicts T vs. n_s for fruit juices, wine, and corresponding concentrates.

Concentration Limits

As freezing progresses and concentration increases, less and less ice forms per unit drop in T. At critical concentrations, solutes may solidify along with ice and are difficult to separate. In binary solutions, this occurs at constant T and constant composition. For example, ice and NaCl crystallize together at −21.1°C when 23.3% aqueous NaCl is cooled. Ice forms alone at lower concentrations; at higher concentrations, NaCl crystallizes by itself. The mixture of crystals is called a eutectic; the temperature and concentration at which it forms, the eutectic point. Food solutions that can form eutectics, e.g., sucrose solutions, sometimes do not, because solute nuclei, seed crystals needed to trigger solute crystal growth, do not form readily during freezing.

When solutions contain more than one solute, behavior is more complex. Water and a solute may solidify simultaneously; but concentrations of remaining solutes continue to increase; so T continues to drop. In multicomponent food solutions, ice formation and concentration increases progressively slow down and ultimately

Table 5.1   Solute Content and Freezing Point Depression of Single-Strength Juices

Juice	n_s	n_m	n_d	$(T_o - T_i)$
Apple A[a]	0.132	0.091	0.019	1.4
Apple B[a]	0.131	0.090	0.011	1.3
Apple C[a]	0.148	0.085	0.036	1.3
Apple D[a]	0.129	0.083	0.016	1.2
Apple E[a]	0.129	0.088	0.010	1.2
Pear A[a]	0.164	0.106	0.003	1.6
Pear B[a]	0.140	0.088	0.016	1.5
Pear C[a]	0.130	0.080	0.008	1.5
Grape[a]	0.160	0.134	0.004	1.95
Cherry A[a]	0.137	—	—	1.5
Cherry B[a]	0.127	—	—	1.4
Raspberry[a]	0.0645	0.029	0.002	0.7
Red currant[a]	0.082	0.049	0.000	0.9
Black currant[a]	0.121	—	—	1.9
Gooseberry[a]	0.070	0.043	0.0003	0.8
Rhubarb[a]	0.0229	0.005	0.000	0.4
Cucumber[b]	0.035	—	—	0.8
Spinach[b]	0.045	—	—	0.5
Tomato[b]	0.045	—	—	0.7
Asparagus[b]	0.065	—	—	0.9
Snapbeans[b]	0.070	—	—	1.1
Raspberry[b]	0.080	—	—	1.3
Strawberry[b]	0.085	—	—	1.0
Peach[b]	0.105	—	—	1.3
Blueberry[b]	0.108	—	—	1.7
Carrot[b]	0.12	—	—	2.0
Pea[b]	0.13	—	—	1.7
Italian prune[b]	0.15	—	—	2.4
Sweet cherry[b]	0.225	0.214[e]	0.0002[e]	3.2
Sweet cherry[d]	0.21	—	—	2.6
Orange[c]	0.117	0.047[e]	0.057[e]	1.1
Orange[d]	0.110	—	—	1.2
Lemon[c]	0.09	0.010[e]	0.070[e]	1.5

[a]Riedel (1949).
[b]Chen (1986).
[c]Gutschmidt (1964).
[d]Maltini (1975).
[e]Van Pelt and Swinkels (1984).

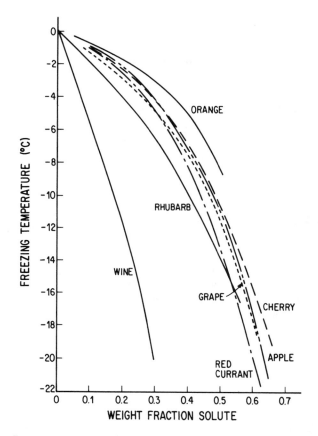

Figure 5.5 Freezing point vs. solute concentration plot for fruit juices and wine.

stop. For juices and beverage extracts, concentrations no longer increase after roughly 65% concentration is reached.

As concentration increases, viscosity increases markedly. Viscosities of 65% fructose, glucose, sucrose, and raffinose, solutions at their freezing points range from 3 to 5 kg/m·s (30–50 poise). Viscosities for 65% fruits juice and beverage extracts are even larger. Ice crystals grow very slowly at such viscosities, and crystals large enough to be cleanly separated become difficult to produce. Separation itself also becomes more difficult, and frictional heating increases. Refrigeration efficiency decreases at T needed for high concentrations, and more energy is expended per unit of water removed. Consequently, practical maximum concentrations for freeze concentration are usually in the 45–55% range.

Freezing Curves

Figure 5.6a is a typical plot of temperature T vs. time for freezing pure water. T drops smoothly until ice nuclei form in large numbers at −5 to −6°C. Then, ice crystals start to grow. T rises to 0°C and stays there until all liquid water changes into ice. Then T drops. ΔT, the temperature driving force for heat transfer, diminishes, so T drops at a progressively decreasing rate and asymptotically approaches the coolant temperature T_e. Figure 5.6b is a freezing curve for a 10% NaCl solution. T drops roughly five degrees below T_i; ice nuclei form, then ice crystal growth starts, and T rises rapidly to −6.1°C, T_i for 10% NaCl. As ice forms, NaCl concentration in the solution rises and T progressively drops, slowly at first, then more and more rapidly. T drops below the eutectic temperature, NaCl nucleates, and NaCl and water solidify together at constant composition, 23.3% NaCl at −21.1°C. When solidification is complete, T drops and asymptotically approaches T_e. Somewhat similar behavior is observed in Fig. 5.6c, a freezing curve for a fruit juice containing 12% solute. Subcooling occurs, nucleation starts at roughly −6.5°C, and T rapidly rises to −1.2°C, T_i for the juice. Concentration increases in the residual solution, T progressively drops and approaches T_e, but no discernable eutectic forms. The residual solution becomes very viscous or glassy, and ice formation radically slows down or stops.

MATERIAL BALANCE CONSIDERATIONS

Residual Water and Ice Contents

Dividing Eq. (5) applied at T_i, where the original water mass fraction is n_{wo}, by Eq. (5) applied at $T < T_i$, where mass fraction of liquid water in the mixture is n_w, we find

$$\frac{(n_w - bn_s)}{(n_{wo} - bn_s)} = \frac{(T_0 - T_i)}{(T_0 - T)} \tag{8}$$

Figure 5.7 depicts how Eq. (8) fits residual water content data for sucrose and glucose solutions during freezing. The weight fraction of ice $n_I = (n_{wo} - bn_s) - (n_w - bn_s)$. Therefore,

$$n_I = (n_{wo} - bn_s) \frac{(T_i - T)}{(T_0 - T)} \tag{9}$$

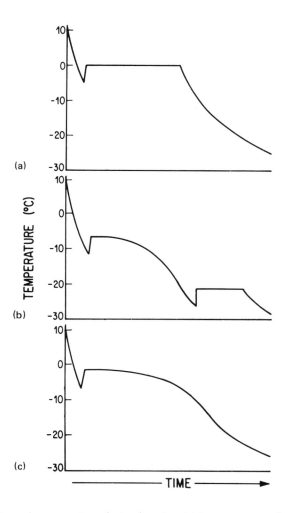

Figure 5.6 Temperature vs. time during freezing: (a) for pure water; (b) for 10% NaCl; (c) for 12% fruit juice.

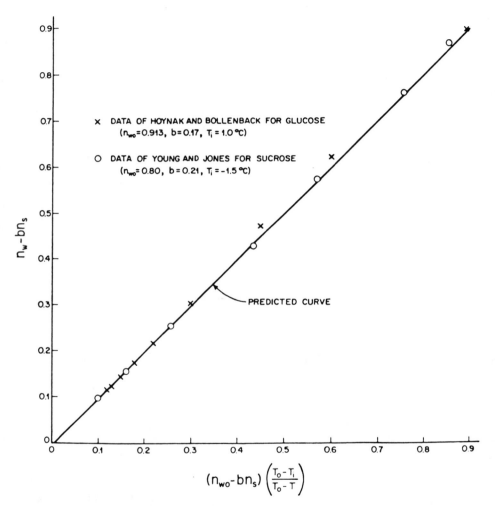

Figure 5.7 Residual freezable water content $(n_w - bn_s)$ vs. $(n_{wo} - bn_s)(T_o - T_i)/(T_o - T)$ for glucose and sucrose.

n_s' the mass fraction of solute in the residual solution = $n_s/(n_w + n_s)$. Hence,

$$n_s' = n_s \frac{(T_o - T)}{[n_{wo} (T_o - T_i) + bn_s (T_i - T) + n_s(T_o - T)]} \tag{10}$$

On the centigrade scale, Eq. (10) becomes

$$\frac{T}{T_i} = \frac{n_s'}{n_s} \frac{[1 - n_s(1 + b)]}{[1 - n_s'(1 + b)]} \qquad (T \text{ in } {}^\circ C) \tag{11}$$

Eqs. (9), (10), and (11) are handy for calculating amounts of ice formed during freeze concentration, solution concentrations at given T, T required for chosen values of n_s', or T during freezing. For example, from Eq. (10), T should be $-10.14°C$ to produce a concentrate containing 50% solute from feed containing 12% solute if $T_i = -1.2°C$ and b = 0.15. The residual solution contains only 0.12 kg of water per 0.12 kg of solute vs. 0.88 kg of water per 0.12 kg of solute in the feed, so $n_I = 0.88 - 0.12 = 0.76$ kg ice/kg feed have to be made. The same answer is obtained from Eq. (9).

Practical Ice Levels

Figure 5.8 depicts how amounts of ice and concentrate change during freezing of a 20% solution of soluble coffee. The equilibrium curve provides n_s' vs. T. The vertical line passes through n_s. Based on the "lever rule," lengths of the labeled line segments are proportional to relative masses of ice and concentrate. At 55% concentration, there is much more ice than concentrate. Slurries that contain too much ice are difficult to pump and stir. To maintain fluidity, concentrate is recirculated in freeze concentration systems to keep ice levels lower than 36% by volume (32% by weight).

HEAT REMOVAL

Energy Requirements

Energy consumption will be analyzed in terms of heat transferred and energy used per kg of feed or per 1000 kg of water removed. Q_b, the basic load, is the heat removed to convert a kg of feed into an appropriate mixture of ice and residual solution. Q_b, Q_p, frictional heat produced by pumping, agitation and scraping, and Q_L, heat inleakage, are components of Q_e, the heat removed from the system by evaporating refrigerant, i.e., $Q_e = Q_b + Q_p + Q_L$. Work W_r is used to drive the

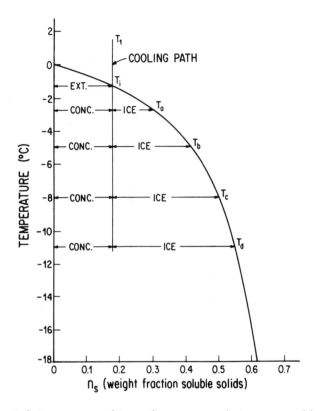

Figure 5.8 Relative amounts of ice and concentrate during course of freezing.

refrigeration system. It can be broken up into components, W_{rb}, W_{rp}, etc., which depend on similarly labeled components of Q_e. Q_p also equals the work used to carry out pumping, agitation, and scraping and other tasks that cause frictional heating of process streams. Work used to drive motors, fans, and devices that do not transfer heat into the system = W_a. The total energy used = $Q_p + W_{rp} + W_{rb} + W_{rL} + W_a$.

The minimum work needed for freeze concentration is W_{rb}. To determine W_{rb}: (A) use Eq. (10) or (11) to calculate the T needed to produce the desired n_s' ; (b) determine Q_b; and (c) from refrigeration cycle calculations determine W_{rb} for cooling load Q_b. $Q_b = Q_o + Q_f$. Q_o is heat removed in reducing feed T from its initial value T_b to T_i, Q_f the heat removed in forming ice and reducing T from T_i to its discharge values. $Q_o = C_o(T_b - T_i)$, where C_o is the heat capacity of unfrozen feed. T_b depends on whether heat exchange between cold product streams and feed is used. Q_f depends on how ice is removed from the system. In cocurrent operation, ice is removed only after the final n_s' is reached. Ice can also be removed at intermediate concentrations. In countercurrent operation, ice flows countercurrent to the concentrate and is discharged at the dilute end of the system.

Expressions for Q_f and examples for these cases are presented below. The freezable water content $(n_{wo} - bn_s) = N$. The examples are based on a kg of feed and conditions previously discussed, i.e., $b = 0.15$, $n_s = 0.12$, $n_{wo} = 0.88$, $T_i = -1.2°C$, $n_s' = 0.5$ and $T = -10.14°C$. Hence, $N = 0.862$; $(T_i - T) = 8.94°C$; $(T_o - T_i) = 1.2°C$; and $(T_o - T) = 10.14°C$. It is assumed that feed temperature $T_b = 10°C$, so $(T_b - T_i) = 11.2°C$.

Heats of solution and dilution can be neglected, and C_o and C_f (the heat capacity in the fully frozen state) can be estimated from C_s, the partial heat capacity of the solute, and C_w and C_I. For sugar-like solutes, $C_s \approx 1.42$ kJ/kg·K, so $C_o \approx (C_w - 2.76$ $n_s) = 3.849$ kJ/kg·K; $C_f \approx [NC_I + n_s(C_s + bC_w)] = 2.047$ kJ/kg·K; and $Q_o = 43.1$ kJ/kg feed.

(a) For cocurrent operation.

$$Q_f = (T_i - T)\left[C_f + \frac{N\Delta H_o}{(T_o - T)}\right] \tag{12}$$

$Q_f = 271.8$ kJ/kg feed. Adding Q_o, $Q_b = 271.8 + 43.1 = 314.9$ kJ/kg feed. Since $1000/0.76$ kg of feed form 1000 kg ice, and 3600 kJ = 1 kWh, Q_f and Q_b should be multiplied by $1000/(0.76 \times 3600) = 0.365$ to express them in kWh/1000 kg ice formed. $Q_f = 99.2$ and $Q_b = 114.9$ kWh/1000 kg ice formed.

(b) If ice is removed in differential increments as soon as it is formed,

$$Q_f = n_s(C_s + bC_w)(T_i - T) + N\left\{C_I(T_o - T_i)\ln\left(\frac{(T_o - T)}{(T_o - T_i)}\right) + \frac{\Delta H_o(T_i - T)}{(T_o - T)}\right\} \tag{13}$$

Substituting appropriate values, $Q_f = 260.4$ and $Q_b = 303.5$ kJ/kg feed, or 95.0 and 110.9 kJ/1000 kg ice formed.

(c) For countercurrent operation

$$Q_f = n_s C_s(T_i - T) + \frac{(T_i - T)}{(T_o - T)}\left[C_w[n_{wo}(T_o - T_i) + bn_s(T_i - T)] + N\Delta H_i\right] \tag{14}$$

where $\Delta H_i = \Delta H_o - (C_w - C_I)(T_o - T_i) = 333.5 - 2.09(T_o - T_i)$. For the feed used, $\Delta H_i = 331.0$. Substituting appropriate values, $Q_f = 257.8$, and $Q_b = 300.9$ kJ/kg feed. These correspond to 94.1 and 109.8 kJ/1000 kg ice formed, respectively.

Freezing in Stages

In freeze concentration systems, ice often forms in increments (Δn_I) in stages, which operate at temperatures T_j. $(\Delta n_I)_j$ = kg ice/kg feed form in the jth stage from the feed end of the system. $(n_w)_j$ is the kg water/kg feed in stage j. By a mass balance,

$$(n_w - bn_s)_j = (n_w - bn_s)_{j-1} - (\Delta n_I)_j \qquad (15)$$

In °C, $T_j = T_i(n_{wo} - bn_s)/(n_w - bn_s)_j$. Assume that three stages are used, $(\Delta n_I)_1 = 0.45$, $(\Delta n_I)_2 = 0.22$, and $(\Delta n_I)_3 = 0.09$. $\Sigma \Delta n_I$ still = 0.76 kg ice/kg feed for the whole system. $(n_w - bn_s)_1 = 0.412$; $(n_w - bn_s)_2 = 0.192$; and $(n_w - bn_s)_3 = 0.102$. $T_1 = -1.2$°C $(0.862/0.412) = -2.51$°C; $T_2 = -5.38$°C; and $T_3 = -10.14$°C.

Eqs. (12) and (14) can be adapted to compute $(Q_f)_j$, the part of Q_f removed in stage j.

For cocurrent operation and j = 1:

$$(Q_f)_1 = (T_i - T_1) \left[C_f + \frac{N\Delta H_o}{(T_o - T_1)} \right] \qquad (16)$$

for j > 1

$$(Q_f)_j = C_f (T_{j-1} - T_j) + N\Delta H_o(T_o - T_i) \left[\frac{1}{(T_o - T_{j-1})} - \frac{1}{(T_o - T_j)} \right] \qquad (17)$$

If ice is withdrawn after each freezing step, $(Q_f)_1$ is still given by Eq. (16). But for j > 1,

$$(Q_f)_j = (T_{j-1} - T_j) \left[C_f + \frac{(n_w - bn_s)_{j-1}\Delta H_o}{(T_o - T_J)} \right] \qquad (18)$$

For countercurrent operation, Eq. (14) applies with $(Q_f)_j$ replacing Q_f, T_j replacing T, and $\Delta H_1 = \Delta H_o - (C_w - C_I)(T_o - T_1)$ replacing ΔH_i.

Table 5.2 lists $(Q_f)_j$ for the three modes of operating using the selected $(\Delta n_I)_j$. Q_o should be added to $(Q_f)_1$ to obtain $(Q_b)_1$.

REFRIGERATION WORK

Refrigerant evaporates at T_e while removing Q_e, and condenses at T_c while transferring heat Q_c to cooling water or ice. $Q_c = Q_e + W_r$. T_e must be lower than

Table 5.2 Ice Formation Levels, Operating Temperatures, and Heat Loads for Different Types of Three-Stage Operation

Stage no.	$(\Delta n_1)_j$ kg ice / kg feed	T_j (°C)	$(Q_f)_j$ for type of operation (kJ/kg feed)		
			Cocurrent	Staged withdrawal	Countercurrent
1	0.45	−2.51	152.7	152.8	152.4
2	0.22	−5.39	79.4	76.3	74.7
3	0.09	−10.14	39.7	33.0	30.7

T during freezing. T_c must be higher than the temperature of the available heat sink. W_r/Q_e increases as $(T_c - T_e)$ increases. As T_e decreases, the density of evaporated vapor decreases; and refrigerant mass-flow rates and refrigeration capacity decrease; but ΔT_e, i.e., $(T - T_e)$, the driving forces for heat removal during freezing, increases; so needed freezer area A decreases.

Scraped surface freezers (SSF) are usually used in FC systems. They cost roughly \$20,000 per square meter of heat-transfer area. Therefore, high ΔT_e (i.e., low T_e) are used to minimize needed SSF area.

$W_r/Q_f = (W_r/Q_e) \cdot (Q_b + Q_p + Q_L)/Q_f$. The smaller W_r/Q_f is, the more efficient freeze concentration is. Q_e and W_r/Q_e increase as T_e drops and ΔT_e increases; but $(Q_b + Q_p + Q_L)/Q_f$ decreases. Because of these opposing effects, W_r/Q_f will be smallest at some ΔT_e, which can be calculated if the T_j, T_c, Q_p per unit area, and heat-transfer coefficients U_f for freezers are known. This ΔT_e, however, is not an overall optimum, because freezer cost and the effect of ΔT_e on ice crystal size must also be considered. The same T_e is often used in all freezers in multistage systems, e.g., all T_e may be set at 253.2 K (−20°C). Then, refrigeration compressor(s) can run at a single suction pressure P_e without loss of efficiency due to refrigerant throttling.

Use of Discharged Ice

$Q_c = Q_f + Q_o + Q_p + Q_L + W_r$. Ice discharged from freeze concentration systems can be used to precool feed and remove part of Q_c. Since only part of one Q_c component, Q_f, causes ice production, there is not enough ice to condense all the compressed vapor. Part has to be condensed by cooling water, which frequently is available at $T_w = 24–27°C$. To avoid excessive water use, water flow is usually regulated so it leaves condensers at 30–45°C. If so, condensation takes place at

35–48°C. Ice-based condensation can take place at 3–5°C. W_r is much smaller when ice-based condensation is used. $W_r/Q_e = 0.0537$ for an imaginary cycle where Q_e is absorbed at $T_e = -10.14$°C with $\Delta T_e = 0$ and Q_c is rejected to ice at 0°C with $\Delta T_c = 0$. When Q_e is absorbed at $T_e = -20$°C and Q_c is rejected to ice at $T_c = 5$°C, $W_r/Q_e = 0.145$; and for absorption at $T_e = -20$°C and rejection to cooling water at $T_c = 35$°C, $W_r/Q_e = 0.375$.

If Q_e is absorbed at -20°C, 75% of Q_c is rejected to ice at 5°C, and 25% is rejected to cooling water at 35°C, $W_r/Q_e = 0.216$. Further, if heat exchange with discharged cold streams is used to precool feed and Q_p, Q_L and Q_o are negligible, W_r would $= 21.4$ kWh/1000 kg of ice for cocurrent operation; W_r would $= 20.6$ kWh/1000 kg ice if ice is removed as it forms; and 20.3 kWh/1000 kg ice for countercurrent operation.

Ice-based condensation can be provided by periodically switching the refrigerant compressor discharge pressure P_c. Condensation at low P_c would be used when ice is available and water-based condensation at high P_c the rest of the time. Alternatively, two sets of compressors, condensers, and condensate receivers could be used. Ice-based condensation would be used in one set, water-based condensation in the other. All the vapor would pass through ice-based set, where part will condense. The water-based set would handle any uncondensed vapor that remains.

THERMODYNAMIC ANALYSIS

Refrigeration Cycles

Let us examine refrigeration cycles for the countercurrent process and feed previously considered. In all cases, refrigerant rejects Q_i to ice and Q_w to cooling water, $Q_c = Q_i + Q_w$, and $Q_i = n_l[\Delta H_o + C_l(T_o - T_i)]$.

Ideal Cycles

W_r is smallest for a Carnot-like cycle in which ΔT_e and ΔT_c are 0, and heat exchange between cold discharged streams and feed precools feed to T_i. $Q_p = 0$. Q_e is accepted at T as T changes during freezing. Q_i is rejected at 0°C, Q_w at 27°C. ΔT_e and $\Delta T_c = 0$.

When mechanical refrigeration is used, Q_w and Q_i are rejected as refrigerant condenses at constant pressures P_c, and Q_e is accepted by evaporating refrigerant at constant pressure(s) P_e. Refrigerant vapor is compressed from P_e to P_c. In ideal and semi-ideal refrigeration evaporation takes place at -2.51°C, -5.39°C, and -10.14°C, T_j for the stages involved; Q_i is rejected at 0°C; Q_w at 27°C; and ΔT_e

and $\Delta T_c = 0$. In ideal refrigeration, compression is isentropic, compression efficiency = 100%; in semi-ideal refrigeration, $W_r = (W_r)_{ideal}/0.75$.

Real Cycles

In real cycles, ΔT_e and $\Delta T_c \neq 0$, and we assume $W_r = (W_r)_{ideal}/0.75$. Effects of different feed temperatures T_b and frictional heat input Q_p are examined. Table 5.3 lists W, W_r/Q_f, Q_c, Q_i/Q_c for ideal cycles and real cycles for three-stage countercurrent operation, for various ΔT_e and ΔT_c, or T_e and T_c. Multiply energy entries in Table 5.3 by 0.365 to convert them from kJ/kg feed to kWh/1000 kg ice removed for the feed and final concentration used.

W_r/Q_f for ideal cycles are much lower than ever attained for real freeze concentration. Low W_r/Q_f were obtained for desalination processes where direct evaporative cooling was used to remove Q_e and direct contact between compressed vapor and ice produced was used to reject Q_c. Q_p, ΔT_e, and ΔT_c were consequently very low. W_r/Q_f was close to W_r/Q_f for the $\Delta T_e = 5°C$ entry in Table 5.3.

Frictional heat input greatly increases W_r. Energy efficiency can be greatly increased by reducing Q_p. Q_p can be broken into two parts, Q_s heat produced by scraping in SSF, and $Q_r = Q_p - Q_s$. In multistage systems where $T_e = -20°C$ in all stages, $Q_p/Q_f \approx 1.25$ and $Q_s/Q_f \approx 0.56$. Q_s is proportional to SSF area A. If Q_s/A and Q_r/Q_f are constant, we can calculate Q_s/Q_f at one T_e from Q_s/Q_f at another ΔT_e from changes in A caused by changes in ΔT_e.

$$(A_2/A_1) = \frac{[1 + (Q_r/Q_f)](\Delta T_{e1}/\Delta T_{e2})}{[1 + (Q_p/Q_f)_1 - (Q_s/Q_f)_1(\Delta T_{e1}/\Delta T_{e2})]} \tag{19}$$

and

$$(Q_p/Q_f)_2 = (Q_r/Q_f) + (Q_s/Q_f)_1(A_2/A_1) \tag{20}$$

Based on Q_p/Q_f and Q_s/Q_f at $T_e = -20°C$, Q_p/Q_f will be 1.0 at $T_e = -28.9°C$. Operation at $T_e = -28.9°C$ with $Q_p/Q_f = 1.0$, and operation at $T_e = -20°C$ with $Q_p/Q_f = 1.25$ (and the same T_b for both cases) are worth comparing; 29% more SSF area is required at $-20°C$, but W_r/Q_f is smaller. When operation at $\Delta T_e = 5°C$ is compared with operation at $T_e = -20°C$, and Q_p is similarly accounted for, much more area is needed at $\Delta T_e = 5°C$. The increase in Q_p/Q_f due to increased A is so large that W_r/Q_f is also much larger at $\Delta T_e = 5°C$ than at $T_e = -20°C$.

Different evaporation temperatures cannot be obtained efficiently using a single compressor operating at a single inlet pressure P_e, but they could be obtained efficiently by using several compressors or by using a multistage compressor with inlet ports between the stages.

Table 5.3 W_r and W_r/Q_f, Q_c, and Q_i/Q_c for different refrigeration cycles and different T_e, T_c, T_b and Q_p/Q_f[a]

Case	Carnot cycle	Ideal mech. refr.	Semi-ideal mech.[b]	Nonideal mechanical refrigeration						
ΔT_e	0	0	0	3	5	15.2[c]	15.2[c]	15.2[c]	24.6[c]	24.6[c]
T_{e1}	T	−2.5	−2.5	−5.5	−7.5	−20	−20	−20	−29	−29
T_{e2}	T	−5.4	−5.4	−8.4	−10.4	−20	−20	−20	−29	−29
T_{e3}	T	−10.1	−10.1	−13.1	−15.1	−20	−20	−20	−29	−29
T_{ci}	0	0	0	3	5	5	5	5	5	5
T_{cw}	27	27	27	35	35	35	35	35	35	35
T_b	−1.2	−1.2	−1.2	6	6	6	6	20	6	6
Q_p/Q_f	0	0	0	0	0	0	1.25	1.25	0	1.0
W_r	3.8	5.2	7.1	24.1	30.7	55.6	176	197	80	200
W_r/Q_f	.015	.020	.028	.093	.119	.216	.684	.762	.310	.777
Q_e	258	258	258	286	286	286	636	662	286	543
Q_c	262	263	265	309	316	341	728	858	416	743

[a]Three-stage countercurrent operation, feed 12% solute, concentrate 50% solute, $T_i = -1.2°C$. Energy values in kJ/kg feed; Temperatures in °C.
[b]In semi-ideal cycles compression efficiency is 75%, ΔT_e and $\Delta T_c = 0$.
[c]Effective $\Delta T_e = \sum Q_{ej}/\sum(Q_{ej}/\Delta T_{ej})$.

ICE GROWTH

Ice

The ice we are familiar with, ice Ih, is a hexagonal, prismatic crystal that forms below 273.16 K (0°C) at atmospheric pressure. Its structure is shown in Fig. 5.9a. Oxygen atoms in constituent water molecules are situated at the vertices of the hexagonal cross-section, but do not lie in a common plane; the hexagon resembles a shallow reclining chair.

Ice Ih grows along four axes: three "a" axes which pass through vertices of the hexagonal cross-section, and a "c" axis normal to the hexagonal cross-section (see Fig. 5.9b). Information about ice Ih is provided by Hobbs (1974), Fletcher (1970), Fennema and Powrie (1964), Fennema et al. (1973), Dorsey (1968), Eisenberg and Kauzman (1969), Horne (1972), and Franks (1982).

Water forms nine other ices. Most exist only at pressures greater than 203 Mpa (2,000 Atm). Ice VI forms at pressures as low as 625 MPa (6,170 Atm) and temperatures as low as 0.16°C and can exist at temperatures up to 81.6°C, where the equilibrium pressure is 2,150 MPa (21,200 Atm); ice VII exists at even higher temperatures and pressures. Freeze concentration based on use of ice VI might greatly reduce W_r and provide low viscosities and high diffusivities that facilitate ice growth and ice separation, but operating at such high pressures would be very difficult.

Radius of Curvature Effects

Freezing temperatures for small crystals are slightly depressed because surface free energy per unit volume is larger for small crystals than for large crystals. Crystals of radius r are in equilibrium when total free energy is minimized. This occurs at solution temperatures $\Delta T'$ lower than equilibrium T for very large crystals.

$$\Delta T' = \frac{2 \sigma T}{\rho_s \Delta Hr} \qquad (21)$$

σ is crystal surface free energy per unit area, T is the equilibrium temperature in K for crystals with infinite r, and ρ_s is the density of the crystal, i.e., 917 kg/m^3 for ice. Reported σ for ice in liquid water range from 1.5 to 3.2×10^{-2} J/m^2 (Hobbs, 1974). In sample computations, we assume $\sigma = 2.0 \times 10^{-2}$ J/m^2. For 2-μm diameter crystals in solution whose normal T is 263.2K (−10°C), $r = 1 \times 10^{-6}$m, $\Delta H = 3.126 \times 10^5$ J/kg, and $\Delta T' = 0.0367$°C. For 10-μm crystals, $\Delta T' = 0.0073$°C. Equilibrium T are plotted vs. diameter D_p for ice in contact with 30% sucrose in Fig. 5.10.

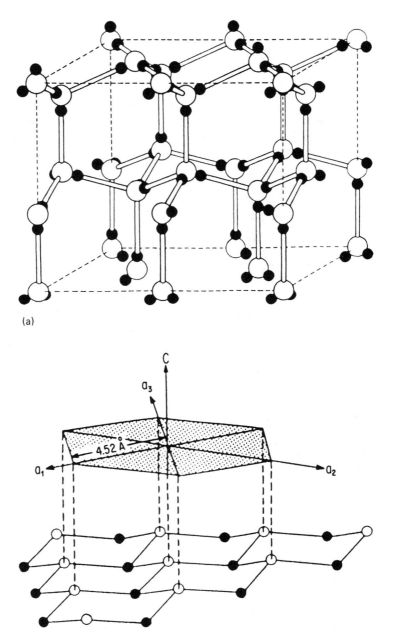

(a)

(b)

Figure 5.9(a) Crystal lattice structure of ice Ih. White circles are oxygen atoms; black circles are hydrogen atoms. *(Reprinted from Fletcher (1970) by courtesy of Cambridge University Press.)* (b) Growth-axes of ice Ih. *(Reprinted from Fennema et al. (1973) by courtesy of Marcel Dekker, Inc.)*

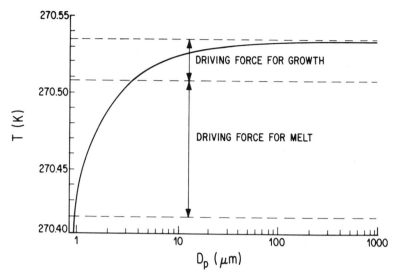

Figure 5.10 Freezing point vs. crystal diameter D_p for 30% sucrose solution, Reprinted courtesy of Grenco Special Applications bv.

Solutions in contact with small crystals may also be regarded as being unsaturated with respect to water concentration c_w. For crystals of radius r,

$$\ln (c_w/c_{we}) = \frac{2 (18.02) \sigma}{\rho_s RTr} \qquad (22)$$

where c_{we} is the concentration in equilibrium with crystals whose radius of curvature is infinite. R = 8.314 kJ/kg mole·K.

Nucleation

Ice-like clusters of water molecules form and break up in cold, ice-free solutions. At some degree of subcooling (or water supersaturation) these clusters have an even chance of surviving. Ice crystal nuclei form, and crystal growth follows. Homogeneous nucleation occurs when small drops of very pure water or particle-free solutions are subcooled to −40°C (Hobbs, 1974). Particles that facilitate nucleation are usually present in larger amounts of solution, so heterogeneous nucleation usually occurs 5–6°C below T_i. Fragments that break off ice crystals often provide secondary nuclei.

Ice Crystal Growth in Solution

Ice growth in solutions is affected by rates at which water diffuses to ice crystal surfaces, how solutes diffuse away from the surfaces, how fast and where water incorporates into ice crystals, radius of curvature effects, rates of removal of heat from the crystal growth surfaces, and interactions between these factors. Overall rates of ice formation are usually regulated by rates of heat removal; the remaining factors determine where local growth occurs and crystal shape and size.

The underlying arrangement of water molecules in the ice lattice remains the same regardless of circumstances of growth, but external shape varies considerably. Branched crystals form in highly subcooled solutions; tree-like crystals (dendrites) form when solutions freeze on chilled surfaces; snowflakes form when ice grows by freezing of water vapor in air. Rapid freezing in the presence of components other than water promotes branching. Slow freezing and low levels of impurities promote smooth crystal formation.

ICE CRYSTAL FORM

Tips that project from ice crystals are exposed to solution whose water content has not been depleted by prior ice formation. Water can diffuse or transfer rapidly to these tips from many directions. Water diffuses more slowly into regions below the tip, particularly stagnant regions between dendrite trunks and branches where c_w are depleted, n_s are higher, and freezing points are lower because of prior ice growth.

Heat is released when water changes into ice, and must be conducted away from growth sites for growth to continue. When crystals grow because of bulk subcooling, heat can transfer away from tips rapidly. On the other hand, radii of curvature are very small at tips of dendrites, so tips T have to be lower to sustain tip growth. Thus, tip growth and dendrite formation occur when bulk subcooling is large. Smooth crystals grow when bulk subcooling is too low to overcome radius of curvature effects at incipient projections.

When suspended, small ice crystals grow in water or aqueous solutions subcooled < 0.9°C, disc-like oblate spheres initially develop. "a" axes grow roughly four times as fast as "c" axes (Huige, 1972). Apparently, screw dislocations or surface imperfections which favor water molecule incorporation are present at the ends of "a" axes, but missing at ends of "c" axes. When the diameters of the growing discs reach roughly 3 mm, notches develop, branching starts, and crystals becomes hexagonal. As growth continues, dendritic branches form. Similar growth occurs at 0.9–2.7°C subcooling; but branching and dendrite formation occur at smaller diameters. More complex crystals form if supercooling > 2.7°C. Since branches retain solution, subcooling > 0.9°C should not be used during final stages of freeze concentration. In practice, ripening-induced subcooling of 0.015–0.02°C is usually used and rounded crystals with final diameters ranging between 200 μm and 500 μm are produced (see Fig. 5.25).

When ripening is used to produce large crystals, rapid growth is obtained if very small crystals are produced initially. To produce such crystals, short holdup time (e.g., 12 sec) and high recirculation rates are used in SSF (Van Pelt and Roodenrijs, 1982) in freeze concentration systems. Only 7.5–8% of the solution entering such SSF change into ice per pass; 4 sec holdup time and 2.2% conversions per pass are cited in earlier work (Thijssen and Huige, 1977). Holdup times and conversions per pass in SSF for ice cream are 24 sec and 32% (Arbuckle, 1986). Cold concentrates are very viscous, so high recirculation rates cause high flow pressure drop and frictional heating. Flow rates through SSF in freeze concentration systems are adjusted to minimize combined disadvantages due to in-freezer ripening and flow friction.

Freezing on Surfaces

The thermal conductivity of ice is roughly 2.25 W/m·K; the thermal conductivity of aqueous solutions and water range between 0.44 and 0.558 W/m·K. Heat is readily conducted along ice trunks away from tips of ice crystals growing on chilled surfaces. Water diffuses down to the base of these crystals very slowly. The same situation prevails for tips and bases of branches that project out from trunks of growing crystals. Thus, trunk and branch tips propagate; and dendrites almost always form when ice is produced from aqueous solutions in SSF. In contrast, when ice forms from pure water on chilled surfaces, adequate water is always available at growth sites, and dense polycrystalline ice forms.

"a" axes grow faster than "c" axes when ice crystals form on a chilled SSF wall. As the crystals grow, solution near the wall becomes slightly lean with respect to water. The "a"-axis tips reach into solution richer in water, where growth is faster; "c"-axis tips lie closer to the wall, where growth is slower. This reinforces "a"-axis predominance, an inference supported by the 60° branching pattern observed in dendrites.

Solutes are rejected when water turns into ice. Therefore, lateral c_w gradients develop as ice grows on chilled surfaces. The gradients are steep when growth is fast and less steep when growth is slow. Lateral gradients in T also develop, but are small because heat transfer is fast compared to mass transfer. T at the surface rises instantaneously as subcooling is relieved when new crystals form, then drops again. As T drops, new crystals form where local c_w are high enough for nucleation to occur. When c_w gradients and rates of temperature drop are small, i.e., when freezing is slow, this occurs far from previously formed crystals. When freezing is fast, nucleation occurs closer to previously formed crystals. This is illustrated in Fig. 5.11.

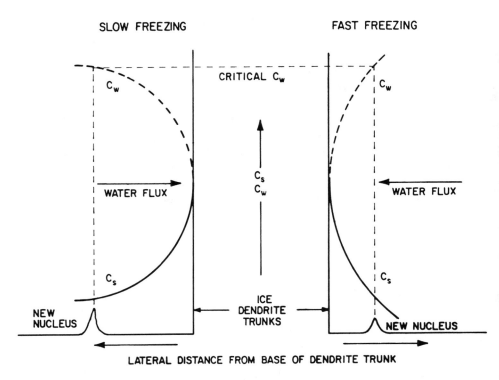

Figure 5.11 Concentration gradients near the base of an ice crystal growing on a freezer wall and their effect on nucleation. C_s = solute concentration. C_w = water concentration.

FREEZERS

In food FC systems, ice is usually produced in cylindrical SSF cooled by evaporating refrigerant, usually R22 or ammonia. Dendrites form on the heat-transfer surface, with their trunks normal to the surface. Fairly regularly spaced branches form at angles of 60° from the trunk (Barnett, 1973); similar sub-branches then form on the branches. Fig. 5.12 depicts ice formation and scraping in SSF (e.g., Votators or Contherms) where rotating scrapers are used. Ice grows rapidly near the outermost tips of the dendrites, and less rapidly on dendrite branches. Heat, removed to sustain growth, has to be conveyed over progressively larger distances as dendrites grow. Therefore, growth progressively slows down. In Fig. 5.13, ice depth at contact time t is plotted vs. $(t)^{1/2}$ for a stainless steel surface maintained at −5°C and dipped into 5% or 10% sugar solutions at T_i (Schwartzberg,

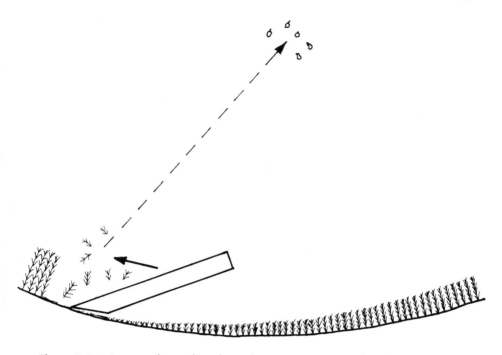

Figure 5.12 Ice crystal growth and transformation in a scraped-surface freezer.

1988). Ice depth and length of the heat-transfer path are proportional to $(t)^{1/2}$. Consequently, ice-layer heat transfer coefficients are proportional to $(t)^{-1/2}$. Time intervals between scraper passes in SSF are inversely proportional to scraping speed. If the ice layer is the dominant heat-transfer resistance, heat-transfer coefficients in SSF should be directly proportional to the square root of scraping speed. However, at scraping speeds and wall thicknesses usually used, tube walls provide the most significant resistance. Scraping removes ice and improves heat transfer but produces heat, which reduces the net amount of ice formed. Scraping speed should be optimized to reduce FC energy consumption.

Lengths of dendrites depend on ΔT_e as well as t. Based on the data for Fig. 5.13, and the 4.5°C ΔT_e used in obtaining that data, I estimate that in a SSF with two blades operating at 120 rpm, a 6-mm thick stainless steel wall and a ΔT_e of 15°C, 25- to 50-µm long dendrites would be produced. Mass- and heat-transfer considerations (Schwartzberg, 1977; Bomben and King, 1982) indicate that spacing between dendrites should be inversely proportional to the square root of the freezing rate. Based on ice growth patterns depicted by Barnett (1973), depths

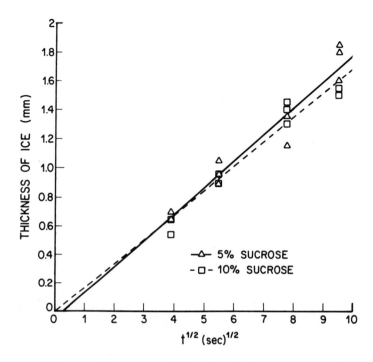

Figure 5.13 Thickness of ice layer vs (t)$^{1/2}$ for freezing from sucrose solutions carried out on a stainless steel surface maintained on −5°C.

of samples, and freezing times be used, I estimate that for SSF operating conditions just cited, the average space between dendrite trunks would be roughly 20–25 µm. The trunks themselves may be roughly 5-µm thick. Thus, thin dendrites that probably break after scraping are produced in SSF. Due to ripening (discussed later), these fragments change into irregular, rounded crystals 5–20 µm in diameter by the time they leave FC SSF (Huige, 1972). Figure 5.14, traced from a photomicrograph taken by Huige, depicts crystals freshly discharged from a SSF.

In SSF used to make ice cream, residence times are longer, and diameters of discharged crystals range from 26 to 58 µm. Overrun, increased fat and nonfat solids content, higher pH, and emulsifiers tend to reduce sizes of these crystals (Arbuckele, 1986).

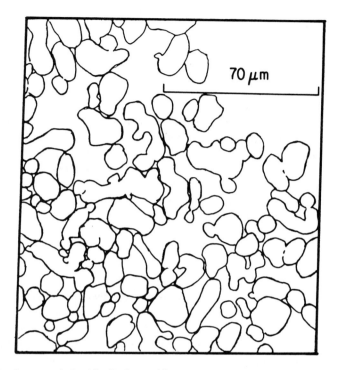

Figure 5.14 Ice crystals freshly discharged from scraped-surface freezer. Ripening has already started. (*Reprinted courtesy of N. Huige.*)

Scraping Power

Five to 12.5 kW of scraper drive power is used per m^2 of heat transfer surface in SSF used to make ice cream slushes. Installed SSF scraper drive power in recently constructed freeze concentration systems corresponds to 3.8 kW/m^2. Correlations for scraper power use nonfreezing service in scraped-surface heat exchangers (Trommelen and Beek, 1971; Harrod, 1986) indicate that only 0.2–0.3 kW/m^2 of the power used during freezing is used for fluid agitation. Apparent, heat-transfer coefficients U of 1100–1400 W/$m^2\cdot$K are obtained in SSF producing 8–10% ice; 600–1100 W/$m^2\cdot$K are obtained with 20–25% ice. No doubt, part of the reduction in U is caused by friction due to the presence of ice. Ice-cream freezers expend power in kneading slush. Since dilute slushes are formed in freeze concentration

SSF, kneading probably does not occur, and scraping ice off walls probably consumes most of the power used.

Reducing Friction

Suitable crystal growth could be obtained with less friction by: (a) using larger ΔT and slower scraping; (b) using types of scraping and coatings that reduce friction; (c) using less intense scraping and longer ripening; (d) using growth produced by bulk subcooling instead of ripening; (e) freezing feed in tubes that are slightly rewarmed at regular intervals to release ice (Thijssen, 1979); (f) using cooling produced by partial evaporation of the feed's water content; (g) using liquid-liquid contact with evaporating refrigerant to carry out freezing; (h) forming gas hydrates instead of ice. Some of these ideas are examined below.

ΔT and Scraping Speed

Scraping interrupts crystal growth and promotes breakage, so faster scraping may produce smaller crystals. Since crystal thickness decreases as freezing rates increase, long, thin crystals that break into small fragments probably can be produced by using higher ΔT at lower scraping speeds. W_r/Q_e would increase; but if suitably low scraping speeds are used, $Q_f/(Q_b + Q_p + Q_L)$ would probably increase to an even greater extent. Therefore, W_r/Q_f might be minimized by simultaneously adjusting ΔT_e and scraping speed. Ice production and freezer utilization might also increase.

Ice Adhesion

Figure 5.15 shows sheer strength as a function of T for ice frozen on stainless steel (S.S.) and polystyrene surfaces (Jellinek, 1959, 1972; Hobbs, 1974). On S.S., adhesive failure occurs between 0 and −13°C; below −13°C, failure occurs in the ice itself. Adhesive shear strength is 40% greater on aluminum than on S.S. Adhesion is smaller for ice formed from sucrose solutions than for ice formed from pure water. This probably occurs because ice trunk coverage decreases as concentration increases. The adhesive strength of ice in shear at −4.5°C is 0.45 kg/cm² on S.S. with a mirrow polish vs 6.0 kg/cm² for S.S. finished on a lathe (Jellinek, 1972). On very smooth surfaces, shear stress peaks just before failure. The peak stress increases as rates of shear increase. Stress levels off to 0.07–0.14 kg/cm² after failure occurs, i.e., while ice is sliding on the surface.

Plastic coatings greatly reduce adhesion (Raraty and Tabor, 1958; Jellinek,

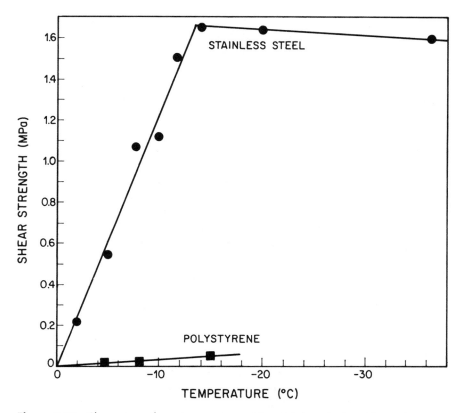

Figure 5.15   Shear strength vs. temperature at ice-stainless steel and ice-polystyrene junctions (after Jellinek, 1959 (*Reprinted courtesy of J. Colloid Science.*)

1959; Landy and Freiberger, 1967) and decrease the lower temperature limit for adhesive failure. Teflon FEP reduces adhesion 90%. Therefore, coating scraped surfaces with Teflon FEP or another suitable plastic might greatly decrease scraping power expenditure. Unfortunately, scraped, thin coatings may not last long; thick coatings will markedly reduce heat-transfer coefficients. Coatings may also affect ice nucleation.

Schoerner (1984) patented use of plastic-coated, shell-and-tube freezers (CSTF) to facilitate ice release. The solution being frozen either flowed through filled tubes or as a falling film down the walls of vertical tubes. Many coatings, polyesters, polytetrafluoroethylene, polyethylene, nylon, etc., are cited as being useful, but no operative examples are provided. Electrically polished chrome plating provided better performance in many cases (Johnson, 1989). CSTF were tested in a desalination plant at Yanbu, Saudi Arabia (Midwest Research Institute, 1987). Freezeup occurred at ΔT_e larger than 4°C or brine concentrations lower than 8%. $Na_2SO_4 \cdot 10H_2O$ precipitated at brine concentrations higher than 11%.

Huige (1972) used a Teflon heat exchanger in freeze concentration tests. Thijssen (1968) used a highly polished heat exchanger, which provided bulk subcooling of 6°C. Wall growth of ice was completely suppressed.

Shuttling Ball Freezer

Johnson (1984) invented a shuttling-ball freezer (SBF) for use in freeze concentration (Concentrex, 1985). SBF cost 80% less per unit of freezing capacity than conventional SSF. They produce slushes containing 10–25% ice, which are pumped through chilled, parallel, 22-mm i.d., S.S. tubes. Valves switch the direction of flow every 7 sec. Plastic balls shuttle back and forth, driven by the flow, scraping ice off the tube walls. Perforated end plates keep the balls inside the tubes. Scraping energy and frictional heat production are quite low at 10% ice content, but increase as ice content increases. Apparent heat-transfer coefficients are 450–570 $W/m^2 \cdot K$ at 10% ice content, but decrease as ice content increased. ΔT_e of 3°C are used. Crystals with an average size of 35–40 μm are produced. Figure 5.16 shows how a SBF might be used as a stage in a multistage system.

Since ΔT_e is less than required for heterogeneous nucleation; ice probably forms on unremoved ice crystal roots and secondary nuclei. When ΔT_e is too large, freezeup occurs, particularly at low solute concentrations and in aluminum tubes, where ice adheres more strongly than on S.S. SBF usually do not produce crystals small enough to ripen rapidly, but do produce ice that separates fairly readily in centrifuges.

Use of Bulk Subcooling

SBF and CSTF might be used to provide ice crystal growth through bulk subcooling. If growth at close to equilibrium conditions occurs and heat flows out through tube walls as fast as it is generated by freezing; subcooling $(T_p - T)$ can be obtained from a heat balance

$$4 U_f (T - T_e)/D_f = 6 h (1 - \varepsilon)(T_p - T)/D_p \qquad (23)$$

where $(1 - \varepsilon)$ is the volume fraction of crystals in the slurry, D_p is the effective crystal diameter, T_p is the equilibrium temperature at D_p, h is the crystal-surface heat-transfer coefficient and U_f is the freezer heat-transfer coefficient based on D_f, the tube i.d.

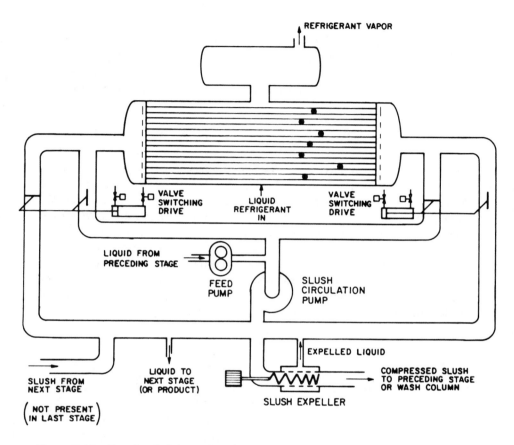

Figure 5.16   Shuttling-ball freezer used as a stage in a multistage freeze concentration system.

From ripening data (Van Pelt and Roodenrijs, 1982), I estimate h is roughly 325 W/m²·K in 50% sucrose and 780 W/m²·K in 29% sucrose. Using these h, $(1 - \varepsilon)$ = 0.1, and other appropriate values in Eq. (23), I calculate $(T_p - T)$ would be 0.08–0.12°C in SBF if ΔT_e = 5°C and heterogeneous nucleation does not occur. These are much larger than the 0.015–0.02°C subcooling usually obtained by ripening. If secondary nucleation and wall growth can be suppressed, SBF or CSTF could replace ripening tanks and standard SSF. Ice concentration would affect both the number of units needed and frictional heating. Mixing energy expenditure and frictional heating should decrease since growth time would be reduced.

Growth was initially rapid in batch crystallization runs where bulk subcooling was used, but leveled off as time increased (Thijssen, 1974). In similar tests where ripening was used, rates of growth remained fairly constant. Because secondary nuclei survive and grow when excessive subcooling is used, rates of growth in size in the subcooling-based runs may have been low because many small crystals were produced. Better results might have been obtained by using less subcooling.

Evaporative Cooling

Evaporative cooling at pressures below 3 mmHg has been used to produce ice in desalination systems (Barduhn, 1967, 1975). Small crystals that ripen rapidly can be produced. Water vapor compression (Johnson, 1965; Johnson et al., 1965a,1965b; Emmerman et al., 1965) and vapor absorption (Barduhn, 1967) have been used in these systems. Ten to twelve kWh of electrical energy were used per 1000 kg of water recovered. When absorption was used, 67 kg steam/1000 kg water were also needed. A desalination system based on use of evaporative freezing induced by water vapor compression is shown in Fig. 5.17. The compressed vapor condenses on ice previously formed by evaporative cooling. This melts the ice. Fresh water is obtained from melted ice and condensed vapor. Correcting for

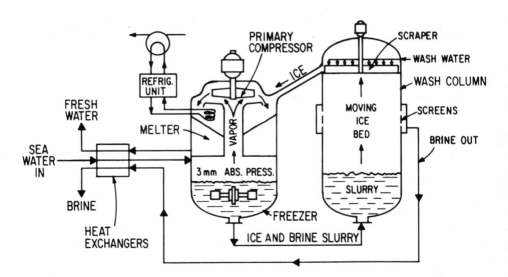

Figure 5.17 Desalination freeze concentration system using freezing produced by mechanically-induced evaporative cooling (Barduhn, 1975). (*Reproduced by permission of the American Institute of Chemical Engineers.*)

differences in Q_f/n_l and freezing T_j distribution, I estimate 14–16 kWh/1000 kg water removed would be needed to concentrate foods by multistage freeze concentration. W_{rf} would be 20% less than for conventional refrigeration, and energy use for scraping and removing scraping heat would be eliminated.

Aromas would be lost nearly completely if freezing caused by water vapor compression was used. Absorption of water vapor in concentrated brine has been used to freeze water by evaporative cooling (Yanniotis, 1980). When LiBr brine is used, water vapor sink temperatures can be as much as 40°C higher than temperature of the water vapor source (the freezing solution). LiBr depresses a_w and water partial pressure, but increases activities and partial pressures of dissolved aroma compounds. Because of vapor source-sink temperature and activity coefficient differences, 80–90% of initial aroma might be retained if evaporation-induced freezing based on use of LiBr is employed in FC systems. We have produced ice in systems like the absorption-driven freezer shown in Fig. 5.18. Large amounts of energy could also be saved by using such freezers instead of SSF.

Gas Hydrates

CO_2, C1 to C3 hydrocarbons, various freons, and other gases and vapors form ice-like gas hydrates (Davidson, 1973), which sequester water in separable solid form at temperatures above 0°C. Gas hydrates have been used for desalination (Donath, 1959; Williams, 1961; Hess and Jones, 1964; Barduhn, 1967) and food freeze concentration (Glew, 1962). Figure 5.19 is a combined P-T phase diagram for CO_2–water and CO_2–45° Brix fruit juice. CO_2 hydrate, whose composition is roughly $CO_2 \cdot 6H_2$), forms along the curve between 1MPa and −1°C and 4.5 MPa and 9.9°C. Quadruple points occur at the lower and upper ends of the curve. CO_2 gas, ice, hydrate, and liquid water are present at the lower end; CO_2 gas, CO_2 liquid, liquid water, and hydrate at the upper end.

Hydrates can be used to raise juice freezing points markedly. Cyclopropane, R22, and ethyl fluoride can provide freezing points of 16–22.8°C at 0.55–0.8 MPa; but they have to be stripped from foods. The maximum freezing point is lower with CO_2; but residual CO_2 can be tolerated. It might be stripped without excessive loss of aroma because of its high relative volatility. CO_2 hydrate slushes have been produced by passing water and CO_2 through SSF (Adler and Timpe, 1965). Rates of formation usually depended on ΔT_e; but when large amounts of hydrate formed, mass-transfer became rate limiting. The hydrate ripened quickly when the slush was held in a stirred autoclave.

CO_2 hydrate has also been produced at rapid rates and high levels of conversion by dispersing water in liquid CO_2 evaporating at the upper quadruple point (Mitchell et al., 1965). Figure 5.20 shows how this method could be used to produce CO_2 hydrate that would ripen rapidly in freeze concentration systems.

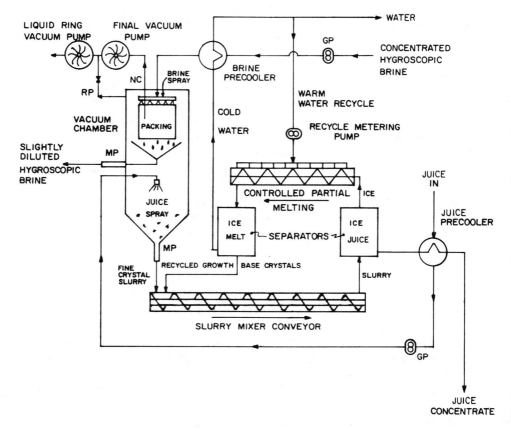

Figure 5.18 Food freeze concentration system in which absorption-induced evaporative cooling and freezing is used. GP = gear pump. MP = mono pump.

Scraping would not be used; therefore, large amounts of energy would be saved. CO_2 evaporated to produce cooling would be condensed and recycled. The hydrate would be separated in a pressurized wash column, and melted under pressure, releasing CO_2, which would be condensed and recycled.

Ripening should be more rapid at 9°C, the upper quadruple point for concentrate and CO_2, than at normal freezing points of concentrate, −2.5 to −10°C. $(T_c - T_e)$ would be small, and small ΔT could be used for heat exchange because direct heat-transfer and evaporation would be used. Therefore, W_r/Q_e should be very low. On the other hand, bubbles produced by outgassing may interfere with separation, some aroma may be lost, and high-pressure equipment would be needed.

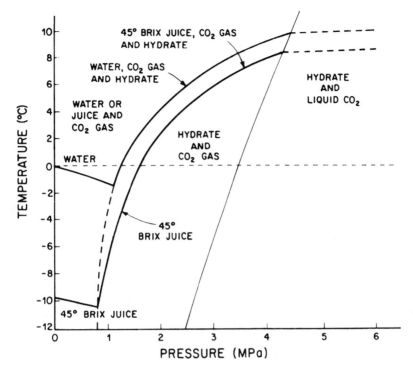

Figure 5.19 P-T phase diagram for CO_2-water and CO_2-fruit juice systems.

Separation difficulties were encountered in systems where gas hydrates were used for desalination (Barduhn, 1967). Though ripening was not used in these systems, and could be used now, use of CO_2 hydrate probably is a risky option for food freeze concentration.

RIPENING

In solutions containing slurried ice crystals of mixed sizes, T adjusts to a value higher than the equilibrium T for small crystals and lower than that for large crystals. The small crystals melt, removing heat from the solution, which, in turn, removes heat from large crystals, which consequently grow. Similar effects occur in different parts of dendrites. Radii of curvature are very small and positive at tips of dendrites and dendrite branches, so they melt. Radii of curvature are very small, but negative, at the base of clefts between dendrite branches, so they fill in. The

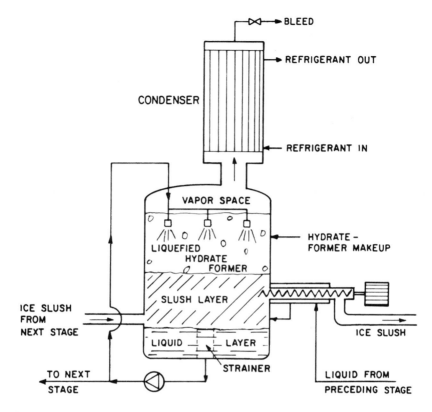

Figure 5.20 Apparatus for producing CO_2-hydrate from water in aqueous food solution using cooling provided by CO_2 evaporation at upper-quadruple point.

process, called Ostwald ripening or simply ripening, is used to produce large, smooth ice crystals which can be readily separated from concentrate.

Ripening Setups

In the Commonwealth Engineering (Olive, 1948), Union Carbide (Smith, 1965), Struthers (Muller, 1966,1967), Daubron (undated) systems and some Phillips Petroleum (McKay and Goard, 1965) systems, freezing and ripening occurred in the same vessel. Ripening times of 1.3–3 hours were used. The Daubron system was particularly simple. Ice crystals were scraped off the walls of a slowly stirred, jacketed tank cooled by evaporating ammonia. Part of the initial crop grew into large crystals. Subsequently formed ice melted due to ripening, which produced growth on crystals from the initial crop. The residence time of the growing crystals was about 3 hours. Sepial (undated) used a system where an initial crop of ice was produced in a first stage and added growth occurred in a second stage (Thijssen, 1974).

In most systems, ice is produced in SSF, and ripening is carried out in tanks containing stirred ice-concentrate slurry. Concentrate is drawn off from the tank through screens. Part is discharged. The rest mixes with entering liquid and is pumped through the SSF, where partial freezing occurs. The dilute ice slurry produced returns to the ripening tank, where it mixes with slurry circulating in the tank. The basic arrangement used is depicted in Fig. 5.21 (Thijssen and Huige, 1977). Similar arrangements are used in Grenco systems, but details differ, e.g., wiped strainers are used. Figure 5.22 is a drawing of a Grenco ripening tank (Van Pelt and Roodenrijs, 1982). Ice-concentrate slurry carried up a draft tube by a rotating auger recirculates to the bottom of the tube through space between the tube and tank wall. Small crystals produced in SSF enter the tank, mix with circulating slurry, and gradually melt, causing ripening-induced growth in larger crystals.

Ice slurry is transferred from a stage to a more dilute stage in a system or to a wash column. Concentrate leaves as product or is transferred to a stage operating at higher concentration. In large installations, two to four SSF are used per ripening

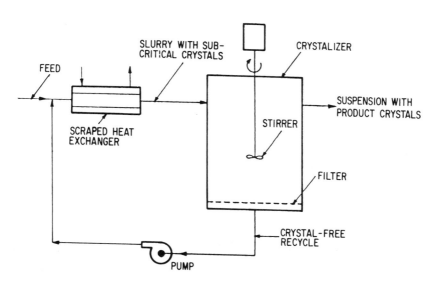

Figure 5.21 Single-stage ripening system (Thijssen, 1974). (*Reprinted by permission of Applied Science Publishers.*)

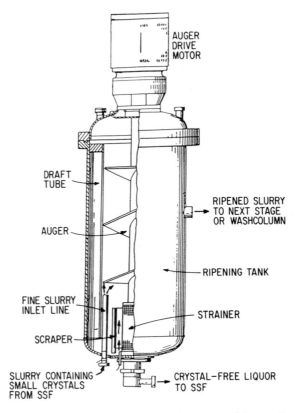

Figure 5.22 Grenco ripening tank. (*Reprinted courtesy of Grenco Special Applications bv.*)

tank. To minimize flow pressure drop, feed for each SSF is drawn off by a separate pump through a separate, wiped strainer.

MULTISTAGE OPERATION

Single-stage freeze concentration still is used occasionally; multistage systems are more efficient. In multistage systems built by Commonwealth Engineering (Olive, 1948), Union Carbide (Smith, 1965), and Struthers Scientific (Struthers, 1965), ice discharged from each stage (Deshpande et al., 1984; Muller, 1966). These systems are no longer made.

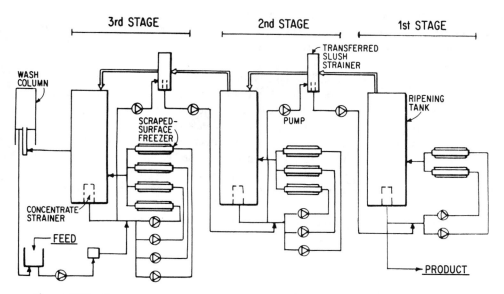

Figure 5.23 Freeze concentration system using ripening and three-stage, countercurrent operation. (*Reprinted courtesy of Grenco Special Applications bv.*)

Grenco countercurrent systems have three to six stages. Figures 5.2 and 5.23 depict three-stage systems. Elements include ripening tanks, associated SSF, wash columns to separate discharged ice, refrigeration to provide cooling, and pumps and other devices to transfer liquid and slurried ice between stages and SSF and ripening tanks. Ice moves in slurry form from stage to stage in the direction of decreasing solute concentration and discharges from the most dilute (last) stage. Solution moves in the opposite direction. Concentrate discharges from the first stage.

The amount of ice transferred increases from the first (most concentrated) stage to the last. Transferred slurries may be compressed and strained (Van Pelt and Roodenrijs, 1982,1984) to minimize carryover of concentrate. Liquid pressed from slurry returns to the ripening tank supplying the slurry.

A small fraction of the crystals produced in first-stage SSF survive and provide bases on which ripening-induced growth occurs. Though the survivors are the largest crystals produced in the SSF, initially they are relatively small, perhaps 20 µm in diameter. In the first ripening tank, surviving crystals grow due to sacrificial melting of smaller crystals produced in the SSF and produce larger crystals (Van Pelt and Roodenrijs, 1982). Slurry containing the larger crystals and entrained smaller crystals is transferred to the second ripening tank where further growth occurs due to melting of small crystals produced in SSF for that stage. These small crystals and small crystals subsequently produced in other SSF virtually completely disappear due to ripening-induced melting (Van Pelt and Roodenrijs, 1982).

Concentrate from the second stage is transferred to the first stage. The process is repeated in successive stages. Ice crystals initially produced in the msot concentrated stage grow as they move toward the least concentrated stage. Solution flows in opposite direction and progressively increases in concentration as it moves from stage to stage. In a three-stage system, effective crystal diameters D_{32} were 135 μm leaving the first stage, 189 μm leaving the second stage, and 230 μm leaving the last stage (Van Pelt and Roodenrijs, 1982). While Figs. 5.2 and 5.23 show one wash column, individual columns are used for each 1900 to 2000 kg/hr of ice separated.

Figure 5.24 depicts stream flow rates, concentrations, and operating temper-

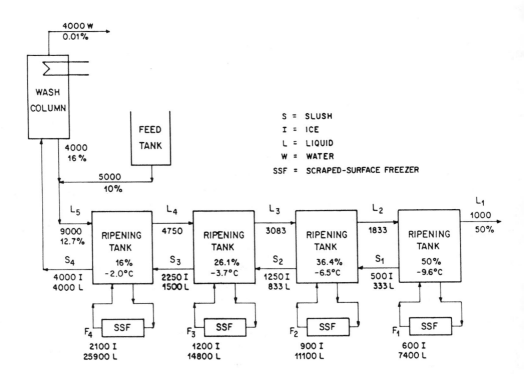

Figure 5.24 Flows, flow rates (kg/hr) concentrations and temperatures in a four-stage freeze-concentrations system.

Table 5.4 Energy Used Per 1000 kg of Water Removed in a Four-Stage Freeze Concentration System

---

Feed: 20°C, 12% s.s. juice (T_i = -1.4°C); b = 0.081
Concentrate: 50% s.s. (T = -9.6°C); Refrigerant: R22
Stage concentrations: 18.2%, 26.7%, 37.5%, 50%
Stage temperatures (°C): -2.3, -3.8, -6.4 and -9.6
Suction temperature: -20°C. Heat rejected at 3°C to ice when available, and at 38°C to cooling tower water at other times.

Refrigeration for freezing and cooling feed: 33.3 kWh (35% of work done and 54% of heat removed when ice is used for condensation)
Mixing, pressing and pumping work: 19.7 kWh
Scraping work: 18.5 kWh
Refrigeration to remove heat due to friction: 16.4 kWh (all done with cooling tower water used for condensation)
Refrigeration to remove heat inleakage 1 kWh
Fan and pump horsepower for evaporative water cooler 4.2 kWh
Motor inefficiency and other power expenditures that do not increase refrigeration load: 6.9 kWh
Total: 100 kWh

---

atures for a four-stage system. Table 5.4 provides a breakdown of energy use in a similar system. Frictional heat input, energy used for refrigeration to remove that heat, and energy used for nonparasitic processes other than freezing are responsible for roughly two-thirds of the energy consumption.

ANALYSIS OF RIPENING

Though local temperatures fluctuate during ripening (Huige, 1972), the average solution temperature does not change; and the amount of ice deposited on large crystals equals the amount melted from small ones. The distribution of crystal diameters D_p can be characterized by function $f(D_p)$, which indicates the number-based fraction of crystals per unit of D_p with diameters between D_p and D_p + dDp. Smith (1984) used Rosin-Rammler distributions to characterize unimodal distributions of ice D_p. Peleg and Normand (1987) provide a function useful for characterizing bimodal distributions that occur when fine crystals from SSF mix with large crystals in ripening tanks.

There is a neutral diameter D_n, whose equilibrium temperature T_n is the same as that of the solution. If ripening is adiabatic and frictional heat production and secondary nucleation are negligibly small, heat taken up by melting crystals with $D_p < D_n$ is generated by growth of crystals with $D_p > D_n$. Equating rates of heat released and absorbed,

$$\int_0^{D_n} hf\ (D_p)\ D_p^2(T_p - T_n)dD_p = \int_{D_n}^{D_{max}} hf(D_p)D_p^2(T_n - T_p)dD_p \tag{24}$$

where h is the heat-transfer coefficient between the particles and surrounding fluid, T_p and T_n are the respective equilibrium temperatures for particle sizes D_p and D_n. $(T_p - T_n) = (T + \Delta T_p' - T + \Delta T_n') = (\Delta T_p' - \Delta T_n')$. From Eq. (21), $(\Delta T_p' - \Delta T_n')$ $= (4\ \sigma\ T/\rho_s\Delta H)\ [(1/D_p) - (1/D_n)]$. With this substitution, all the variables in the integral are functions of D_p or constant. Smith (1984) found that mass-transfer coefficients in stirred slurries containing ice crystals were only weakly affected by D_p. Therefore, it is likely that h is also reasonably constant. If so, h can be taken out of the integral sign and cancelled from both sides, as can $(4\ \sigma\ T/\rho_s\Delta H)$. If this is done, Eq. (24) can be rearranged to provide

$$D_n = \int_0^{D_{max}} f\ (D_p)D_p^2dD_p\ /\ \int_0^{D_{max}} f(D_p)D_p dD_p = <D_N^2>/<D_N> \tag{25}$$

Where $<D_N>$ is the number-based mean value of D_p, and $<D_N^2>$ is the number-based mean value of D_p^2. Analysis of mass transfer in ripening (Smith and Schwartzberg, 1985, Smith, 1984) leads to the same result. Once D_n is known, rates of change of D_p with respect to time t can be found for each D_p, i.e.,

$$\frac{dD_p}{dt} = \frac{8\ \sigma\ h\ T}{(\rho_s\Delta H)^2}\left[\frac{1}{D_n} - \frac{1}{D_p}\right] \tag{26}$$

If $D_p < D_n$, D_p decreases with time, and if $D_p > D_n$, D_p increases. D_n can also be calculated using size ranges. If the mean size in range i is D_{pi}, the number fraction N_i and the mass fraction W_i,

$$D_n = \sum_i(N_iD_{pi}^2)\ /\ \sum_i(N_iD_{pi}) = \frac{<D_N^2>}{<D_N>} \tag{27}$$

or, in terms of W_i,

$$D_n = \sum_i(W_i/D_{pi})\ /\ \sum_i(W_i/D_{pi}^2) \tag{28}$$

For example, for 5 wt% 5-μm crystals and 95 wt% 200-μm crystals, $D_n = 7.29\ \mu$m. $(T_n - T_p) = 0.0096°$ for the 200-μm crystals if T = 268.2 K (−5°C). Similarly, for the 5-μm crystals $(T_p - T_n) = 0.0045°$. In ripening tanks, moderately small weight fractions of small crystals produced in SSF are mixed with large crystals. The previous example shows D_n will be fairly close to D_p for the small crystals, and that there will be an appreciable driving force for growth of the large crystals.

If h varies with D_p, but the nature of that variation is known, D_n can still be determined if h is placed inside the integrals in Eq. (25) or summations in Eqs. (27) and (28). Values of h appropriate for particular D_p also have to be used in Eq. (26).

As large and small crystals mix, D_p decreases for small crystals and increases for large ones. Suppose D_p for the 5-μm crystals in the previous example decreases to 2.5 μm, and the mass of ice remains constant. Since the masses of individual crystals are proportional to D_{pi}^3, W_i for the smaller crystals $= 0.05(2.5/5.0)^3 = 0.00625$. Hence for the larger crystals, $W_i = 1 - 0.00625 = 0.99375$, and $D_{pi} = 200 \cdot (0.99375/0.95)^{1/3} = 203$ μm. D_n equals 7.22 μm for the new W_i and D_{pi}, and $(T_n - T_p)$ for the large crystals now equals 0.0097°. Thus, for large crystals, $(T_n - T_p)$ and growth rates scarcely changes at all. For the small crystals, $(T_p - T_n)$ now equals 0.0189°, and their rate of shrinkage markedly increases.

In continuous ripening tanks, freshly introduced small crystals travel up the draft tube with recirculating crystals and then mix with more freshly introduced small crystals. Nevertheless, if the tank operates at steady state, D_n will be constant, and small crystals will still shrink at progressively more rapid rates as their size decreases.

Ice Concentration

Ripening speed increases when the volume fraction occupied by ice increases and stirring speed increases (Huige, 1972), but Q_p also increases. Huige obtained good ripening at ice volume fractions of 0.08–0.36 and stirring speeds of 20–45 rpm. Smith (1984) used 150 rpm at an ice volume fraction of 15% without excessive melting. Spheroidal crystals as opposed to disc-like ones are obtained at high ice volume fractions. This may be due to abrasion, which creates favorable growth sites on "c" axis surfaces (Huige, 1972). Figure 5.25 depicts large crystals obtained by ripening small crystals, like those shown in Fig. 5.14.

Batch Ripening Tests

Useful data about heat and mass transfer during ripening can be obtained by analyzing samples periodically withdrawn from batches of ice crystals and precooled solution mixed at adiabatic conditions. $f(D_p)$ are determined from D_p measurements and particle counts taken from photographs of the samples. Figure 5.26 shows how $f(D_p)$ vs. D_p changed with time during batch ripening tests with

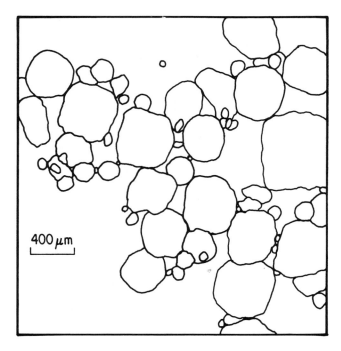

Figure 5.25 Large crystals produced by ripening. (*Reprinted courtesy of N. Huige.*)

ice crystals when $f(D_p)$ was unimodal (i.e., there was only one $f(D_p)$ peak). As time increased, distributions became wider and $<D_N>$ and D_n increased.

By using sequential analysis, a computational technique (Smith and Schwartzberg, 1985; Smith, 1984), one can determine sizes D'_p of crystals after a time step from sizes D_p before the step and $f(D_p)$ just before and just after the step. At D_n, $D'_p = D_p$. Figure 5.27 shows how sequential analysis was to determine how D_p changes. As expected, growth occurs when $D_p > D_n$, shrinkage when $D_p < D_n$. Though D_p remains almost constant at D_n for small time intervals, D_n progressively changes as $f(D_p)$ changes with time.

$dD_p/dt = (D'_p - D_p)/\Delta t$. h vs. D_p values were determined for $D_p > D_n$ and $D_p < D_n$ by substituting appropriate dD_p/dt, D_p, and D_n in Eq. (26). h did not vary significantly with D_p and was roughly the same for crystals that grew and crystals that melted.

Effect of Solutes on Ripening

Ripening rates and apparent h were four times greater at 10% sucrose concentration than at 42%. Huige (1972) found ice crystal growth in dextrose solutions was

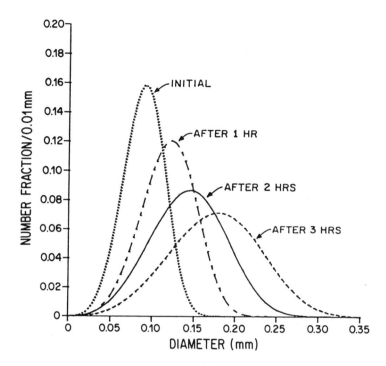

Figure 5.26 Charge in particle size distribution during ripening of a mixture of ice crystals with a unimodal distribution of sizes. (Smith and Schwartzberg, 1985). *(Reproduced by permission of the American Institute of Chemical Engineers.)*

20 times faster at 10% concentration than at 50%. Figure 5.28 shows $f(D_p)$ for monodisperse ice crystals after 5 hours of ripening in 10, 15, 22, and 42% sucrose solutions. $f(D_p)$ spreads more rapidly at higher ripening rates.

Heat- and mass-transfer coefficients and rates were calculated from standard correlations using thermal conductivities, viscosities, and water diffusivities corresponding to different sugar solution concentrations and freezing points. Calculated mass-transfer coefficients and rates decreased as concentration increased. Heat-transfer coefficients also decreased, but less markedly. Mass-transfer rate ratios at different concentrations agreed fairly closely with corresponding observed crystal growth and ripening rate ratios; heat-transfer rate ratios do not agree nearly as well. Therefore, mass transfer limits ripening-induced growth more than heat transfer does.

Solutes carried to ice crystal surfaces by water transfer do not incorporate in ice and accumulate at ice surfaces until back-diffusion counterbalances water-flow–

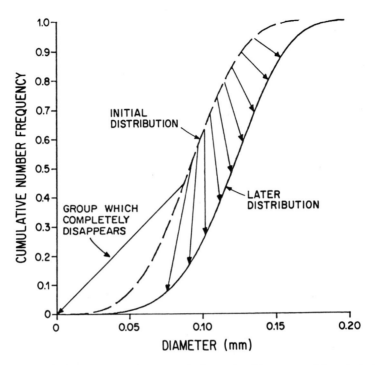

Figure 5.27 Changes in sizes of ice crystals determined by sequential analysis. (Smith and Schwartzberg, 1985). (*Reproduced by permission of the American Institute of Chemical Engineers.*)

induced solute transfer. Viscosity increases, and local freezing points, heat-transfer ΔT, heat- and mass-transfer coefficients decrease at the surface because of the higher surface concentration. Solutes probably also absorb at growth sites on crystal surfaces (Huige, 1972). These factors, combined, decrease growth rates and apparent h.

Figure 5.29 shows $f(D_p)$ vs. D_p after 5 hours of ripening for 10% sucrose solutions containing small percentages of gelatin. Diffusivities of gelatin and other macromolecules are very low. Macromolecular solutes accumulate at ice surfaces and restrict crystal growth and are used to limit crystal growth in manufacturing frozen confections. Pectin found in fruit juices greatly increases sugar solution viscosities at low temperatures (Bellows and King, 1973). Soluble coffee contains oligosaccharides created by hydrolysis of coffee grounds. Ripening rates for soluble coffee and orange juice are very much smaller than for sucrose (Van Pelt,

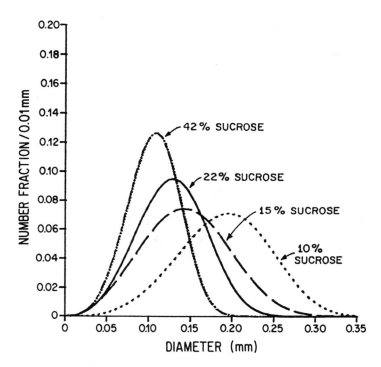

Figure 5.28 Ice crystal size distribution after five hours of ripening in sucrose solution of different concentrations. (Smith and Schwartzberg, 1985). (*Reproduced by permission of the American Institute of Chemical Engineers.*)

1975) even though the solutes involved behave like sucrose in terms of freezing point depression. Removal of macromolecules by ultrafiltration (Gobel et al., 1985) has substantially reduced viscosities of juices subjected to freeze concentration. For example, at 40% concentration and T_i, the viscosity of strawberry juice was 0.057 kg/m.sec, that of ultrafiltered juice, 0.015 kg/m.sec.

RIPENING IN MULTISTAGE SYSTEMS

Continuous Ripening

Ripening tanks operate continuously at steady state in multistage freeze concentration systems. D_p population balances for these tanks can be written in

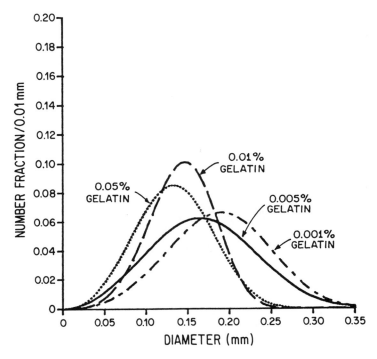

Figure 5.29 Ice crystal size distributions after five hours of ripening in 10% sucrose solution containing different concentrations of geletin. (Smith and Schwartzberg, 1985). *(Reproduced by permission of the American Institute of Chemical Engineers.)*

terms of population density distributions $N(D_p)$ that specify numbers of crystals per unit volume per unit diameter in range dD_p centered about D_p. $N(D_p) = f(D_p) \cdot N_t$, where N_t is the total number of crystals per unit volume. A balance used by Huige (1972) can be extended and used for well-mixed tanks operating at steady state in multistage systems.

$$V_r \frac{d[N_r(D_p)\,(dD_p/dt)]}{dD_p} = -V_o' N_r(D_p) + \sum_i V_i' N_i(D_p) \tag{29}$$

where V_r is the volume of the ripening tank, $N_r(D_p)$ is the population density distribution in the tank, $N_i(D_p)$ the distribution in the ith stream entering the tank, V_i' is the volume flow rate of that stream, V_o' is the volume flow rate of the stream transferring ice out of the tank. $N(D_p)$ for that stream is the same as $N_r(D_p)$ in the tank. If secondary nucleation does not occur, dD_p/dt is given by the RHS of Eq. (26). Frictional heating causes melting in ripening tanks. Because of this, Q'_r/N_tV_r should be added to the right hand side of Eq. (24), and $Q'_r\rho_s/4\sigma hTN_tV_r$ to the right hand side of Eqs. (25), (27), and (28), where Q'_r/V_r is the amount of heat generated or transferred into a tank per unit volume per unit time. If ripening tanks are cooled, negative Q'_r/V_r values should be used. If secondary nucleation does not occur, Eq. (26) is unaffected once effects of heating and cooling on D_n are accounted for through Eqs. (25), (27), or (28). Based on patent data (Van Pelt and Roodenrijs, 1982,1984), 23–28% of ice produced in first stage SSF melt because of frictional heating in the ripening tank for that stage, 14–21.5% of ice produced in intermediate stages melts, and 12–14% melts in the most dilute stage.

Only crystals from SSF enter first-stage ripening tanks. In subsequent stages, ripening tanks contain crystals produced in SSF and crystals transferred from the preceding stage. To solve Eq. (29) for $N_r(D_p)$ vs. D_p, one has to know how $N_i(D_p)$ and dD_p/dt vary vs. D_p. Finding dD_p/dt requires knowing D_n, which in turn, through $f(D_p)$, depends on $N_r(D_p)$, which is being sought. Trial and error calculations are required. These are beyond the scope of this work. Therefore, multistage ripening will be discussed mainly in qualitative terms.

Figure 5.30 qualitatively depicts how size distributions change in different stages of a multistage system. At all D_p, some crystals leave before they can grow larger or smaller. Therefore, $f(D_p)$ and $N(D_p)$ progressively decrease as D_p increases for $D_p > D_n$. For $D_p \gg D_n$, dD_p/dt increases very slightly as D_p increases, so $d(dD_p/dt)/dD_p$ approaches zero. Since large crystals do not enter the first stage, $N_i(D_p) = 0$ for large D_p, so $dN_r(D_p)/dD_p = -(V'_o/V_r)N_r(D_p)/(dD_p/dt)$. When integrated between a suitably large reference diameter D_m and D_p, this yields

$$N_r(D_p) \approx N_r(D_m) \exp[-k(D_p - D_m)] \qquad \text{(for } D_p > D_m \text{ in stage 1)} \qquad (30)$$

where $k \approx V'_o/[V_r(dD_p/dt)]$ and (dD_p/dt) is evaluated at $(D_p + D_m)/2$. For large D_p, $dD_p/dt \approx (8\sigma hT/D_n)/(\rho_s\Delta H)^2$; and $k \approx V'_oD_n(\rho_s\Delta H)^2/(8V_r \sigma hT)$. $f(D_p)$ in the ripening tank can be used in place of $N_r(D_p)$ in Eq. (30).

$N_r(D_p)$ for large D_p in subsequent stages are obtained from similar derivations. $N_r(D_p)$ in the nth tank is given by the sum of n terms, whose form is $C_j\exp[-k_j (D_p - D_m)]$. $k_j = V'_oD_n(\rho_s\Delta H)^2/(8V_r \sigma hT)$ based on D_n, V'_o, V_r, h, and T for the jth stage. C_j is a function of k_j and k_{j-1} and $(V'_o)_j$ $(V'_o)_{j-1}$.

Change in Effective Size

The effective D_p in terms of separation efficiency is D_{32}.

$$D_{32} = \sum_i N_i(D_{pi})^3 / \sum_i N_i(D_{pi})^2 \qquad (31)$$

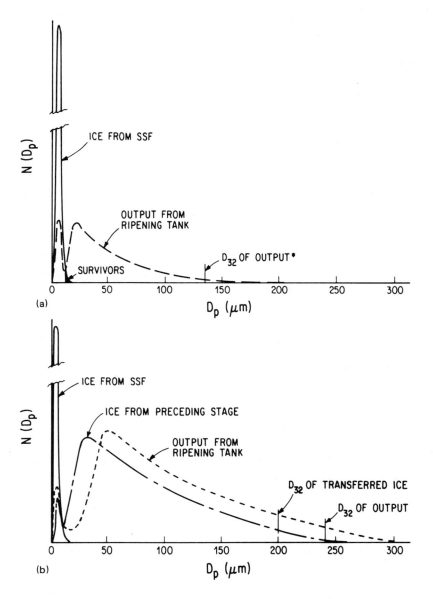

Figure 5.30 Crystal size distributions in different stages of a multistage freeze concentration system. (a) First stage. (b) Later stage.

D_n is often roughly 1.5 times $<D_N>_s$ the number-based mean size of crystals entering from SSF. The presence of small crystals decreases D_{32} markedly. If $f(D_p)$ decays exponentially as D_p increases and all crystals smaller than D_n completely melt, $D_{32} \approx 0.75 \cdot \Sigma(W_i \cdot D_{pi})$. However, if discharged slush contains 5 wt% crystals smaller than D_n, $D_{32} \approx 0.25 \cdot \Sigma(W_i \cdot D_{pi})$. If 5 minutes holdup time is provided for slush between last ripening stages and ice separation systems, crystals with $D_p < D_n$ should completely melt, D_{32} should markedly increase, and improved separation should be obtained.

Eqs. (32) and (33) can be used to roughly predict changes in D_{32} in ripening tanks and calculate ripening tanks volumes needed to obtain desired changes in D_{32}.

For the first stage,

$$(D_{32})_o = 2<D_N>_s + \frac{16\ V_r o h T}{V'_o(\rho_s \Delta H)^2} \left[\frac{1}{3<D_N>_s} - \frac{1}{(D_{32})_o + 2<D_N>_s} \right] \quad (32)$$

For subsequent stages,

$$(D_{32})_o = (D_{32})_i + \frac{16\ V_r o h T}{V'_o(\rho_s \Delta H)^2} \left[\frac{1}{3\ <D_N>_s} - \frac{1}{(D_{32})_o + (D_{32})_i} \right] \quad (33)$$

$(D_{32})_o$ and $(D_{32})_i$ are the respective D_{32} for ice leaving the tank and for transferred ice entering the stage. A method for determining D_{32} experimentally is presented in the appendix. V_r/V'_o (i.e., ice holdup times) for three-stage plants producing 50% concentrate from 10% feed are roughly 2.6 hr in the first (most concentrated) stage, 1.0 hr in the second stage, and 0.5 hr in the last stage. Solute holdup times (i.e., masses of solute in stages divided by the solute mass flow rate leaving the stage) are roughly 4.3 hr for the first stage, 2.3 hr for the second stage, and 1.3 hr for the last stage, a total of 7.9 hr. Van Pelt and Swinkels (1984) and Van Pelt and Jansen (1988) list much shorter total residence times for four- and five-stage systems.

Crystal Size Uniformity

Part of the small crystals produced in first-stage SSF and intermediate size crystals produced from them have to survive and grow to produce larger crystals. Small crystals produced in subsequent stages tend to completely melt. Incompletely melted crystals leave these stages; most subsequently melt. Since growth occurs on transferred larger crystals in all stages but the first, and residence times become more uniform when stirred stages are used in series, multistage operation improves crystal size uniformity.

Ripening also occurred in large freezers used in some systems (Daubron, undated; Smith, 1965). Van Pelt and Jansen (1988) claim this provided smaller product crystals. Ripening also occurs in SBF and SSF. SSF in which large fractions of feed froze in a single pass were used in the Votator (1963) and Sepial (undated) freeze concentration processes (Thijssen, 1974). Product D_p were less than 50 μm.

Improving Size Uniformity

Multistage operation has improved size uniformity, but crystals still cover ranges of sizes because intermediate-size crystals produced to make larger crystals in the first stage continuously leave the stage along with larger crystals. Greater uniformity could be obtained by feeding both small crystals that solely melt, and medium-size crystals that solely provide bases for growth into first-stage ripening tanks. Medium-size crystals could be produced by operating SSF at high holdup and low scraping speed. Twenty-four to 56 μm ice crystals are regularly made using 24 sec holdup in SSF that produce ice cream (Arbuckle, 1986). I have produced ice with much larger D_p in SSF that provided long holdup times and operated at slow scraping speed.

Uniform, large D_p crystals could be obtained by ripening mixtures of growth-base crystals and small crystals in plug-flow conveyors, as shown in Fig. 5.31. For example, 200-μm crystals could be produced from 1.6 wt% of 50-μm growth-base crystals and 98.4 wt%, 5-μm crystals. Small crystals should completely disappear, and growth-base crystals should grow uniformly if good radial mixing and adequate, uniform holdup are provided. Growth-base crystals would be fed into the first stage ripener only; transferred crystals and small crystals from SSF would be fed into ripeners in all other stages. Combining plug-flow, first-stage ripening with standard ripening tanks in other stages might be cost-effective. The combination could be used to retrofit existing multistage freeze concentration.

Small crystals produced in SSF cover a range of sizes. If D_n is too small, the largest crystals produced in the SSF will grow instead of melting. Weight fractions of crystals from SSF present in mixtures of small crystals and large crystals have to be kept within an acceptably low range to prevent this. This could be done by progressively adding small crystals from SSF along the length of conveyors instead of all at once at conveyer inlets when plug-flow ripening conveyers are used.

Ice Recycling

Very large growth-base and product crystals could be obtained by partially melting discharged crystals (Smith and Schwartzberg, 1985). The unmelted cores of

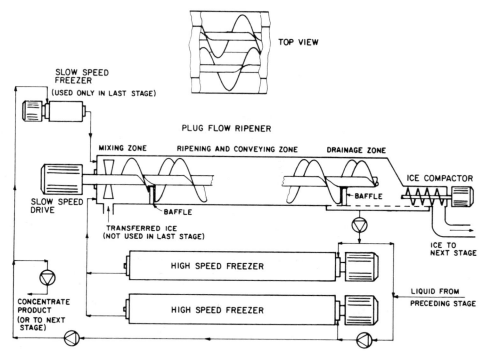

Figure 5.31 Ripening of a mixture growth-base crystals and small crystals in a plug-flow, twin-screw ripening conveyor.

partially melted crystals would be spun dry and recycled through ripeners. The cores might be washed with liquid of increasing concentration prior to ripening to minimize dilution. Figure 5.32 shows a recycling arrangement for single-stage freeze concentration. In multistage operation, recycled crystals and small crystals from SSF would feed into the first-stage ripener; transferred crystals and fresh, small crystals from SSF would enter ripeners in other stages.

Recycling should increase ripening rates, but also increase ice flow. Amounts of growth needed for clean separation would decrease. Effects of these factors on needed system volume depend on how much melting is used and are difficult to predict. Breaking discharged ice into separate crystals and controlled melting may be difficult to arrange. De novo production of moderately large growth-base crystals would be simpler, but recycling should yield larger crystals.

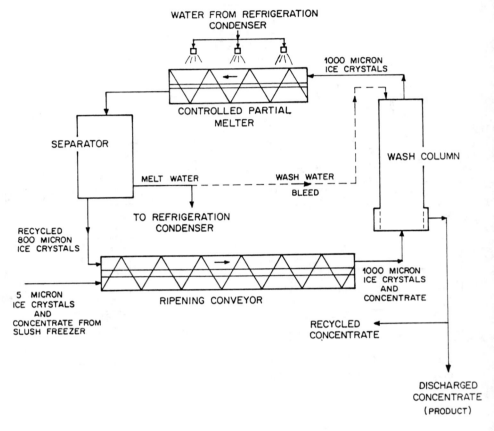

Figure 3.32 Recycling of partially-melted discharged ice to provide growth-base crystals.

SEPARATION

Separation Devices

Simple drainage, pressing, centrifugation, filtration, countercurrent washing, and displacement in wash columns have been used to separate ice from concentrate. Cutting away or drainage of exuded concentrate has also been used. Since the volume of ice is 8.5% greater than the water it replaces, concentrate exudes to a certain extent when solutions are frozen in static layers or containers. Concentrate can be recovered by scraping exudate off as solutions freeze or by cutting out or draining cores from material frozen in cylindrical vessels. Though yields are very low, commercial use has been investigated (Shaw, 1951). The method was used on farms to concentrate apple jack.

Pressing

Concentrate can be pressed out of ice-concentrate slurries (Joslyn, 1961). Porous cakes form. As compaction proceeds, porosity decreases and filtration resistance increases. Flow pressure drop and stress transmission in presscakes cause buildup of filtration resistance near outflow surfaces (Schwartzberg, 1983). Ultimately, resistance becomes so great that further expulsion of fluid becomes impractical. 0.6 kg concentrate/kg ice remain in ice presscakes at 10 MPa (10 atm) pressure; at 100 MPa; 0.05 kg/kg remains (Thijssen, 1974). For countercurrent operation and typical feeds, the latter figure corresponds to 4.8% solute loss. This normally would be intolerable. Nevertheless, pressing has been used in freeze concentration systems (Daubron, undated; Muller, 1967), and patents have been issued for processes based on use of pressing (Doat, 1976; Vella, 1984). Vella used pressing in a process to reduce the alcohol content of wine. Clean separation was not needed.

Centrifugation

Ice and concentrate have been and occasionally still are separated by centrifugation (Votator, 1963; Smith, 1965; Gasquet, undated; Muller, 1967; Gottesman and Pascal, 1972; Dwyer, 1974), but solute losses are usually undesirably high, 2% or more (Thijssen, 1974). One to 3% loss occurs even when washing is used (Dwyer, 1974). Oxidation and aroma loss can occur during centrifugation, but can be reduced by using inert atmospheres, fully enclosed centrifuges, and countercurrent operation.

Other Separation Systems

Olson (1985) used rotary vacuum filtration to separate concentrate from ice. Based on typical filter performance, solute losses for most freeze concentrated products would be excessively high. Countercurrent, multistage, mixing, and centrifugation have been used to produce concentrated juice with only moderate solute losses (Joslyn, 1971). Other forms of countercurrent contact have been used to reduce solute carryover with ice (Wrobel and Fraser, 1983; Merle et al., 1984). Reverse osmosis can be used to recover and partially concentrate solute from melted discharged ice when imperfect forms of separation are used.

Wash Columns

Separations of solution from ice are usually carried out in wash columns (McKay and Goard, 1965; Probstein, 1971; Thijssen, 1973,1974; Grossman, 1976; Korzonas, 1985). Ice mixed with solution is pumped into a column. A bed of ice forms and part of the solution drains. The bed moves upward through wash water, which displaces residual, entrained solution. Very large columns have been used for desalination (Korzonas, 1985). Gravity-based displacement has been used in open columns. Bouyancy and flow-induced drag caused ice to rise, and displaced solution drained through perforated zones in standpipes or side walls. Capacities of 4,500 kg/m²·hr have been obtained in open columns; 27,000 kg/m²·hr has been obtained in small, pressurized columns (Grossman, 1976).

Less ice forms per unit of feed during desalination than for freeze concentrating foods. Concentrations and viscosities are also lower, and more solute carryover can be tolerated. Nevertheless, fresh water produced in "successful" desalination wash columns contained 400 ppm salt vs. 35,000 ppm in the feed (Barduhn, 1975). More than 1% of feed solute was carried over with ice. Flow nonuniformity and cake cracking often occur in large columns (Grossman, 1976). Separation has been poor when large, open columns were used for separating ice from liquid foods; therefore, open columns are usually avoided when working with foods.

Parallel, modular, small (e.g., 0.6-m diameter) pressurized wash columns usually are used in food freeze concentration systems. These are automatically controlled and provide mechanically assisted ice movement. Thijssen and van der Malen (1982) developed a widely used efficient column (Fig. 5.33). Liquid in entering slush drains or is expelled through screens mounted on rotating vanes, which force ice upward as a uniform, annular bed. The bed passes through a stationary layer of melt water in the top half of the column. The water displaces solution from pores in the bed. Ice is scraped off at the top of the column and melted, using heat provided by condensing compressed refrigerant vapor. Some of the melt is used as makeup displacement water in the wash column. Since the water layer is stationary, very little makeup is needed. Sensors (photocells or thermocouples) detect the position of the water-solution interface. If the interface advances towards the top of the column, more pressure is applied to the wash water to force the interface back to the desired level. Reciprocating pistons have been used to propel ice upward in similar columns (Thijssen, 1973,1974,1975; Van Pelt, 1975).

If the entering ice is uniformly distributed and its upward velocity is small enough, smooth, clean displacement occurs and solute losses are usually less than 0.01%. Ice-handling capacities of 9000–18000 kg/m²·hr have been obtained in efficient, small columns (Van Pelt and Swinkels, 1984). Capacities in recently installed plants average 8000–9000 kg ice/m²·hr.

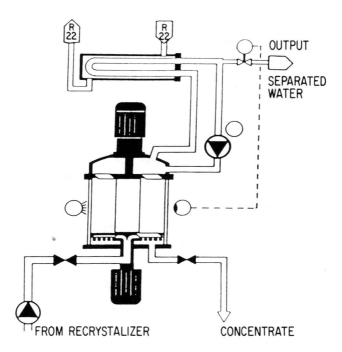

Figure 5.33 Grenco wash column. (*Reprinted courtesy of Grenco Special Applications bv.*)

FLOW THROUGH BEDS OF ICE CRYSTALS

Permeability

Cleanness of separation is strongly influenced by ice bed permeability K.

$$K = \frac{\mu U}{\alpha P/dZ} \approx \frac{\mu U}{\Delta P/\Delta Z} \tag{34}$$

where U is the superficial permeation velocity, (dP/dZ) the pressure gradient, and μ viscosity. K can be estimated from the viscous term in the Ergun (1952) equation.

$$\frac{\Delta P}{\Delta Z} \approx \frac{dP}{dZ} = \frac{150 \, U\mu(1 - \varepsilon)^2}{\varepsilon^3(6 \, V_p/S_p)^2} = \frac{U\mu}{K} \tag{35}$$

where ε, the bed porosity, is usually ≈ 0.4, S_p is the mean surface area of the particles, and V_p is their mean volume. For spherical particles of mixed size, (6 $V_p/S_p) = D_{32}$. Hence,

$$K = \frac{(D_{32})^2 \; \varepsilon^3}{150 \; (1 - \varepsilon)^2} \tag{36}$$

K and D_{32} can be determined from permeability tests described in the appendix. D_{32} can also be determined from $f(D_p)$ computed from D_p vs. number count measurements taken from photographs, a laborious, time-consuming procedure. $f(D_p)$ could be obtained more efficiently by using a suitably chilled Coulter counter. Cold stage microscopes with TV recorders have been used to take pictures of ice crystal slurries, which could be used with image analysis programs to obtain $f(D_p)$, D_{32} and other size and shape information.

ΔP in Wash Columns

Suppose 50% sucrose solution at its freezing point is being displaced from ice whose $D_{32} = 100$ μm in a wash column processing 8500 kg ice/m²·hr, depth ΔZ of the displaced fluid = 0.3 m, and $\varepsilon = 0.4$. The solution viscosity is 0.052 kg/m·s (52 cps) (Thijssen, 1974). Based on Eq. (36), K = 1.19×10^{-11}m². Since $\varepsilon = 0.4$, ice occupies 0.6 of the bed cross-section, and its velocity is (8500 kg ice/m²·hr)·(hr/3600s)·(m³/917 kg ice)·(m³bed/0.6 m³ice) = 4.29×10^{-3} m/s. It moves through stationary displaced concentrate, whose superficial velocity U relative to the ice is $4.29 \times 10^{-3} \times 0.4 = 1.72 \times 10^{-3}$ m/s. Applying Eq. (34), $\Delta P = 2.25$ MPa (327 psi). For $D_{32} = 50$ μm, $\Delta P = 9.0$ MPa (1,300 psi); and for $D_{32} = 250$ μm, $\Delta P = 0.36$ MPa (52 psi). The ΔP calculated for 250-μm crystals is close to measured ΔP for crystals of similar size (Vorstman and Thijssen, 1973). If 15% sucrose at its freezing point, with viscosity 0.0035 kg/m.s (3.5 cps), is displaced instead of 50% sucrose, the ΔP would only be 3.5/52 = 0.067 as great. The effect of D_{32} on ΔP illustrates an incentive for producing large crystals, the effect of μ on ΔP, an incentive for countercurrent operation.

ΔP Across Strainers

What are ΔP across strainers used in lines leading from ripening tanks to SSF? From Van Pelt and Roodenrij's (1982) data and drawings, I estimate U = 4.5×10^{-3} m/s and $\Delta Z = 0.025$ m for flow through strainers in the most concentrated stage

Table 5.5 Effect of Ice Crystal Particle Size on Drainage

D_{32} (μm)	K (m^2)	G	B_d	S_o	Z_d (m)
100	1.19×10^{-11}	g	2×10^{-6}	0.075	0.47
100	1.19×10^{-11}	500 g	0.001	0.075	0.009
1000	1.19×10^{-9}	g	2×10^{-4}	0.075	0.047
1000	1.19×10^{-9}	500 g	0.1	0.018	1×10^{-5}

of a multistage system. For $D_{32} = 135$ μm and 50% sucrose, $\Delta P = 270$ kPa (39 psi). At the flow rates involved, 0.8 kWh of frictional heat is generated per 1000 kg ice produced in the stage. In the most dilute stage, flow and ice deposition rates are twice is great, but viscosity is only 0.067 as large, $\Delta P = 70$kPa, and 0.27 kWh of heat is generated per 1000 kg ice produced.

DRAINAGE

In drained porous beds there often are a saturated layer where pores are completely filled, a "funicular" layer where connected liquid partly fills pores, and a "pendular" layer where disconnected pools exist and the fractional saturation is S_o. The combined liquid content of the funicular and saturated layers corresponds to an effective fully saturated depth Z_d. The average saturation $S = [Z_d + (Z - Z_d)S_o]/Z$, where Z is the total bed depth. S_o and Z_d depend on the drainage number $B_d = KG\rho_L/\sigma_L$, hence they depend on permeability K. For gravitational drainage $G = g$; in centrifugal fields, $G = \omega^2 R$. ρ_L and σ_L are the density and surface tension of the draining liquid, g is gravitational acceleration, ω angular velocity, and R radius of rotation. $S_o = 0.075$ for $B_d \leq 0.02$, $S_o = 0.0018/B_d$ for $B_d > 0.02$, and $Z_d = 0.275$ $K^{1/2}/B_d$ (Dombrowski and Brownell, 1964). Table 5.5 illustrates the effects of ice crystal particle size on drainage when $\sigma_L = 0.07$ kg/s^2 (70 dynes/cm) and $\rho_L = 1200$ kg/m^3.

Even for 1000-μm crystals drained at 500 g in a centrifuge, significant loss, $S_o\varepsilon/(1 - \varepsilon) = 0.018(0.4/0.6) = 0.012$, still occurs due to pendular cling. This roughly corresponds to loss data for 500 g centrifugal separation of freeze concentrated coffee from ice (Dwyer, 1974).

DISPLACEMENT STABILITY

Very clean separation can be obtained if the water-concentrate interface is stable when concentrate is displaced from ice by water in wash columns. Unstable displacement often occurs when viscous, dense fluid (concentrate) is displaced by less viscous, less dense fluid (water). Upflow displacement is always unstable in such cases. Downflow can be stable at proper conditions. Projections of less viscous invading fluid tend to move faster than more viscous displaced fluid. In downflow, these projections are surrounded by denser fluid, and the interface acts like a manometer which tends to level out. Stable displacement occurs when the displacement velocity U is less than a critical value U_c. If $U > U_c$, fingers of invading fluid pass through the displaced fluid, and displacement is highly imperfect and incomplete. Based on the pressure gradients involved, Hill (1962) showed that

$$U_c = \frac{Kg(\rho_2 - \rho_1)}{(\mu_2 - \mu_1)} \tag{37}$$

where the subscript 2 indicates the displaced fluid and 1 indicates the invading fluid. Dumoré (1964) proposed a more stringent criterion based on $d\rho/d\mu$ at the displacement interface.

$$U_c = Kg \frac{d\rho}{d\mu} \tag{38}$$

Displacement of low temperature concentrate by 0°C water causes dendritic growth on ice crystals, which changes K. Therefore, Thijssen proposed that

$$U_c = \frac{g(\rho_2 - \rho_1)}{(\mu_2/K_2 - \mu_1/K_1)} \tag{39}$$

where K_2 is the permeability in the concentrate layer and K_1 the permeability in the water layer. Table 5.6 lists critical displacement velocities calculated from Eq. (37) for two sucrose solution concentrations for various sizes of ice crystals using ice water as the displacing fluid.

The effect of concentration on U_c shows why countercurrent operation is beneficial. The effect of D_{32} on U_c shows why large crystals are needed for rapid separation.

These U_c values are all much smaller than the U of 1.72×10^{-3} m/sec previously calculated for typical wash column operation in food freeze concentration systems. Why does this discrepancy occur? Changes in K caused by displacement, i.e., applying Eq. (39) instead of Eq. (37), raise U_c, but probably not by amounts needed. On the other hand, water that starts to project through the displacement interface

Table 5.6 Calculated Critical Displacement Velocities for Various Ice Crystal Particle Sizes and Displaced Sucrose Solution Concentrations

D_{32} (μm)	Conc (w%)	ρ_2 (kg/m^3)	μ_2 (kg/m·s)	U_c (m/s)
100	50	1233	0.052	5.4×10^{-7}
250	50	1233	0.052	3.4×10^{-6}
1000	50	1233	0.052	5.4×10^{-5}
100	15	1060	0.0035	4.1×10^{-6}
250	15	1060	0.0035	2.5×10^{-5}
1000	15	1060	0.0035	4.1×10^{-4}

Viscosity of wash water = 0.00179 kg/m·s; density of water = 999.9 kg/m^3.

is surrounded by colder ice at the freezing point of concentrate. Consequently, small projections freeze, and growth of projections stops. Longitudinal variation in K caused by propulsion-induced bed compaction also helps stabilize displacement (Vorstman and Thijssen, 1973). In contrast to small projections, cracks in beds or bed nonuniformity cause melting and faster breakthrough (Grossman, 1976). Uniform, fault-free deposition and propulsion of ice is necessary for efficient wash column operation. Based on known wash column performance, Eq. (37) should be used with its RHS multiplied by 65. In terms of ice handling capacity, IHC in kg/m^2·hr, this corresponds to

$$IHC = \frac{3.87 \times 10^7 \ (D_{32})^2(\rho_2 - 999.9)}{(\mu_2 - 0.00179)} \tag{40}$$

with D_{32} in m, ρ_2 in kg/m^3, and μ_2 in kg/m.s.

CONCLUSIONS

Freeze concentration provides superior liquid food concentrates. It is superficially a simple process, but subtle problems have to be resolved to provide successful operation. Plausible improvement schemes often have failed because unanticipated difficulties were encountered, development resources were inadequate, or market potentials were unrealistically assessed. Improvements have been made by following consistent, theoretically sound approaches (use of ripening, wash columns, and multistage operation) and by progressively improving equipment and operating conditions. Ideas presented in this article build on those approaches. New approaches are also suggested.

Based on potential benefits and relative ease of development for the ideas presented, I recommend investigating: (a) growing ice by direct subcooling at nonnucleating conditions in freezers that provide large surface-to-volume ratios; (b) use of ultrafiltration to temporarily remove high molecular weight solutes from feeds; (c) use of separate freezers to produce growth-base crystals in addition to small crystals in first stages of multistage systems; (d) optimizing ΔT_e and scraping speeds in SSF; (e) use of cheaper, ice-separation methods combined with reverse osmosis for solute recovery; and (f) providing 5 minutes holdup time between last-stage ripening tanks and wash columns. Using separate refrigeration systems and low ΔT_e for stages in multistage systems could be beneficial, but only if frictional heat Q_p is greatly reduced and much cheaper freezers become available. Developing methods for recycling growth-base crystals produced by partially melting discharged ice would probably take much effort, but could produce great benefits. Using plug-flow ripeners fed with small crystals and growth-base crystals in first stages of multistage systems would also be worth pursuing.

Coatings that prevent or reduce surface nucleation and ice adhesion may prove useful in SBF or other heat exchangers used to grow crystals by subcooling. Disadvantages and risks offset benefits for other new ideas examined in this chapter.

APPENDIX

Permeability Test

Figure 5.34 shows a cell used by Huige (1972) to measure K for ice samples withdrawn from freeze concentration process vessels. Ice-free concentrate from a chosen vessel is pumped through the cell jacket surrounding it to prechill it. Then the cell is filled with ice-concentrate slurry from the vessel. The slurry is allowed to settle until clear concentrate fills the bottom of the cell and ice fills its top. A filter chamber is passed through the ice several times to form a test bed of set depth ΔZ. The bed is compacted using a standard pressure, e.g., 34 kPa (5 psi). Then, by applying measured vacuum ΔP, clear concentrate from the bottom of the cell is drawn up through the bed. The flow rate is determined by measuring volumes of discharged concentrate vs. time. K is calculated from Eq. (34), the measured flow rate, ΔP, ΔZ, and μ.

The bed is removed and allowed to melt, and ε, the porosity of the bed, is calculated from n_b, the solute concentration in the melted bed, and n_s', that of the filtrate, ρ_s, the density of ice, and ρ_f, the density of the filtrate, i.e.,

$$\varepsilon = \frac{n_b \rho_s}{\rho_f(n_s' - n_b) + n_b \rho_s} \tag{41}$$

Then, D_{32} is determined from K and ε using Eq. (36).

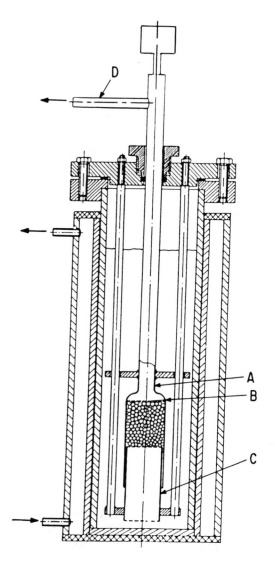

Figure 5.34 Cell for measuring permeability of ice crystal bed. (*Reprinted courtesy of N. Huige.*)

SYMBOLS

A	heat-transfer area in SSF (m^2)
a_w	water activity
B_d	drainage number, $KG\rho_L/\sigma_L$
b	water binding factor (kg bound water/kg solute)
c_w	water concentration in equilibrium with ice crystals of radius $r(kg/m^3)$
c_{we}	water concentration in equilibrium with ice crystals of infinite radius (kg/m^3)
C_o	heat capacity of feed in thawed state (kJ/kg·K)
C_f	heat capacity of feed in fully frozen state
C_I	heat capacity of ice (2.09 kJ/kg·K)
C_w	heat capacity of water (4.18 kJ/kg·K)
C_s	partial heat capacity of solute (kJ/kg·K)
D_f	diameter of freezer tube (m)
$<D_N>$	number-based mean diameter of crystals (m)
$<D_N>_s$	$<D_N>$ of crystals from SSF (m)
$<D_N^2>$	number-based mean value of D_p^2 (m^2)
D_n	diameter of crystal in equilibrium at bulk temperature of solution
D_p	diameter of crystal (m)
D_{pi}	mean diameter of crystals in size class i (m)
D_{32}	effective diameter of crystals (m) — see Eq. (31)
$(D_{32})_o$	D_{32} of crystals leaving ripening tank (m)
$(D_{32})_i$	D_{32} of transferred crystals entering ripening tank (m)
E	$18.02/M_s$
$f(D_p)$	number fraction of crystals in diameter range between D_p and $D_p + dD_p$
g	acceleration of gravity (9.806 m/s^2)
G	acceleration factor for draining bed, g for gravity drainage, $\omega^2 R$ for drainage in centrifuges
h	heat-transfer coefficient at surface of crystal (W/m^2·K)
ΔH	latent heat of fusion of water at T (kJ/kg)
ΔH_m	mean latent heat of fusion of water between 273.16 K and T (kJ/kg)
ΔH_o	latent heat of fusion of water at 273.16 K (333.5 kJ/kg)
IHC	ice handling capacity of wash column (kg/m^2·hr)
K	permeability of bed (m^2)
K_1	K in water-filled portion of bed (m^2)
K_2	K in concentrate-filled portion of bed (m^2)
M_i	molecular weight of ith solute component

M_s effective molecular weight of solutes

n_b weight fraction of solute in saturated bed

n_d weight fraction of disaccharides

n_i weight fraction of ith solute component

n_I weight fraction of ice in mixture

$(\Delta n_I)_j$ weight fraction of ice created in jth freezing stage (kg ice/kg feed)

n_m weight fraction of monosaccharides

n_s weight fraction of solute in feed

n'_s weight fraction of solute in concentrate

n_w weight fraction of water in concentrate

n_{wo} weight fraction of water in feed

N $(n_{wo} - bn_s)$ weight fraction of freezable water

N_i number fraction of crystals in class size i

$N(D_p)$ population density of crystals in size range between D_p and D_p + $dD_p(1/m^3)$

$N_r(D_p)$ $N(D_p)$ in ripening tank

$N_i(D_p)$ $N(D_p)$ in ith stream entering ripening tank

$N_o(D_p)$ $N(D_p)$ in slush leaving ripening tank

N_t total number of crystal per unit volume $(1/m^3)$

ΔP pressure drop across bed (Mpa or kPa)

P_c pressure in refrigeration system condenser (kPa)

P_e pressure in refrigeration system evaporator (kPa)

Q_b $Q_o + Q_f$ (kJ/kg feed)

Q_c heat rejected in condenser (kJ/kg feed)

Q_e heat absorbed in evaporator (kJ/kg feed)

Q_f heat removal to reduce feed temperature from T_i to T (kJ/kg feed)

$(Q_f)_j$ portion of Q_f removed in jth freezing stage (kJ/kg feed)

Q_i portion of Q_c supplied by melting of ice (kJ/kg feed)

Q_L heat inleakage (kj/kg feed)

Q_o heat removal required to reduce feed temperature from T_b to T_i (kJ/kg feed)

Q_p frictional heat production (kJ/kg feed)

Q_r $Q_p - Q_s$ (kJ/kg feed)

Q_s frictional heat produced due to scraping in SSF (kJ/kg feed)

Q_w portion of Q_c supplied by heating of cooling water (kJ/kg feed)

R radius of centrifuge (m)

R perfect gas law constant (8.314 kJ/kg mole·K)

r radius of curvature of crystal or region of crystal (m)

S saturation, fraction of pore volume filled by liquid

S_o saturation in pendular region

S_p mean surface area of particles or crystals (m^2)

t time, contact time, ripening time, freezing time (s)

Δt change in time (s)

T   temperature, freezing temperature (K) (°C sometimes)

T_c condensation temperature of refrigerant (K)

$\Delta T'$ freezing point depression caused by radius of curvature effects (K or°C)

ΔT_c heat-transfer temperature difference for condenser (K or °C)

T_e evaporation temperature of refrigerant, coolant temperature (K)

ΔT_e heat-transfer temperature difference for evaporator (K or °C)

T_i initial freezing point of solution (K)

T_o freezing point of pure water

T_p equilibrium freezing temperature at diameter D_p (K)

U   superficial velocity of fluid flowing through bed (m/s)

U_c value of U at which unstable displacement starts (m/s)

U_f heat-transfer coefficient of freezer (W/m²·K)

V'_i volume flow rate of ith stream entering ripening tank (m²/s)

V'_o volume flow rate of slush leaving ripening tank (m³/s)

V_p mean volume of particles or crystals (m³)

V_r volume of ripening tank (m³)

W_a work that does not produce frictional heat in process stream (kJ/kg feed)

W_i weight fraction of crystals in class range i

W_p work that produces frictional heat in process stream (kJ/kg feed)

W_r work done for refrigeration (kJ/kg feed)

W_{rf} portion of refrigeration work done in removing Q_f (kJ/kg feed)

W_{rb} portion of refrigeration work done in removing Q_b (kJ/kg feed)

X_w mole fraction of water

ΔZ depth of filter bed (m)

Z   distance in direction of flow, bed depth (m)

Z_d effective height of completely saturated portion of bed (m)

Greek Letters

ε volume fraction of bed or slurry occupied by solution

μ viscosity (kg/m·s)

μ_1 viscosity of invading fluid (kg/m·s)

μ_2 viscosity of displaced fluid (kg/m·s)

ρ_s density of ice (kg/m³)

ρ_1 density of invading fluid (kg/m³)

ρ_2 density of displaced fluid (kg/m³)

σ surface free energy of crystal (J/m²)

σ_L surface tension of liquid (J/m²)

ω   angular velocity (radians/s)

ACKNOWLEDGMENT

I wish to acknowledge contributions by Po-Ching Wu, who carried out surface freezing and ice-adhesion experiments, Dr. Carol Smith, who carried out ripening experiments, and Richard Grant, who drew many of the figures for this chapter.

REFERENCES

Adler, I.L. and Timpe, W.C. 1965. Carbonated ice process. U.S. patent 3,220,204. Nov. 30.

Arbuckle, W.S. 1986. *Ice Cream*, p. 319. AVI Publishing Co., Westport, CT.

Barduhn, A.J. 1967. Desalination by crystallization processes. *CEP* 63(1): 98.

Barduhn, A.J. 1975. Status of freeze-desalination. *CEP* 71(11): 80.

Barnett, S. 1973. Freezing of coffee extract to produce a dark colored freeze-dried product. In *Engineering of Food Preservation and Biochemical Processes.* Vol. 69, C.J. King (Ed.). p. 26. AIChE Symposium Series 132.

Bellows, R.J. and King, C.J. 1973. Product collapse during freeze drying of liquid foods. In *Engineering of Food Preservation and Biochemical Processes.* Vol. 69, C.J. King (Ed.). p. 33. AIChE Symposium Series 132.

Bomben, J.L. and King, C.J. 1982. Heat and mass transport in the freezing of apple tissue. *J. Food Technol.* 17: 615.

Chen, C.S. 1982. Evaporation in the citrus processing industry. Presented at AIChE National Meeting, Orlando, Florida, Feb. 28—Mar. 3.

Chen, C.S. 1986. Effective molecular weight of aqueous solutions and liquid foods calculated from freezing point depression. *J. Food Sci.* 51(6): 1537.

Chen, C.S., Carter, R.D., and Buslig, B.S. 1979. Energy requirements for the TASTE citrus juice evaporator. In *Changing Energy Use Futures,* Vol. 6, R.A. Fazzolare and C.B. Smith (Ed.) p. 1641. Pergamon Press, New York.

Chowdhury, J., Short, H., Hunter, D., Johnson, E., and Ushio, S. 1988. CPI warms up to freeze concentration. *Chemical Engineering* 95 (April 25): 24.

Concentrex, 1985. A major advance in freeze concentration technology. Bulletin CS-10,000.

Cottle, J.E. 1967. Fractional crystallization, U.S. patent 3,339,372. Sept. 6.

Daubron, undated. Special brochure No. 23,687, Daubron S.A., France.

Davidson, D.W. 1973. Clathrate hydrates. In *Water, A Comprehensive Treatise,* Vol. 2, F. Franks (Ed.), p. 230. Plenum Press, New York.

Deshpande, S.S., Cheryan, M., Sathe, S.K. and Salunkhe, D.K. 1984. *CRC Crit. Rev. in Food Sci. and Nutr.* 20(3): 175.

Doat, J. 1976, Preparation of concentrated extracts of instant coffee. U.S. patent

3,986,346. Oct. 19.

Dombrowski, H.S. and Brownell, L.E. 1964. Residual equilibrium saturation of porous beds. *Ind. Eng. Chem.* 46: 1207.

Donath, W.E. 1959. Apparatus and method for producing purified water from aqueous saline solutions. U.S. patent 2,904,511. Sept. 15.

Dorsey, N.E. 1968. *Properties of Ordinary Water-Substance.* Hafner Publishing Co., New York.

Dumore, J.M. 1964. Stability considerations in downward miscible displacement. Soc. Petr. Eng. of the *AIME J.* 4(4): 356.

Dwyer, D. E. Jr. 1974. Freeze concentration of instant coffee. U.S. patent 3,845,230. Oct. 29.

Eizenberg, D. and Kauzman, W. 1969. *The Structure and Properties of Water,* p. 1. Oxford University Press, New York.

Emmerman, D.K., Davids, J.H., Johnson, W.E., and Zarchin, A. 1965. Compressor construction. U.S. patent 3,255,602. June 14.

Ergun, S. 1952. Fluid flow through packed columns. *CEP* 48: 89.

Fennema, O.R. and Powrie, W.D. 1964. Low temperature food preservation. *Adv. Food. Res.*13: 219.

Fennema, O.R., Powrie, W.D., and Marth, E.H. 1973. *Low Temperature Preservation of Foods and Living Matter.* Marcel Dekker, New York.

Fletcher, N.H. 1970. *The Chemical Physics of Ice,* p. 23. Cambridge University Press, Cambridge, England.

Franks, F. 1982. *Water, A Comprehensive Treatise,* Vol. 7, p. 215. Plenum Press, New York.

Gasquet. undated. Catalogue No. 1685, Bordeaux France.

Glew, D.N. 1962. Solution treatment. U.S. patent 3,058,832. Oct. 16.

Gobel, G., Behr, N., van der Mel, H., Garduck, F.-J., and von Ettinghausen, O. 1985. Process for concentrating aqueous solutions having temperature-sensitive components. U.S. patent 4,491,600. Jan. 1.

Gottesman, M. and Pascal, F.D. 1972. Method of reducing solid loss in freeze concentration of coffee extract. U.S. patent 3,684,532. Aug. 15.

Grossman, G. 1976. Melting, freezing, and channeling phenomena in ice counterwashers. *AIChE J.* 22(6): 1033.

Gutschmidt, J. 1964. cited in *Cooling Technology in the Food Industry,* A Ciobanu et al. (Ed.). Abacus Press, Tunbridge Wells, Kent, England.

Harrod, M.J. 1986. Scraped surface heat exchangers. *Food Proc. Eng.* 9(1): 1.

Helferich, J. 1988. Industrial scale-up of freeze concentration. Presented at the AIChE National Meeting, Denver, Colorado, Aug. 21–24.

Hess, M. and Jones, G.E. Jr. 1964. Process and apparatus for separating water from an aqueous system. U.S. patent 3,119,772. Jan. 28.

Hill, M.A. 1962. Channeling in packed columns. *Chem. Eng. Sci.* 1: 247.

Hobbs, P.V. 1974. *Ice Physics,* p. 408. Oxford University Press, London.

Horne, R.A. 1972. *Water and Aqueous Solutions.* Wiley-Interscience, New York.

Huige, N.J.J. 1972. Nucleation and growth of ice crystals from water and sugar solutions in continuous stirred tank crystallizers. Ph.D. thesis, Eindhoven Univ. of Technology, The Netherlands.

Jellinek, H.H.G. 1959. The adhesive properties of ice. *J. Colloid Sci.* 14: 268.

Jellinek, H.H.G. 1972. The ice interface. In *Water and Aqueous Solutions,* R.A. Horne (Ed.), p. 65. Wiley-Interscience, New York.

Johnson, W.E. 1965. Evaporating and condensing chamber apparatus. U.S. patent 3,255,605. June 14.

Johnson, W.E. 1984. Freeze crystallization subassembly, U.S. patent 4,468,930. Sept. 4.

Johnson, W.E. 1986. Energy use in Concentrex system. Private communication. Concentrex Inc., Andover, MA.

Johnson, W.E. 1989. Platings and coatings for freezers. Private communication. Topsfield, MA.

Johnson, W.E., Davids, J.H. and Emmerman, D.K. 1965a. Freeze concentration apparatus for separating a solvent. U.S. patent 3,255,603. June 14.

Johnson, W.E., Davids, J.H. and Emmerman, D.K. 1965b. Evaporating and condensing chamber and compressor apparatus. U.S. patent 3,255,604. June 14.

Joslyn, M.G. 1961. Concentration by freezing. Ch. 10. In *Fruit and Vegetable Juice Processing Technology,* D.K. Tressler and M.A. Johnson (Ed.). AVI Publishing Co., Westport, CT.

Karnofsky, G. 1976. Method and apparatus for melting ice in saline water conversion systems. U.S. patent 3,992,170. Nov. 16.

Karnofsky, G. and Steinhoff, P.F. 1960. Use of butane evaporation to induce freezing by direct contact heat transfer during desalination. U.S. Office of Saline Water Progress Report 40.

Korzonas, R. 1985. Apparatus for separating ice from a slurry and washing the ice. U.S. patent 4,517,806. May 21.

Krauss Maffei. undated. Gefrierkonzentration G.K. pamphlet B4721, 3rd ed.

Landy, M., and Freiberger, A. 1967. Studies of ice adhesion I. Adhesion of ice to plastics. *J. Colloid Interfacial Sci.* 25: 231.

Maltini, E. 1975. Possibili modifcazioni de fase durante la liofilizzazione dei succhi di frutta e dei produtti fluidi in genere. Instituto Sperimentale per la Valorizzazione Technologica dei Prodotti Agricoli, Annale VI: 5.

McKay, D.L. and H.W. Goard. 1965. Continuous fractional crystallization. *CEP* 61(11): 99.

Midwest Research Institute. 1987. Solar-powered water desalination project at Yanbu. Report MRI/SOL 0415 published for SOLERAS, the United States–Saudi American joint program for cooperation in the field of solar energy.

Mitchell, W.A., Barnes, H.M., and Schwartzberg, H.G. 1965. Process for making carbonated ice. U.S. Patent no. 3,255,600. June 14.

Merle, R., Decker, M., and Kozlik, R. 1984. Freeze concentration apparatus. U.S. Patent 4,438,634. March 27.

Muller, J.G. 1966. Recent developments in crystallization of ice as applied to freeze concentration. *Bull. Int. Inst. du Froid. Annexe* 3: 15.

Muller, J.G. 1967. Freeze concentration of food liquids: Theory, practice and economics. *Food Technol.* 21(4): 49.

Olive T.R. 1948. Freeze concentration becomes practical. *Chem. Eng.* 55(Oct.): 118.

Olson, J. 1985. Method and apparatus for obtaining a concentrated extract of liquid smoke. U.S. patent 4,551,339. Nov. 5.

Omram, A.M. and King, C.J. 1974. Kinetics of ice crystallization in sugar solutions and fruit juices. *AIChE J.* 20(4): 795.

Peleg, M. and Normand, M.D. (1987). Computer simulation of the attrition patterns of particulated and agglomerated foods. *J. Food Sc.* 52: 943.

Probstein, R.F. 1971. Method of separating solid particles from a slurry with wash column separators. U.S. patent 3,587,859. June 28.

Raraty, L.E. and Tabor, D. 1958. The adhesion and strength properties of ice. *Proc. R. Soc.* A245: 184.

Riedel, L. 1949. Brechungsvermogen und Gefriertemperaturen von Fruchtsaften in Abhangigkeit von der Konzentration. *Zeit. für Lebensmittel-Untersuchung und Forschung* 89(3): 289.

Riedel, L. 1951. The refrigerating effect required to freeze fruits and vegetables. *Refrig. Eng.* 59: 670.

Schoerner, W.S. 1984. Heat exchanger with polymeric-covered cooling surface and crystallization method. U.S. patent 4,452,302. June 5.

Schwartzberg, H.G. 1976. Effective heat capacities for the freezing and thawing of foods. *J. Food Sci.* 41: 152.

Schwartzberg, H.G. 1977. Lightness and darkness—the reflectance of dried porous foods. In *Drying—Principles and Technology.* Vol. 23, p. 157. AIChE. Symp. Ser. 63.

Schwartzberg, H.G. 1981. Mathematical analysis of freezing and thawing. Presented at the AiChE National Meeting, Detroit, Michigan. Aug. 16–19.

Schwartzberg, H.G. 1983. Expression related properties. In *Physical Properties of Foods.* M. Peleg and E. Bagley (Ed.), p. 423, AVI Publishing Co., Westport, CT.

Schwartzberg, H.G. 1988. Potential improvements in freeze concentration. Presented at the AIChE National Meeting, Denver, CO, Aug. 21–24.

Sepial. undated. Societe d'Etudes et d'Exploitation du Procedes pour l'Industrie Alimentaire, Clichy, France.

Shaw, C.L. 1951. Canadian packer to freeze apple juice concentrate. *Western Canner & Packer* 43(4): 19.

Smith, C.E. 1984. Ice crystal growth during the ripening stage of freeze concentration. Ph.D. thesis, University of Massachusetts, Amherst.

Smith, C.E. and Schwartzberg, H.G. 1985. Ice crystal size changes during ripening in freeze concentration. *Biotech. Prog.* 1: 11.

Smith, P.L. 1965. Freeze concentration of fruit juices and beer. *ASHRAE J.* June: 87.

Struthers Scientific and International Corp. 1965. A new way to concentrate liquids, Bulletin Fc 65.

Thijssen, H.A.C. 1968. Freeze concentration of liquid foods. *Dechema-Monografien* 63: 153.

Thijssen, H.A.C. 1974. Freeze concentration. In *Advances in Preconcentration and Dehydration of Foods,* A. Spicer (Ed.), p. 115, John Wiley and Sons, New York.

Thijssen, H.A.C. 1973. Apparatus for the separation and treatment of solid particles from a suspension. U.S. patent 3,872,009. March 18.

Thijssen, H.A.C. 1975. Current developments in the freeze concentration of liquid foods. In *Freeze Drying and Advanced Food Technology,* S.A. Goldblith, L. Rey and W.W. Rothmayr (Ed.), p. 481. Academic Press, London.

Thijssen, H.A.C. 1979. Process and apparatus for separating crystallizable material from a multicomponent mixture. U.S. patent 4,143,524. March 13.

Thijssen, H.A.C. 1981. Process for concentration of alcoholic beverages. U.S. patent 4,265,920. May 5.

Thijssen, H.A.C. and Huige, N.J.J. 1977. Two stage concentration process and apparatus for crystallization. U.S. patent 4,004,886. Jan. 25.

Thijssen, H.A.C. and van der Malen, B.G.M. 1982. Continuous packed column wash column. U.S. patent 4,332,599. June 1.

Trommelen, A.M. and Beek, W.J. 1971. The mechanism of power consumption in a Votator-type scraped-surface heat exchanger. *Chem. Eng. Sci.* 26: 1977.

Van Pelt, W. 1975. Freeze concentration of vegetable juices. In *Freeze Drying and Advanced Food Technology,* S.A. Goldblith, L. Rey and W.W. Rothmayr (eds.), p. 549. Academic Press, London.

Van Pelt, W.H.J.M. and Jansen, H.A. 1988. Freeze concentration economics and applications. In *Preconcentration and Drying of Food Materials,* S. Bruin (Ed.), p. 77–86. Elsevier Science Publishers, B.V., Amsterdam.

Van Pelt, W. and Roodenrijs, J. 1982. Multi-stage countercurrent concentration method. U.S. patent 4,316,368. Feb. 23.

Van Pelt, W. and Roodenrijs, J. 1984. Multi-stage countercurrent concentration system and method and separator. U.S. patent 4,430,104. Feb. 7.

Van Pelt, W.H.J. and Swinkels, W.J.M. 1984. New developments in freeze concentration. Presented at the AIChE Summer National Meeting, Philadelphia, PA. Aug. 19–22.

Vella, P.J. 1984. Process for the reduction of the alcohol content of wine. U.S. patent 4,468,407. Aug. 28.

Vorstman, M.A.G. and Thijssen, H.A.C. 1973. Stability of the displacement of viscous aqueous solutions by water in a packed bed of ice crystals. International Symposium on Heat and Mass Transfer Problems in Food Engineering, Wageningen, The Netherlands, Oct.

Votator. 1963. Votator continuous freeze concentration process. Process Bulletin 153-345C 02.1265.

Williams, V.C. 1961. Hydrate-forming saline water conversion process. U.S. patent 2,974,102. March 7.

Wrobel, P.J. and Fraser, J.H. 1983. Apparatus for and method of preparing crystals for washing. U.S. patent 4,406,679. Sept. 27.

Yanniotis, S. 1980. Absorption-driven evaporative cooling using hygroscopic brines concentrated by solar-energy-induced evaporation. M.S. thesis, University of Massachusetts, Amherst.

6

Supercritical Fluid Extraction

Mark A. McHugh

The Johns Hopkins University
Baltimore, Maryland

If a gas or liquid at ambient conditions is heated and compressed to conditions above its critical point, it becomes a supercritical fluid. In the critical region this "gas" exhibits a liquid-like density and a much increased solvent capacity. Supercritical fluids have a unique combination of solvent and transport properties which make them useful for purifying fine chemicals, foods, pharmaceuticals, and other specialty chemicals. They are especially attractive solvent media since it is possible to operate at low to moderate temperatures when processing thermally labile materials, to fine-tune solvent capacity by varying pressure, and to effectively remove all traces of residual solvent in the resultant material by decreasing the system pressure. Although many technical articles in trade journals have described the advantages of this "new" technology in a broad range of applications, such as coffee decaffeination, spice extraction, and lipids purification, the processing principles have been long well known and practiced in a number of industries.

INTRODUCTION

Within the past 10 years numerous accounts of the processing potential of supercritical fluid solvents have appeared in trade and business journals (e.g., *Chemical Engineering,* March 1979; *Business Week,* July 1981; *Food Development,* August 1981; *Chemical Processing,* January 1985; *Chemical Engineering,* 1985). These articles extol the almost mystical properties of supercritical fluids by describing them as offering the promise of quick extraction, improved separations, lower operating costs, and broad industrial applications to name just a few. In retrospect, it is obvious that no single solvent or class of solvents could possibly live up to such a broad billing. Yet during this period in which the American trade journals talked about the potential of supercritical fluid technology, European companies were constructing and operating plants to decaffeinate coffee and tea, to process hops, and to extract essential oils and flavors. Recently, two plants have been constructed in the United States. One for decaffeination of coffee is in Houston, Texas (B. Perkins, *The Cincinnati Enquirer,* January 10, 1987) and the other for hops extraction is in Syndey, Nebraska (announcement by C. B. Cookson at the Spring Technical Symp. of Bulk Pharmaceutical Chemicals, Newport, May 1987). These successes demonstrate that supercritical fluid processing can be applied to foods and natural products (*Chemical Processing,* May 1987; *Chemical Week,* June, 1987; *Food Processing,* July 1987).

The areas of potential application of supercritical processing include the extraction of flavors, aromas, and colorants (Krukonis, 1988; Coenan et al., 1983; Shultz et al., 1984), concentration of eicosapentaenoic acid (EPA) from fish oils (Eisenbach, 1984, 1985; Krukonis, 1984; Suzuki et al., 1986; Nilsson et al., 1986), extraction of cholesterol from butter, lard, tallow, and eggs (*Newsweek,* 1986, July 14, p. 5; *Ind. Chem. News,* 1986, September p. 11), decaffeination of coffee (Zosel, 1974), and extraction of hops (Laws et al., 1980).

The objective of this chapter is to describe briefly the thermodynamic and process factors which make supercritical fluids attractive candidate solvents for extraction and purification processes. A few specific applications are described in some detail to highlight the characteristics of this emerging technology.

SOLUBILITY AND PHASE-BEHAVIOR CONSIDERATIONS FOR SUPERCRITICAL FLUID PROCESSING

As the critical point of a gas such as carbon dioxide is approached, its isothermal compressibility tends to infinity. This means that the density of the gas changes quite rapidly near its critical point. As the density increases, the gas, now a supercritical solvent, begins to take on the solvent characteristics of a liquid. Therefore, the forces of attraction between solvent and solute increase with increasing density, an observation made repeatedly about supercritical fluids. Let's briefly consider the forces of attraction which become more apparent at liquid-like densities.

Prausnitz et al. (1986) describe how the solubility of one component in another depends on the intermolecular forces between these components which include attractive (induction, dipole-dipole, and dipole-induced dipole), repulsive (hard sphere type-electron overlap), and chemical (donor-acceptor complexing and hydrogen bonding) forces. Each of these forces act over a distance; for example, dispersion and dipole forces are proportional to $1/r^6$, where r is the distance between molecules. Therefore, it is reasonable to expect the solvent power of a supercritical fluid to increase as the system density increases.

Dipolar forces, which can be very strong, are also inversely proportional to temperature. This temperature dependence is very important when considering candidate supercritical fluids. Remembering that "like dissolves like," it is tempting to choose ammonia to extract a polar substrate since ammonia has a dipole moment of 1.47 debye. However, since the critical temperature of ammonia is relatively high, 132°C, the polar nature of ammonia is greatly diminished in the process of making ammonia supercritical. For example, Dhalewadikar and coworkers (1987) show that supercritical ammonia displays very little selectivity for an aromatic compound relative to a parafinnic compound even though at room temperatures liquid ammonia shows very high selectivity for the aromatic. The loss in the selectivity of supercritical ammonia can be partly explained by the decrease in the attraction between the quadrapolar aromatic and the dipolar ammonia at high temperature, and the decrease in the acceptor-donor interactions between these two species (Anderson et al., 1962). Most supercritical fluids face this properties dilemma—that is, the more polar the substance at room conditions, the higher its critical temperature (T_c) and the more likely the polar nature will diminish as the substance is heated to temperatures above T_c. Spiking a nonpolar supercritical fluid with a polar cosolvent offers one way to overcome this polarity-temperature dilemma. The area of cosolvents is discussed elsewhere in the literature and is not considered here (see, for example, Wong and Johnston, 1986).

When assessing the use of supercritical solvent extraction, it is also very important to consider the effect of polarity of the substance being extracted. For example, in many polymer extractions it is necessary to operate to very high temperatures to obtain an appreciable solubility of the polymer in the supercritical fluid-rich phase (McHugh and Krukonis, 1989). High temperatures are needed to reduce polymer-polymer interactions which occur in the polymer-rich phase. These interactions are either polar in nature, hydrogen bonding, or strong dispersion forces magnified by packing arrangements in the solid phase. Since elevated operating temperatures are used in these cases, high pressures are needed to maintain the supercritical fluid solvent at a reasonable density to endow it with the ability to solvate the condensed-phase species when they become accessible.

Condensed-phase interactions can be especially important in the food industry where often the substance of interest is entrenched in some kind of strong relationship with its nearest neighbors making it inaccessible to extraction.

A lack of understanding of the variations in phase behavior which can occur at high pressures is another potential pitfall that awaits the practitioner of supercritical processing. Once any amount of a substance is dissolved into a supercritical fluid, that fluid becomes a mixture which may or may not be supercritical any longer. Consider first the types of behavior exhibited at high pressures (Scott and van Konynenberg, 1970; McHugh and Krukonis, 1986). These classes of behavior cover mixtures ranging from methane–ethane, to toluene–polystyrene, to carbon dioxide–soybean oil. A detailed description of this phase behavior is beyond the scope of this chapter; the reader is referred to McHugh and Krukonis (1986) for more details. Suffice it to say that in many instances supercritical fluid extraction entails more than loading a high pressure vessel and passing carbon dioxide through the vessel at some given pressure and temperature. For example, it is possible for solids to melt at tens of degrees below the normal melting temperature due to the interactions with the high pressure supercritical fluid. It is possible to have a single liquid phase split into two liquid phases when it is contacted with a supercritical fluid. This commonly occurs with alcohol–water–supercritical fluid systems, and it can also occur with supercritical fluid–oil or polymer systems. Finally, in our laboratories we have seen many instances where the supercritical fluid becomes more dense than the condensed phase, thus causing a phase inversion to occur (McHugh et al., 1984, 1988). This means that in continuous flow, countercurrent operation there are operating conditions where the liquid phase, which starts out as the heavier phase and is fed at the top of the column, is pushed out of the column by the supercritical fluid solvent, which was the lighter phase at lower pressures and was fed to the bottom of the column. This inversion phenomenon need not be catastrophic; as long as it is recognized as a possibility in certain classes of extractions (especially oils), the plumbing of the extraction equipment can be designed to allow the flow pattern of the fluid and liquid to be switched.

SUPERCRITICAL FLUID EXTRACTION AND FRACTIONATION OF NATURAL MATERIALS

The development of new information on the processing of natural materials with supercritical fluids continues at a rapid pace. At the recent International Symposium on Supercritical Fluids held in Nice, France, October 1988, 20 of the 134 papers dealt specifically with processing natural materials. These papers, which are found in the reference "Proceedings of the International Symposium on Supercritical Fluids, Societe Francaise De Chimie, Paris, 1988," include the extraction of fatty acids of various oils from natural triglycerides (Wu, A. H., Stammer, A., Prausnitz, J. M., p. 107; Ashour, I., Wennersten, R., p. 115), the extraction of limonen and cineole with carbon dioxide (Azevedo, E. G., Matos, H. A., Ponte, N. M., Simoes, P. C., p. 135), the concentrating eicosapentaenoic acid in fish oil esters (Krukonis, p. 545, Eisenbach, W. O. p. 719; Castera, A., Morin, O., Coustille, J. L. p. 727; Ikushima, Y., Hatakeda, K., Saito, N., Ito, S., Asano, T., Goto, T., p. 735; Bertucco, A., Guarise, G. B., Dahir, M., Navazio, G., p. 791), the removal of cholesterol from butter, beef tallow, and egg yolk (Krukonis, p. 547), the extraction of ginger, rosemary, and sage (Pellerin, p. 677; Sankar, K. U., Manohar, B., p. 807), the extraction of primrose seed (Tolboe, Hansen, and Shukla, p. 685), the extraction of pepper with cosolvents added to carbon dioxide (Duboc, S., Laugier, S., Richon, D., Renon, H., and Mizandjian, J. L., p. 759), the processing of fermentation broths with supercritical carbon dioxide (van Eijs, A.M.M., Wokke, J. M. P., Ten Brinke, B., and Dekker, K. A., p. 799), the extraction of flavor compounds (Lorne, J. L., Adda, J., p. 815), food colors (Jay, A. J., Smith, T. W., and Richmond, P., p. 821), enzymatic reactions in carbon dioxide (Nakamura, K., Min Chi, Y., and Yano, T., p. 925), and coffee decaffeination (Brunner, p. 691). This information complements the already large body of experimental data that has been developed on the solubility and extractability of natural products such as steroids, alkoloids, anticancer agents, oils from seeds, and caffeine from coffee beans using various SCF solvents such as CO_2, ethane, ethylene, and N_2O. Carbon dioxide is probably the most widely investigated SCF solvent since it is nontoxic, nonflammable, and environmentally acceptable. Also, it possesses a modest critical temperature ($T_c = 31.0°C$), which makes it an ideal solvent for extracting materials that are thermally labile.

Stahl and coworkers have compiled a large body of information concerning the solubility of numerous compounds in supercritical fluids. To obtain solubility information, Stahl uses a microextraction apparatus directly coupled to a thin-layer chromatograph (Stahl and Quirin, 1983; Stahl et al., 1980,1984). From these studies Stahl has developed several heurisitics which describe the variables that control the solubility of natural products in supercritical fluids, especially CO_2. These heuristics follow the molecular thermodynamic principles described earlier in this chapter. For example, they report that low molecular weight hydrocarbons and lipophilic organic compounds such as esters, ethers, and lactones are easily extracted with CO_2. This is not surprising considering that CO_2 acts like a lipophilic solvent (Hyatt, 1984; Kim and Johnston, 1987). Carbon dioxide also possesses a quadrapole moment which enhances its interaction with polar compounds. And, CO_2 exhibits Lewis acid-base characteristics, which promotes complexing with many compounds (Hyatt, 1984). Stahl finds that as the number of hydroxyl and carboxcylic acid groups on the mixture constituents increases, the solubility of the substance in CO_2 decreases. This effect is a direct manifestation of strong solute–solute interactions which cannot be broken by carbon dioxide. Not surprisingly, sugars and amino acids are not extracted by supercritical CO_2 probably due to the large amount of intramolecular hydrogen bonding exhibited by these compounds. To gain a better appreciation of the solvent characteristics of carbon dioxide, the reader is directed to the excellent papers by Francis (1954) and Dandge et al. (1985).

In the next section, the SCF processing of coffee and edible oils is briefly described to illustrate the principles employed with supercritical fluid technology.

Coffee Decaffeination

Decaffeination of coffee with supercritical carbon dioxide as well as the extraction of stimulants from tea and cocoa has received a large amount of research and development effort over the past few years (Paulaitis et al, 1983; McHugh and Krukonis, 1986). To explain this process it is necessary to first address the effect of water on the extractability of caffeine by "selective" supercritical carbon dioxide. Dry carbon dioxide cannot extract caffeine from dry coffee, either green or roasted, but moist carbon dioxide can. This same moist-dry effect is experienced if, for example, methylene chloride is used as the extraction solvent. Caffeine is chemically bound in a cholorogenic acid structure present in the coffee bean. Thus, water acts as a chemical agent that frees the bound caffeine from the coffee matrix, whether carbon dioxide or methylene chloride is used in the decaffeination process.

The selectivity of carbon dioxide for the aromas and oils relative to caffeine can be understood by comparing the extraction results with roasted and green coffee beans. With moist carbon dioxide it is possible to extract the caffeine from green coffee beans. However, there is nothing else present in green coffee beans that dissolves in carbon dioxide. Aroma oils are not present in green coffee beans since they are generated during the roasting process. If moist carbon dioxide is used to extract roasted coffee beans, the aroma oils are extracted along with the caffeine. If dry carbon dioxide is used to extract roasted coffee beans, only the aroma oils are extracted. In spite of the confusion about this point in the literature, the supercritical carbon dioxide extraction of coffee is a sound and clever process. The process currently operates in Bremen, West Germany, at 60,000,000 pounds per year. Moreover, it is the first example of a supercritical fluid process that has reached the commercial processing level and whose primary step is, indeed, supercritical.

Another feature of the decaffienation process is that a large excess of carbon dioxide is required to decaffeinate coffee because caffeine does not dissolve to its neat solubility level during the extraction process. Instead, the concentration achieved in the carbon dioxide phase is governed by an equilibrium interaction which is present in the carbon dioxide–water–caffeine–coffee bean system. It is speculated that the moisture-sensitive chemical binding between caffeine and coffee bean lowers the activity of caffeine so that it also does not dissolve to its neat solubility level (McHugh and Krukonis, 1986).

The separation of caffeine from the carbon dioxide phase leaving the extractor is also more complicated than using a simple pressure reduction. In the coffee decaffeination process, because caffeine dissolves to well below its solubility limit, an extremely large pressure reduction step is necessary to precipitate the caffeine from the carbon dioxide phase. Efficient processing would require that the recycled carbon dioxide be essentially free of caffeine. Two alternative schemes are availble for recovering the caffeine from the loaded carbon dioxide phase: stripping with water or adsorption onto activated carbon (McHugh and Krukonis, 1986).

To analyze the stripping process, distribution coefficients for the ternary system carbon dioxide–caffeine–water are needed. At conditions of about 80°C and 310 bar, the distribution coefficient is about 0.03–0.04 (weight basis) (Krukonis, 1981). Although the distribution coefficient is rather small for caffeine extraction from coffee solution using carbon dioxide, it is excellent for the reverse process of removing caffeine from carbon dioxide with a water wash. A U.S. patent discusses in more detail the water washing of recycled carbon dioxide (Prasad et al., 1981).

The other recovery concept uses activated carbon to remove caffeine from the carbon dioxide before recycle (Zosel, 1981). The adsorption behavior of caffeine from a loaded carbon dioxide phase indicates that it is possible to adsorb readily onto activated carbon (Krukonis, 1983).

Edible-Oils Extraction

The process development work being carried out at the USDA's Northern Regional Research Center on the extraction of vegetable oils with supercritical fluid solvents has been reported in the literature (Friedrich et al, 1982). Supercritical carbon dioxide is being considered as a replacement for hexane, which is now used in soybean oil extraction.

The supercritical fluid–soybean extraction process would operate in the following manner. The soybean, which is first flaked to break the cell membrane walls to make the oil accessible to the solvent, is charged to an extraction vessel. Carbon dioxide is passed through the bed, and the CO_2–oil solution leaving the extractor is expanded to a lower pressure to precipitate the oil. The carbon dioxide is then recompressed and recycled to the vessel. One of the factors that affect the economics of the SCF–soybean oil process is the development of a method for continuously feeding and removing soybean solids.

The extraction and oil composition data on soybean oil have been described in a number of journals (see, for example, Friedrich and Pryde, 1984). A number of U.S. and European researchers have reported triglyceride solubility data (Christianson et al., 1984; List et al., 1984; Stahl et al., 1980, 1984). At conditions of 80°C and 825 bar, carbon dioxide and soybean triglycerides become miscible. The recovery of the oil from a 825 bar carbon dioxide–oil stream can be carried out by dropping the pressure by only 70–140 bar at 80°C. While these are high pressures, the recompression costs would be minimized since carbon dioxide is essentially incompressible at these pressures.

Carbon dioxide has been tested with other materials, such as corn and wheat germ, sunflower and safflower seeds, and peanuts. To a very good first approximation, the solubility of all vegetable triglycerides is identical, and all these seeds can be extracted completely if the cells are macerated to make the oil accessible.

Some work on the supercritical carbon dioxide concentration of active fatty acid fractions of fish oils has also been reported (Krukonis, 1984). Evidence has accumulated that suggests that certain polyunsaturated fatty acids comprising fish oil triglycerides have a therapeutic effect on the cardiovascular system. Increasing attention is being directed to clinical studies of fish oils in the human diet and to the development of new refining methods for purifying and concentrating the active components, such as eicosapentaenoic acid (EPA). The active components possess a high degree of unsaturation, and conventional processing by high vacuum and molecular distillation can cause degradative reactions at the high temperature levels required to separate the fish oils from free fatty acids, protein residues, and polychlorinated biphenyls.

Because the active fatty acid moieties are in relatively low concentration in fish oils, because they are dispersed on the triglyceride chains with other fatty acids, and because there are many fatty acids with identical carbon number but with varying unsaturation, it is difficult to concentrate the C20:5 component by distillation or by supercritical fluid extraction. If, however, the triglycerides are first transesterified to methyl (or ethyl) esters, the EPA fraction can be concentrated using supercritical carbon dioxide fractionation. If a reflux stage is added to the process with a two-pass processing sequence, the C20:5 fatty acid concentration can be increased to 95 wt% in a selected fraction (Eisenbach, 1984).

REFERENCES

Anderson, R., Cambio, R. and Prausnitz, J.M. 1962. Physical and chemical forces in solvent selectivity for hydrocarbons. *AIChE J.* 8: 66.

Christianson, D.D., Friedrich, J.P., List, G.R., Warner, K., Bagley, E.B., Stringfellow, A.C., and Inglett, G.E. 1984. Supercritical extraction of dry milled corn germ with carbon dioxide. *J. Food Sci.* 49: 229.

Coenan, H., Hagen, R., and Knuth, M. 1983. U.S. Patent #4,400,398.

Dandge, D.K., Heller, J.P. and Wilson, K.V. 1985. Structure solubility correlations: Organic compounds and dense carbon dioxide binary systems. *Ind. Eng. Chem. Prod. Res. Dev.* 24: 162.

Dhalewadikar, S.V., Seckner, A.J., McHugh, M.A., and Guckes, T.L. 1987. Separation of dodecane-biphenyl mixtures with supercritical carbon dioxide, ethane, and ammonia. *Ind. Eng. Chem. Res.* 26: 976.

Eisenbach, W. 1985. Supercritical extraction of ethyl esters of fish oils. 190th ACS Meeting, Chicago, Sept.

Francis, A.W. 1954. Ternary systems of liquid carbon dioxide. *J. Phys. Chem.* 58: 1099.

Friedrich, J.P. and Pryde, E.H. 1984. Supercritical CO_2 extraction of lipid-bearing materials and characterization of the products. *J. Am. Oil Chem. Soc.* 61: 223.

Friedrich, J.P., List, G.R., and Heaking, A.J. 1982. Petroleum-free extraction of oil from soybeans with supercritical CO_2. *J. Am. Oil Chem. Soc.* 59: 288.

Hyatt, J.A. 1984. Liquid and supercritical CO_2 as organic solvents. *J. Org. Chem.* 49: 5097.

Kim, S. and Johnston, K.P. 1987. Molecular interactions in dilute supercritical fluid solutions. *Ind. Eng. Chem. Res.* 26: 1206.

Krukonis, V. 1984. 75th Am. Oil Chem. Soc. Mtg., Dallas, May.

Krukonis, V.J. 1988. Processing with supercritical fluids: Overview and applications. In *Supercritical Fluid Extraction & Chromatography: Techniques & applications.* ACS Symposium Series 366.

Krukonis, V.J. 1981. Adsorption isotherm for the caffeine-coffee-carbon dioxide system. Unpublished data.

Krukonis, V.J. 1983. Adsorption isotherm for the carbon dioxide-caffeine-coffee-activated carbon system. Unpublished data.

Krukonis, V.J. 1984. Supercritical fluid fractionation of fish oils. Concentration of eicosapentaenoic acid. Paper presented at the 75th Annual Am. Oil Chem. Soc. Meeting, Dallas, TX.

Laws, D.R.J., Bath, N.A., Ennis, C.S., and Wheldon, A.G. U.S. patent 4,218,491.

List, G.R., Friedrich, J.P., and Pominski, J. 1984. Characterization and processing of cottonseed oil obtained by extraction with supercritical carbon dioxide. *J. Am. Oil Chem. Soc.* 61: 1847.

McHugh, M.A. and Krukonis, V.J. 1986. *Supercritical Fluid Extraction: Principles and Practice.* Butterworths, Stoneham, MA.

McHugh, M.A., Seckner, A.J., and Yogan, T.J. 1984. High pressure phase behavior of binary mixtures of octacosane and carbon dioxide. *Ind. Eng. Chem Fund.* 23: 493.

McHugh, M.A., Watkins, J.J., Doyle, B., and Krukonis, V.J. 1988. High pressure napthalene-xenon behavior. *Ind. Eng. Chem. Res.* 27: 1025.

McHugh, M.A. and Krukonis, V.J. 1989. Processing polymers and monomers with supercritical fluids. *Encyclo. Polym. Sci. & Eng.* 16: 368.

Nilsson, W.B., Stout, V.F., Hudson, J.K., Spinelli, J., and Gauglitz, E.J. 1986.

Supercritical extraction of methyl esters of menhaden oil. 77th Am. Oil Chem. Soc. Mtg., Honolulu, May.

Paulaitis, M.E., Penninger, J.M.L., Gray, R.D., and Davidson, P. 1983. *Chemical Engineering at Supercritical Fluid Conditions.* Ann Arbor Science, Ann Arbor, MI.

Prasad, R.M., Gotesman, M. and Scaralla, R.A. 1981. U.S. patent 4,246,291.

Prausnitz, J.M., Lichtenthaler, R.N., and Azevedo, E.G. 1986. *Molecular Thermodynamics of Fluid Phase Equilibria.* Prentice-Hall, Englewood-Cliffs, NJ.

Proceedings of the International Symposium on Supercritical Fluids. 1988. Societe Francaise De Chimie, Paris.

Scott, R.L. and van Konynenberg, P.B. 1970. Static properties of solutions—Van der Waals and related models for hydrocarbon mixtures. *Discuss. Faraday Soc.* 49: 97.

Shultz, E., Vollbrecht, H.R., Sandner, K., Sand, T., and Muhlnickel, P. 1984. U.S. Patent #4,470,927.

Stahl, E. and Quirin, K.W. 1983. Dense gas extraction on a laboratory scale: A survey of some recent results. *Fluid Phase Equil.* 10: 269.

Stahl, E., Quirin, K.W., Glatz, A., Gerard, D., and Rau, G. 1984. New developments in the field of high pressure extraction of natural products with dense gases. *Ber. Bunsenges. Phys. Chem.* 9: 900.

Stahl, E., Schliz, W., Schutz, W., and Willing, E. 1980. A quick method for the microanalytical evaluation of the dissolving power of supercritical gases. In *Extraction with Supercritical Gases.* (Ed.) Schneider, G.M., Stahl, E. and Wilke, G., p. 93. Verlag Chemie, Dearfield Beach, FL.

Suzuki, Y., Shimazu, M., Arai, K., and Saito, S. 1986. Supercritical extraction of fish oil and fractionation of methyl esters with CO_2. Annual AIChE Mtg., MiamiBeach, Nov.

Wong, J.M. and Johnston, K.P. 1986. Solubilization of biomolecules in carbon dioxide-based supercritical fludis. *Biotech. Prog.* 2: 29.

Zosel, K. 1974. U.S. patent 3,806,619.

Zosel, K. 1981. U.S. patent 4,260,639.

7

Drying of Foods

Enrique Rotstein

The Pillsbury Company
Minneapolis, Minnesota

INTRODUCTION

It is hard to decide where to focus in reviewing the issues of food drying. Many authors have devoted significant time and effort to covering the fundamentals of drying of foods, studying issues pertaining to the structure of food materials, equilibrium, and its influence on drying. The field of food dehydration concerns not only this "microscopic" analysis; there are a number of issues related to the design and operation of dryers. They belong to the realm of engineering practice. We have put together in this chapter a number of definitions and elements representing the state of the art in this particular field, including a lumped parameter approach to design. In this way we have preferred to present material that is, we hope, useful and practical. This selection leaves aside other subjects that are equally important and relevant: specific drying systems, operational and safety issues, drying of specific products, and many more. The reason is obvious—there is no way we could cover all issues in a comprehensive manner without going well beyond the space allocated.

We start with general definitions and then review the properties of water–air mixtures. With this information we put together the balances of mass, energy, entropy, and exergy. Then we cover heat and mass transfer coefficients and discuss drier design, transport phenomena analysis, and equilibrium. We end with a brief review of drying of cellular tissue.

GENERAL DEFINITIONS

Systems and Properties

System. In studying a drying operation, we need to define precisely to what volume in space, quantity of matter, or section of a piece of equipment the analysis is to be addressed. This is the system. Everything else around the system are the surroundings. The decision is ours entirely; it is made trying to simplify the analysis as much as possible.

Proper definition of the system to be studied is an important tool in problem solving. Although physical reality does not change by the approach used to study it, the amount of work in solving the same problem can be considerably decreased by selecting an adequate definition of the system to be studied.

Given a point inside a system, the set of values of properties needed to be able to reproduce it entirely is defined as state of the system.

Intensive and Extensive Properties. In designing, optimizing, or operating driers, we need to measure or quantify properties that are related to the different exchanges of mass and energy. They fall into two categories. Those properties for which the value is proportional to the amount of matter involved are called extensive properties. Typical examples are mass, volume, energy, and money. Extensive properties are additive, and as such they are subject to balance. When the property value is independent of the amount of mass involved, it is an intensive property. This is the case of properties such as temperature and pressure. Intensive properties are not additive, and they are not subject to balance. We can perform a mass or energy balance, not a temperature or pressure balance. From another standpoint, we can talk about thermodynamic equilibrium between intensive properties but not between extensive properties. It is possible to exchange extensive properties, not intensive properties. Thus, energy may be exchanged, temperature may not be exchanged.

An extensive property may be converted into an intensive one by associating it to a specific amount of matter. This is the case of specific volume, concentration, etc.

WATER–AIR MIXTURES

Since most drying operations take place between air and the material being dried, it is convenient to summarize the most frequently used psychrometric properties and their calculation.

Absolute Humidity

In a drying operation we typically use air as a carrier for moisture. It is convenient to refer the mass of moisture being carried to the unit mass of dry air associated with it. This is Y, the absolute humidity.

$$Y = \frac{m_w}{m_g} \tag{1}$$

At the pressure prevailing in usual drying operations, we can use the ideal gas law to express absolute humidity in terms of partial pressures of water and dry air. Thus:

$$p_w V = \frac{m_w}{M_w} RT \tag{2}$$

$$p_w V = \frac{m_g}{M_g} RT \tag{3}$$

Using Eqs. (2) and (3) in Eq. (1):

$$Y = \frac{p_w}{p_g} \quad \frac{M_w}{M_g} \tag{4}$$

We often find it convenient to refer to the total pressure p, rather than the dry air pressure:

$$p = p_g + p_w \tag{5}$$

Using Eq. (5) to replace p_g in Eq. (4):

$$Y = \frac{p_w}{p - p_w} \quad \frac{M_w}{M_g} \tag{6}$$

In some applications we need to use mole fractions rather than partial pressures. By definition, the water mole fraction in moist air is:

$$y_w = \frac{n_w}{n_g + n_w} \tag{7}$$

since:

$$n_w = \frac{m_w}{M_w} \tag{8}$$

$$n_g = \frac{m_g}{M_g} \tag{9}$$

it follows from Eqs. (2), (3), and (6) that the absolute humidity is related to y_w by:

$$Y = \frac{y_w}{1 - y_w} \frac{M_w}{M_g} \tag{10}$$

Accurate knowledge of the vapor pressure of water at saturation is critical. Correlations suitable to calculate this and other properties useful for psychrometric calculations have been recommended by the American Society of Agricultural Engineers (Agricultural Engineers Yearbook, 1986). The cited publication provides all constants given below. We can estimate the saturated vapor pressure by means of the following equation:

$$\ln \frac{p_{ws}}{R} = \frac{A + BT + CT^2 + DT^3 + ET^4}{FT - GT^2} \tag{11}$$

For quick estimations we recommend the approximation (Smith and Van Ness, 1975):

$$\frac{d \ln p_{ws}}{d \left(\frac{1}{T}\right)} = - \frac{\Delta H_v}{R} \tag{12}$$

Eq. (12) can be integrated between 30°C and the desired value of T, using the mean value of ΔH_v between 30 and T. After integrating and rearranging, we obtain for the range 30°C < T < 80°C:

$$\ln p_{ws} = 8.3506 - 5130.2833 \left(\frac{1}{T} - \frac{1}{303.16}\right) \tag{13}$$

Eq. (13) gives p_{ws} in Pascals for T in degrees Kelvin. Starting from Eq. (12), we can find analogous equations for other temperature ranges or other system of units.

If we need to calculate the saturation temperature, given p_{ws}, Eq. (14) applies (Agricultural Engineers Yearbook, 1986):

$$T = T^* + \sum_{i=0}^{8} A_i[\ln(0.00145\, p_{ws})]^i \tag{14}$$

for 62 kPa $< p_{ws} <$ 4,688 kPa, the constants T^* and A_i are given in the reference.

Fractional Saturation

It is convenient to refer to the water content of moist air in terms of how far it is from saturation. One way of doing this is in terms of the absolute humidity ratio, Y/Y_s. Using Eq. (6), we obtain the relationship between this ratio and the water vapor partial pressure:

$$\frac{Y}{Y_s} = \frac{p_w}{p_{ws}} \frac{(p - p_{ws})}{(p - p_w)} \tag{15}$$

We define relative humidity as:

$$\phi = \frac{p_w}{p_{ws}} \tag{16}$$

Although Eq. (15) is a more accurate description of the distance to saturation, the relative humidity is of more customary use than the absolute humidity ratio.

Replacing Eq. (16) in Eq. (15), we find the relationship between relative humidity and absolute humidity ratio:

$$\frac{Y}{Y_s} = \phi \frac{(p - p_{ws})}{(p - p_w)} \tag{17}$$

Thus, ϕ will depart more from Y/Y_s at lower ϕ values.

Thermodynamically, the relative humidity is directly associated with the water chemical potential of the moist air. By definition:

$$\mu_w - \mu_w^o = R\, T \ln \frac{f_w}{f_w^o} \tag{18}$$

where μ_w and μ_w^o are chemical potentials in the actual and reference states, respectively;

f_w and f_w^o are fugacities in the same states. At the low pressures associated with drying processes, we can consider air as an ideal gas, in which case we can replace the fugacity ratio by the partial pressure ratio p_w/p_{ws}, and Eq. (18) becomes:

$$\mu_w - \mu_w^o = R T \ln \phi \tag{19}$$

It is possible to obtain relative humidities from absolute humidities. Replacing in Eq. (6) p_w by ϕp_{ws} and rearranging:

$$\phi = \frac{Yp}{[(M_w/M_g) + Y] \, p_{ws}} \tag{20}$$

at saturation:

$$1 = \frac{Y_s \, p}{[(M_w/M_g) + Y_s] \, p_{ws}} \tag{21}$$

Dividing Eq. (20) by Eq. (21), it follows:

$$\phi = \frac{Y \, [(M_w/M_g) + Y_s]}{Y_s \, [(M_w/M_g) + Y]} \tag{22}$$

Moist Air Volume

The *humid volume* v_H is the volume of unit mass of dry gas plus its accompanying vapor, at the prevailing temperature and pressure. As indicated above, we can consider moist air an ideal gas in most practical situations. Thus its molar volume at T and p is:

$$V = R \, \frac{T}{p}$$

where $V = 22.416$ m^3/kmol when $p = 1$ and $T = 273$ K. The number of kmoles in the moist air is:

$$n_H = \frac{1}{M_g} + \frac{Y}{M_w}$$

The humid volume is:

$$v_H = N_H V$$

For the SI system, with Y in kg/kg dry air, $M_g = 28.966 \times 10^{-3}$ kg/mol, pressure in Pa and v_H in m^3/kg dry air:

$$v_H = 8{,}314 \left(\frac{1}{28.966} + \frac{Y}{18.016} \right) \frac{T}{p} \qquad (23)$$

An alternative approach is to calculate v_H as a function of p_w and T, instead of Y and T:

$$v_H = \frac{R}{M_g} \frac{T}{(p - p_w)} \qquad (24)$$

where $\dfrac{R}{M_g} = 287.04 \dfrac{J}{kgK}$ in the Si system.

We often use dry air specific volume v_g as an approximation for v_H:

$$v_g = \frac{287.04\ T}{p} \qquad (25)$$

Humid Heat

We call humid heat, c_H, the specific heat of unit mass of dry air with its associated water vapor. For a moist air of absolute humidity Y:

$$c_H = c_g + c_w\ Y \qquad (26)$$

Air and water specific heat show a slight dependence on temperature. For all practical purposes they are considered temperature independent. The SI value is $c_g = 1.0069$ kJ/kgK and $c_w = 1.8757$ kJ/kgK.

Enthalpy

Enthalpy is the sum of internal energy and pressure-volume product:

$$H = U + pV \qquad (27)$$

Enthalpy is the natural thermodynamic property for energy accounting in open systems. Since driers are generally open systems, it becomes important to evaluate the enthalpy of moist air.

The exchange in enthalpy with temperature depends on whether the substance is in the liquid, vapor, or gas state. Figure 7.1 shows a typical representation of an enthalpy–temperature diagram for a pure fluid. Line AC is the saturated liquid line,

C is the critical point, line CB corresponds to saturated vapor. The area under ACB is the two-phase region. At temperatures and pressures higher than the critical values, the fluid phase is named a gas. At each temperature T, the difference between the enthalpy of the saturated vapor and that of the saturated liquid is the heat of vaporization ΔH_{vT}. When a gas mixture is cooled at constant pressure from, say, point 1, it will reach the saturated vapor line at a temperature T_{dp}. We call this temperature the *dew point*. Points above the saturated vapor line correspond to the so-called superheated condition. For a liquid or vapor, the slope of the constant pressure lines is the heat capacity. Vapors at low pressure approach ideal behavior, in which case H becomes independent of pressure and is a function of temperature only.

For the purpose of practical calculations, enthalpy values are property changes with respect to a specified reference state. In the case of air and water vapor, gaseous air and saturated liquid water at 0°C (32°F) are often used as the reference state. We build the equation to evaluate enthalpy changes on the basis of the dry air contribution plus the vapor contribution. This is better understood observing Fig. 7.1, where the reference state for the vapor corresponds to H_o, T_o. On the basis of unit mass of dry air:

$$H_1 - H_o = c_g (T_1 - T_o) + [C_w (T_1 - T_{dp}) + \Delta H_{vdp} + c_{lw}(T_{dp} - T_o)] Y_1 \qquad (28)$$

where c_{lw} = 4.1868 kJ/kg K. On the right hand side of Eq. (28), the first term is the air contribution, and the three terms between brackets describe the contribution of water vapor, heat of vaporization, and liquid water, respectively. Because of the low operating pressures of conventional driers, points 1 and 1′ in Fig. 7.1 tend to coincide. As a result, we can approximate Eq. (28) by an equation describing the vaporization of water at T_o followed by constant pressure heating up to point 1′:

$$H_{1'} - H_o = c_g (T_1 - T_0) + [c_w (T_1 - T_0) + \Delta H_{vo}] Y_1 \qquad (29)$$

Recalling the definition of humid heat, Eq. (2), Eq. (29) can be rearranged and written as:

$$H_{1'} - H_0 = c_H (T_1 - T_0) + \Delta H_{vo} Y_1 \qquad (30)$$

Heat of Vaporization Estimates

We estimate the heat of vaporization using the equation suggested in the Agricultural Engineers Yearbook (1986), which provides the corresponding constants and temperature ranges:

$$\Delta H_v = h_o - h_1 (T - T_b) \qquad (31)$$

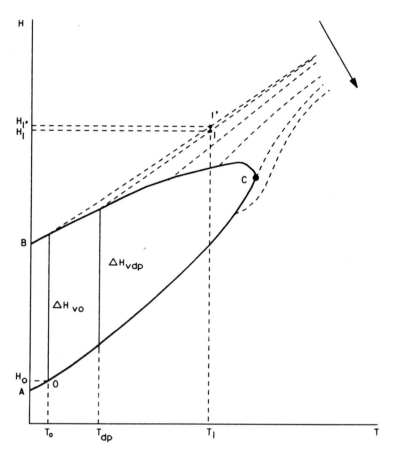

Figure 7.1   Enthalpy–temperature diagram for a pure fluid.

Dew Point

The dew point temperature of an air-vapor mixture, air being noncondensable at usual drying operating conditions, is the śame as that for vater vapor alone. Thus, if we cool moist air at constant total pressure out of contact with liquid water, the temperature when it reaches saturation is T_{dp}, the dew point temperature (see Fig. 7.1). In terms of absolute humidity, this means that given a starting value Y at T > T_{dp}, when T = T_{dp} the same value of Y becomes Y_s. We can use Eq. (14) to evaluate the dew point temperature if we know p_w at the initial conditions.

Practical Evaluation of Psychrometric Properties

We can evaluate the properties of moist air discussed so far using a psychrometric chart, available in handbooks such as the *Chemical Engineers' Handbook* (Perry and Green, 1984). More accurate is to use a computer package such as PSYCHR (Ratti et al., 1989a), with the added advantage of providing a listing of relevant data and the potential of interfacing with other programs.

BALANCES

Mass Balance

Usual drying situations involve an entering wet solid or liquid stream which leaves as a solid with a specified moisture content. We dry this stream by means of dry air, which picks up the moisture as it goes through the drier. The air stream may or may not be mixed with an additional air stream somewhere along the equipment. This case is shown in Fig. 7.2, where G_1, G_2, and G_3 are air flow rates in terms of mass of dry air per unit time, and L_1, L_2 represent masses of bone dry material per unit time. It is convenient to represent water concentration in the solid material as X mass of water per unit mass of bone dry solid. Thus, we relate the total mass flow rate of moist air through any port q, $w_{q,A}$, to the air flow rate by:

$$w_{q,A} = G_q (1 + Y_q) \tag{32}$$

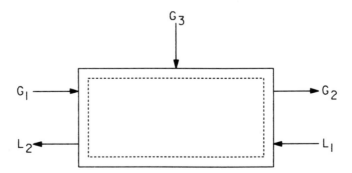

Figure 7.2 A typical drying situation: The dotted line indicates the system.

In the same manner, we relate the bone dry material flow rate through a suitable port q to the corresponding total mass flow rate $w_{q,D}$, by the following equation:

$$w_{q,D} = L_q (1 + X_q) \tag{33}$$

If we intend to make a water balance, it is convenient to split the system in Fig. 7.2 into two subsystems: air and drying material, as shown in Fig. 7.3. We represent the water flowing as vapor through any air port q by:

$$w_{q,w} = G_q Y_q \tag{34}$$

In the case of the water flow through any wet material port q:

$$w_{q,w} = L_q X_q \tag{35}$$

and there is an additional port which represents the water vapor leaving the material stream and entering the air stream, w_w.

For the system in Fig. 7.3, operating in steady state, the material balance is:

$$G_1 (1 + Y_1) + G_3 (1 + Y_3) - G_2 (1 + Y_2) \\ + L_1 (1 + X_1) - L_2 (1 + X_2) = 0 \tag{36}$$

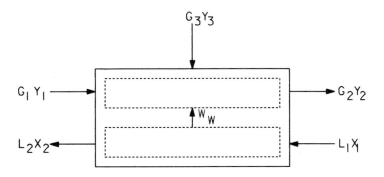

Figure 7.3 Drier illustrated in Fig. 7.2 split into two subsystems, as indicated by dotted lines.

and the water balance:

$$G_1 Y_1 + G_3 Y_3 - G_2 Y_2 + w_w = 0 \tag{37}$$

$$L_1 X_1 - L_2 X_2 - w_w = 0 \tag{38}$$

The dry air balance is:

$$G_1 + G_3 - G_2 = 0 \tag{39}$$

and the drying material dry matter mass balance is:

$$L_1 - L_2 = 0 \tag{40}$$

Notice that Eq. (36) is the sum of the water dry matter and dry air balances, i.e., Eqs. (37) through (40).

Illustration

We set up a drier in the manner shown in Fig. 7.3, to dry a solid food material. The product feed and output conditions are $L_1 = 400$ kg bone dry/hr, $X_1 = 7$ kg/kg, $X_2 = 0.3$ kg/kg. We specify the following conditions for air streams: $Y_1 = 0.01 = Y_3$, $G_1 = 10,000$ kg/hr, $G_2 = 13,000$ kg/hr. Find the amount of secondary air, G_3, output air absolute humidity Y_2, evaporation rate w_w, and product output (dry basis) L_2.

We find L_2 from Eq. (40), w_w from Eq. (38), Eq. (39) gives G_3 and Eq. (37) yields Y_3. From Eq. (40):

$$L_2 = L_1 = 400 \text{ kg bone dry/hr}$$

From Eq. (38):

$$\begin{aligned} w_w &= L_2 X_1 - L_2 X_2 \\ &= 400 (7 - 0.3) \\ &= 2,680 \text{ kg } H_2O/hr \end{aligned}$$

Using Eq. (39):

$$\begin{aligned} G_3 &= G_2 - G_1 \\ &= 13,000 - 10,000 \\ &= 3,000 \text{ kg/hr} \end{aligned}$$

Finally, Eq. (37):

$$Y_2 = \frac{G_1 Y_1 + G_3 Y_3 + w_w}{G_2}$$

$$= \frac{10,000 \times 0.01 + 3,000 + 0.01 + 2,680}{13,000}$$

$$= 0.22 \text{ kg/kg}$$

Lever Rule

When two streams are mixed to yield a third one, the so-called lever rule applies. Consider, for instance, two air streams that mix, as shown in Fig. 7.4. The water balance is:

$$G_1 Y_1 + G_2 Y_2 = G_3 Y_3 \tag{42}$$

and the air balance:

$$G_1 + G_2 = G_3 \tag{42}$$

Using Eqs. (42) in (41) and rearranging:

$$G_1(Y_3 - Y_1) = G_2 (Y_2 - Y_3) \tag{43}$$

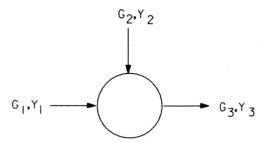

Figure 7.4 Two air streams mix to yield a third.

We call this mixture property the lever rule, in the sense that Eq. (43) is analogous to the moment equation if G_1 and G_2 are thought of as if they were forces and the $\Delta Y'$s as distances.

Illlustration

We use a countercurrent tunnel drier to dry 4,642 kg/day of fresh 85% moisture apple rings to a 22% moisture content, wet basis. The drying set-up is sketched in Fig. 7.5. The absolute humidities of air entering and leaving the tunnel are $Y_1 = 0.01895$ kg/kg, $Y_2 = 0.03072$ kg/kg, and $Y_5 = 0.00822$ kg/kg. $G_3 = 0$. Find G_1, G_2, G_4, G_5, G_6, L_1, L_2, w_w, Y_4, Y_6, X_1, and X_2.

Equations (37) through (40) are still valid, with $G_3 = 0 = Y_3$. As a result of mixing air streams 4 and 5 and splitting air stream 2, six new equations are added. From a water balance:

$$G_5\, Y_5 + G_4\, Y_4 - G_1\, Y_1 = 0$$

$$G_2\, Y_2 + G_4\, Y_4 - G_6\, Y_6 = 0$$

From a dry air balance:

$$G_4 + G_4 - G_1 = 0$$

$$G_2 - G_4 - G_6 = 0$$

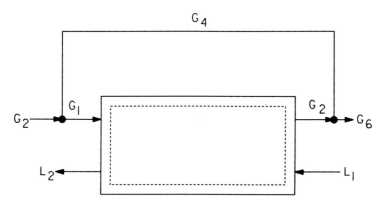

Figure 7.5   Sketch of a tunnel drier with air recirculation.

By inspection

$$Y_6 = Y_4 = Y_2 = 0.03072 \text{ kg/kg dry air}$$

so that there is only one independent equation corresponding to the splitting of stream 6.

$$X_1 = \frac{85}{15} = 5.667 \text{ kg water/kg dry matter}$$

$$X_2 = \frac{22}{78} = 0.282 \text{ kg water/kg dry matter}$$

$$L_1 = \frac{4{,}642}{24} \times 0.15 = 29 \text{ kg dry matter/hr}$$

From Eq. (40):

$$L_2 = L_1 = 29 \text{ kg/hr}$$

Using Eq. (38)

$$\begin{aligned} w_w &= L_1 (X_1 - X_2) \\ &= 29 \times (5.667 - 0.282) \\ &= 156.17 \text{ kg/hr} \end{aligned}$$

From Eq. (39), G_3 being equal to zero:

$$G_1 = G_2$$

thus, we can use Eq. (37) to find G_1:

$$G_1 (Y_1 - Y_2) + w_w = 0$$

$$\begin{aligned} G_1 &= \frac{w_w}{Y_2 - Y_1} \\ &= \frac{156.17}{(0.03072 - 0.01895)} \\ &= 13{,}268.48 \text{ kg dry air/hr} \end{aligned}$$

We now use the equations from mixing streams to find G_4 and G_5:

$$0.00822 \ G_5 + 0.03072 \ G_4 = 13{,}268.48 \times 0.01895$$
$$G_5 + \ G_4 = 13{,}268.48$$

Solving this system of two equations and two unknowns, we find:

$$G_4 = 430.49 \text{ kg dry air/hr}$$
$$G_5 = 12{,}837.99 \text{ kg dry air/hr}$$

which allows calculation of G_6:

$$G_6 = G_2 - G_4$$
$$= 13{,}268.48 - 430.49$$
$$= 12{,}837.99 \text{ kg dry air/hr}$$

The Energy Balance

The general equation for energy balance in an open system operating at steady state is:

$$\sum_{q_{out}} (G_q \hat{H}_q + L_q \hat{H}_q) - \sum_{q_{in}} (G_q \hat{H}_q + L_q \hat{H}_q) = \sum_{l_{in}} (\dot{Q}_l + \dot{W}_{ul}) - \sum_{l_{out}} (\dot{Q}_l + \dot{W}_{ul}) \qquad (44)$$

Illustration

To the drier of the previous illustration we feed fresh apples entering the drier at 18°C. The dried rings leave at 60°C. Figure 7.5 shows a sketch of the drier, identifying the different streams. Stream flow rates and concentrations corresponding to steady state operation are shown in Table 7.1. Air flow rates and humidities were shown in the previous illustration. Considering the system made of the drier, air streams 1 and 2, and apple streams 1 and 2, find the heat losses.

Using Eq. (44), identifying air enthalpies with subscript g and apple enthalpies with subscript d:

$$G_2 H_{2g} + L_2 H_{2d} - G_1 H_{1g} - L_1 H_{1d} = Q_{out}$$

Thus:

$$Q_{out} = 13{,}268.48 \ (120.37 - 116.20) - 29.00 \ (457.22 - 171.32)$$
$$= 63{,}620.67 \text{ kJ/hr}$$

Table 7.1 Stream Flow Rates and Concentrations in Apple Drier

Stream	Temperature (°C)	Flow Rate kg dry/hr)	Concentration (kg water/kg dry)	Enthalpy (kJ/kg dry)
Air,1	70	13,268.48	0.01895	120.37
Apple,1	18	29.00	5.667	457.22
Air,2	37	13,268.48	0.03072	116.20
Apple,2	60	29.00	0.282	171.32

The Entropy Balance

This balance equation has a unique feature which makes it quite different from the energy balance equation: entropy is not necessarily conserved. According to the Clausius inequality, the production of entropy $\dot{R}_s$ cannot be negative:

$$\dot{R}_s \geq 0 \qquad (45)$$

Equation (45) means that when a process is reversible, the production of entropy is nil. A reversible process is one which we carry out with infinitesimal driving forces, i.e., with practically no departure from equilibrium; this would be the case, for instance, if two streams—one at T and other at T + dT—exchange heat. A reversible process is an idealization; it can be approached but it cannot be accomplished in reality. When the process is irreversible and all real processes are irreversible, the production of entropy is greater than zero.

For steady state with q = 1, 2, 3 ..., entrance and exit ports, and l = 1, 2, 3 ..., heat exchange surfaces, the entropy balance is:

$$\dot{R}_s = \sum_{q_{out}} (G_q \hat{S}_q + L_q \hat{S}_q) - \sum_{q_{in}} (G_q \hat{S}_q + L_q \hat{S}_q) + \sum_{l_{out}} \frac{\dot{Q}_l}{T_l} - \sum_{l_{out}} \frac{\dot{Q}_l}{T_l} \qquad (46)$$

The Exergy Balance

The exergy balance combines the energy and entropy balances. Exergy of a stream q is defined as:

$$B_q = H_q - T_a S_q \qquad (47)$$

In a steady state open system, with q exit and entrance ports and l work exchange surfaces, the exergy balance is:

$$\sum_{q_{out}} (G_q \hat{B}_q + L_q \hat{B}_q) - \sum_{q_{in}} (G_q \hat{B}_q + L_q \hat{B}_q) = \sum_{l_{out}} \dot{W}_{u,l} - \sum_{l_{in}} \dot{W}_{u,l} - T_a \dot{R}_s \qquad (48)$$

HEAT AND MASS TRANSFER COEFFICIENTS

Mass Transfer Coefficients

Consider a slab with a wet surface. From this surface water evaporates at a rate which, at a given temperature, will depend on the absolute humidity and velocity of the drying air which flows past it. If the mechanism by which water migrates into the air were purely a diffusional one, the diffusion equations would be enough to obtain a quantitative description of the phenomenon. This is rarely the case, because many practical situations involve turbulant flow. On the other hand, in a situation such as the one described, there will be a velocity distribution in the air. It will decelerate as it approaches the surface, and, if the surface has a null velocity in the direction of the air flow, the air velocity will be zero at the interphase. Thus, most of the change in concentration from the interphase to the bulk of the air will occur near the interphase, where a diffusional mechanism is more likely to occur, at least partly. As a result we define the mass transfer coefficients on the basis of diffusion equations.

In the one-dimensional case (Treybal, 1980):

$$N_w = C D_{w,nw} \frac{\partial y_w}{\partial z} + y_w (N_w + N_{nw}) \qquad (49)$$

we define the fluxes N_w and N_{nw} as fluxes at the phase interface. Two important cases arise: one corresponds to the convective term $y_w(N_w + N_{nw})$ being negligible; the other to the opposite case. Notice that this term may be negligible either because fluxes are low or because y_w is low. When the convective term is neglected, integration of Eq. (49) yields:

$$N_w = - C \frac{D_{w,nw}}{(z_1 - z_0)} (y_{w1} - y_{wo}) \qquad (50)$$

Actual calculations would require knowledge of the film thickness $(z_1 - z_0)$ over which the purely diffusional mechanism occurs. This is not an easy task and in practice $C D_{w,nw}/(z_1 - z_0)$ is replaced by a mass transfer coefficient K_y. Thus, we replace Eq. (50) by:

$$N_w = -K_y (y_{w1} - y_{w0}) \qquad (51)$$

If we multiply the mole fractions by the total gas concentration:

$$N_w = -K_c (C_{w1} - C_{w0})$$
(52)

Conversions from one coefficient to another results from referring to the starting equation, Eq. (50). From it:

$$K_y = C \frac{D_{w,nw}}{(z_1 - z_0)}$$
(53)

and

$$K_c = \frac{D_{w,nw}}{(z_1 - z_0)}$$
(54)

Thus, assuming ideal gas behavior:

$$K_y = C K_c = \frac{p}{RT} k_c$$
(55)

In a similar manner, replacing C by p/RT in Eq. (49) and using partial pressures as driving forces:

$$N_w = -K_G (p_{w1} - p_{w0})$$
(56)

and

$$K_G = \frac{K_c}{RT}$$
(57)

When the convective term in Eq. (49) is not negligible, two particular cases are important from a drying standpoint: flow of water vapor through stagnant air and equimolar counterdiffusion. The latter corresponds to migration of water from one enclosure to another, both being at the same total pressure. The need to keep the same pressure implies that the interconnected gas phases must exchange an equal number of moles. Integration of Eq. (49) for $N_{nw} = 0$ results in:

$$N_w = C \frac{D_{w,nw}}{z_1 - z_0} \ln \frac{1 - y_w}{1 - y_{w0}}$$
(58)

Defining the logarithmic mean of two values b_0 and b_1 as:

$$b_{lm} = \frac{b_0 - b_1}{\ln \dfrac{b_0}{b_1}} \qquad (59)$$

We can write Eq. (58) in terms of the driving force $(y_{w0} - y_{w1})$ by using Eq. (59) and recalling that:

$$y_w + y_{nw} = 1$$

Then

$$N_w = C \, \frac{D_{w,nw}}{z_1 - z_0} \, \frac{1}{(y_{nw})_{lm}} \, (y_{w0} - y_{w1})$$

which we can write:

$$N_w = K_y^s \, (y_{w0} - y_{w1}) \qquad (60)$$

If we need to use the total moisture content Y as the driving force, Eq. (10) can be used in Eq. (58) to obtain:

$$N_w = K_y \, \ln \frac{(M_w/M_g) + Y_0}{(M_w/M_g) + Y_1} \qquad (61)$$

Keey (1978) suggested the development of a more conventional shape for Eq. (61). To do this, we multiply the equation by M_w to obtain n_w, then by

$$\frac{M_g}{M_g} \, \frac{(M_w/M_g + Y_0)}{(M_w/M_g + Y_1)} \, \frac{(Y_0 - Y_1)}{(Y_0 - Y_1)}$$

and we rearrange, to obtain

$$n_w = K_y \, M_g \, F_1 \, F_2 \, (Y_0 - Y_1) \qquad (62)$$

where

$$F_1 = \frac{(M_w/M_g)}{(M_w/M_g) + Y_0}$$

$$F_2 = \frac{(M_w/M_g) + Y_0}{(Y_0 - Y_1)} \, \ln \frac{(M_w/M_g) + Y_0}{(M_w/M_g) + Y_1}$$

For small $(Y_1 - Y_0)$, F_2 approached unity. For $Y_0 << M_w/M_g$, F_1 also approached unity. Otherwise it may be handier to use either Eq. (60) or Eq. (61). A more compact expression of Eq. (62) is:

$$n_w = K_y F(Y_0 - Y_1) \qquad (63)$$

where:

$$K_y = K_y M_g$$

$$F = F_1 F_2$$

From Eqs. (60) and (53), it follows that:

$$K_y^s = \frac{K_y}{(y_{nw})_{lm}} \qquad (64)$$

The case of equimolar counterdiffusion corresponds to $N_w = -N_{nw}$, thus the equations are formally the same as those corresponding to a negligible convective term.

Heat Transfer Coefficients

Consider a surface at $z = 0$ subject to heat exchange with a fluid flowing past it. Say the fluid flows along the x direction (Fig. 7.6). At any point at a distance x from the origin, the surface will have a temperature T_0 lower than the corresponding fluid bulk temperature T_b. We define the local heat transfer coefficient h_{Loc} by the equation:

$$dQ = h_{Loc} \, l \, dx \, (T_b - T_0) \qquad (65)$$

where l is the width of the table in the y direction.

If the surface temperature distribution is known and the fluid properties do not change appreciably along the surface, we may define a heat transfer coefficient for the entire process:

$$Q = h A (T_b - T_0)_{mean} \qquad (66)$$

The value of h will depend on the specification for A and the mean temperature.

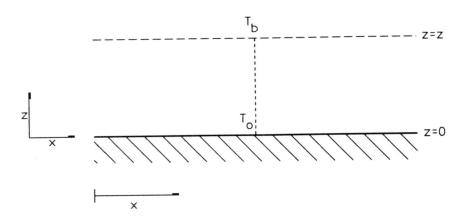

Figure 7.6 A surface subject to heat exchange.

Wet Bulb Temperature

Consider a liquid film in contact with a rapidly moving unsaturated vapor–air mixture. The incoming mixture properties are constant, its temperature measured with a dry bulb thermometer being T_{db}. Assume the liquid film is at a temperature $T_l < T_{db}$; as water evaporates from the liquid, the required vaporization energy is provided by the liquid film, which as a result cools down. When $T_{dp} < T_l < T_{db}$, the gas stream can provide additional energy to the vaporization process. Eventually the liquid film will cool down to a constant equilibrium temperature, T_{wb}. This is the wet bulb temperature.

The fact that air is flowing rapidly over the liquid film surface allows us to assume that the temperature and water partial pressure drops occur primarily across a thin air film in contact with the liquid film (Fig. 7.7). This film theory provides a simple but effective picture of the pheneomenum, on which basis predictions can be made.

We can estimate transfer from gas to liquid using:

$$q = h (T_{db} - T_{wb}) \tag{67}$$

where h is the gas side convective heat transfer coefficient. We can approximate the mass transferred into the air stream using Eq. (63) with $K'_y = K_y F$:

$$n_w = K'_y (Y_{wb} - Y) \tag{68}$$

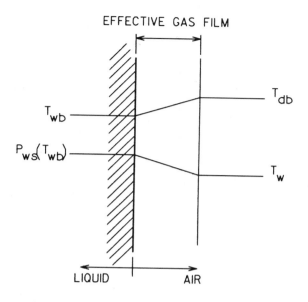

Figure 7.7 Wet bulb temperature corresponding to an air stream at T_{db}, p_w, and Y.

Where K'_y is the gas side mass transfer coefficient and Y_{wb} is the saturated absolute humidity corresponding to the liquid at T_{wb}. The equilibrium condition implies:

$$q_g + \Delta H_{vwb} \cdot n_w = 0 \tag{69}$$

Replacing Eqs. (67) and (68) into (69) and rearranging:

$$T_{db} - T_{wb} = \frac{\Delta H_{vmb} (Y_{wb} - Y)}{(h_g/K'_y)} \tag{70}$$

Equation (70) relates the wet bulb depression ($T_{db} - T_{wb}$) to the increase in humidity from the actual air condition to the saturated value Y_{wb}.

Experimental results show that we can replace the psychrometric ratio (h_g/K'_y) by humid heat. Using this fact in Eq. (70) and rearranging:

$$Y = \frac{Y_{wb} - [(T_{db} - T_{wb})/\Delta H_{vwb}]c_g}{1 + [(T_{db} - T_{wb})/\Delta H_{vwb}]c_w} \tag{71}$$

DRYING KINETICS AND DRIER DESIGN

A Practical Lumped Parameter Approach

Dependable drying models and the required data to fit them to practical situations are scarce. Much work remains to be done in that direction. A practical suggestion was put forward by Van Meel (1958). Assuming it was possible to obtain a normalized drying curve for each material, Van Meel developed the elements for a lumped parameter approach. This curve was called the characteristic drying curve (CDC). Keey (1972,1978) discussed applications of the concept, and Keey and Suzuki (1974) examined its theoretical foundations.

The concept of a characteristic drying curve is based on normalizing the water flux n_w at any stage of the drying dividing it by the maximum evaporation rate n_w^m. The ratio is the relative evaporation rate f:

$$f = \frac{n_w}{n_w^m} \tag{72}$$

It is assumed that the normalization renders f independent of external process conditions and dependent only on the structure of the material. For a certain drying process it is reasonable to assume that the structure of the material will depend only on its normalized moisture content $\Phi = (X - X_e)/(X_{cr} - X_e)$:

$$f = f(\Phi) \tag{73}$$

Two concepts need further clarification—the maximum evaporative flux and the critical moisture content. The maximum evaporative flux is that corresponding to unhindered evaporation, the flux that would be obtained in free evaporation from a liquid surface. We can calculate this using Eq. (63):

$$n_w^m = K_Y F (Y_0 - Y_1) \tag{74}$$

The above is the flux that can be obtained in certain materials at the initial stage of drying. When this is the case, n_w^m remains constant for the duration of the stage, and the time span is termed the constant rate period. For this period to exist, the surface of the material must be fully wet. Afterwards the flux falls below the value given by Eq. (74) and the falling rate period begins. The moisture content at which the first period ends and the second starts is called the critical moisture content, X_{cr}.

Equation (72) is, in reality, a first approximation. As shown by Keey and Suzuki (1974), f is also a function of temperature, the balance of internal and external resistances to moisture transfer and the intensity of drying. Nevertheless, in many practical applications the single lumped parameter f has allowed successful interpolations, comparison of drying alternatives and description of processes where internal resistances prevail.

The water flux at any time is obtained from experimental data, $X = X(t)$ or from the theoretical predictions, if they are available. The moisture content loss data are transformed into rates and the following equation provides the value of n_w:

$$n_w = \frac{d \, (\rho_{nw} \, X/a)}{dt} \tag{75}$$

where ρ_{nw} is the bulk density of nonaqueous matter and a is the evaporation area per unit volume of material being dried. In general, both ρ_{nw} and a are functions of moisture content. The difficulty in applying the normalization procedure to foodstuffs is that in many cases the constant rate period is nonexistent. Fornell et al. (1980) suggested that this difficulty could be overcome by using a calculated n_w^m;

$$n_w^m = \frac{h \, (T_{db} - T_{wb})}{\Delta H_{vwb}} \tag{76}$$

Using findings from Loncin (1976), the authors indicate that $h/\Delta H_{vwb}$ is proportional to the square root of the air velocity. Thus:

$$n_w^m = \beta \, u^{1/2} \, (T_{db} - T_{wb}) \tag{77}$$

where β is a proportionality constant. Using Eqs. (77) and (75) in Eq. (72) and taking into account Eq. (73):

$$\frac{-d \, (\rho_{nw} \, X/a)/dt}{\beta \, u^{1/2} \, (T_{db} - T_{wb})} = f \, (\Phi) \tag{78}$$

it follows that a plot of $[d(\rho_{nw} \, X/a)/dt]/[u^{1/2}(T_{db} - T_{wb})]$ vs. Φ will result in a single CDC for the material being studied. Actually, Fornell et al. (1980) plotted $[(dX/dt)]/[u^{1/2}(T_{db} - T_{wb})]$ vs. X for several materials (apple, carrot, and potato dices) and succeeded in getting a characteristic curve for each. The result implies that, to the extent that a CDC exists, a does not change with time, a property which has not yet been proved.

For design, extrapolation, comparison, or other purposes, it is often desirable to find the drying time. This is possible using the above ideas and integrating Eq. (75):

$$dt = \frac{d(\rho_{nw} X/a)}{n_w} \tag{79}$$

$$t = - \int_{X_i}^{X} \frac{d(\rho_{nw} X/a)}{n_w} \tag{80}$$

If the material is such that ρ_{nw}/a is independent of moisture content, Eq. (80) simplifies to:

$$t = - \frac{\rho_s}{a} \int_{X_i}^{X} \frac{dX}{n_w} \tag{81}$$

We carry out the integration numerically, starting at X_i, finding Φ and, through the corresponding $f(\Phi_i)$, the value of n_w from Eq. (78). In the case of data such as that provided by Fornell et al. (1980), the calculation is simpler. A value:

$$\frac{dX/dt}{u^{1/2}(T_{db} - T_{wb})} = f'(X) \tag{82}$$

is obtained for each value of X. Then:

$$t = - \frac{1}{u^{1/2}} \int_{X_i}^{X} \frac{dX}{(T_{db} - T_{wb}) f'(X)} \tag{83}$$

Illustration

We need to design a countercurrent drier for the drying of apples. The required output is L = 29 kg dry matter/hr with raw material and product moisture contents $X_{in} = 5.67$ and $X_{out} = 0.28$. The incoming air has $T_{db} = 70°C$, $T_{wb} = 35°C$, and the air leaving the drier has $T_{db} = 37°C$ and the same T_{wb}. The air velocity is u = 3.72 m/s. Find the drying time.

Figure 7.8 shows a flow chart of the calculation procedure. We obtain the value of f' from the general plot of Fornell et al. (1980), as indicated above. Using the psychrometric computer package PSYCHR (Ratti et al., 1989a), $Y_{wb} = 0.03445$ kg/kg, $Y_{in} - 0.01$ kg/kg, $Y_{out} = 0.0331$ kg/kg and $\Delta H_{wb} = 2421.4$ kJ/kg. We calculate the values of $(T_{db} - T_{wb})$ at each iteration using Eqs. (67) through (69).

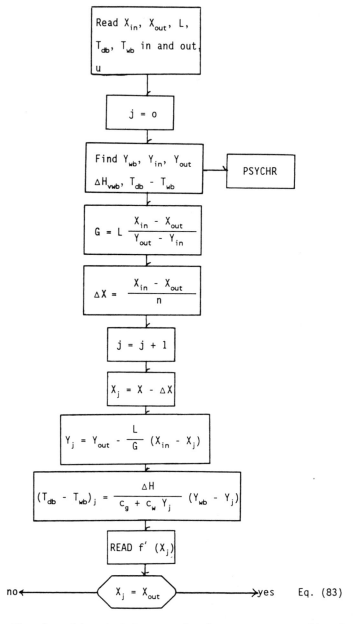

Figure 7.8 Flow chart of the calculation procedure for a countercurrent drier of apples.

Table 7.2   A Ten-Step Discretization to Calculate Drying Time

j	X_j	Y_j	$(T_{dbj} - T_{wb})$, (°C)	10^6 f'(X_j), $(s°C)^{-1}(m/s)^{½}$	$\dfrac{-\Delta X}{(T_{dbj} - T_{wb})}$ f'(X_i), hr (m/s)^{½}
0	5.67	0.0331	3.0	111.4	
					0.317
1	5.131	0.0317	6.2	93.9	
					0.221
2	4.592	0.0303	9.4	79.5	
					0.185
3	4.053	0.0289	12.7	66.8	
					0.169
4	3.514	0.0275	15.9	57.3	
					0.161
5	2.975	0.0261	19.1	49.3	
					0.161
6	2.436	0.0248	22.2	40.7	
					0.173
7	1.897	0.0234	25.5	31.8	
					0.193
8	1.358	0.0220	28.8	25.4	
					0.238
9	0.819	0.0206	32.1	15.9	
					0.370
10	0.280	0.0192	36.0	7.9	
					Total 2.188

We show in Table 7.2 a 10-step discretization and the corresponding result. From it and Eq. (84):

$$t = \frac{2.188}{1.93} = 1.57 \text{ hr}$$

Transport Phenomena Analysis

A much broader understanding, with the corresponding implications in terms of product and process optimization, is obtained from transport phenomena analysis.

Basically we need to pose the mass and energy equations and solve them. A typical mass transfer equation will be of the shape:

$$\rho_{nw} \frac{\partial X}{\partial t} + \rho_{nw} u_{sh} \frac{\partial X}{\partial z} = \frac{\partial(\rho_{nw} D_{eff} \, \partial X/\partial z)}{\partial z} \tag{84}$$

where:

$$D_{eff} = f\,(\varepsilon, X) \tag{85}$$

$$\rho_{nw} = \frac{\rho_b}{1 + X} \tag{86}$$

Similarly, a representative energy equation is:

$$\rho_{nw} \, \Delta H_{sp} \frac{\partial X}{\partial t} + \rho c \left(\frac{\partial T}{\partial t} + u_s \frac{\partial T}{\partial z} \right) = \frac{\partial(k_{eff} \, \partial T/\partial z)}{\partial z} \tag{87}$$

where:

$$k_{eff} = f\,(\varepsilon, X) \tag{88}$$

Solving this set of nonlinear partial differential equations is not a simple task, but we can do it via numerical solutions. The real challenge is that for each product we need densities, porosities, shrinkage data, effective diffusivities, specific heats, and thermal conductivities, as they change with moisture content and drying method. The literature on these data is still scarce. We need both hard data and predictive procedures.

EQUILIBRIUM

To solve the transport equations, Eqs. (84) and (87), we usually use the assumption of local equilibrium. To this effect we need dependable equilibrium data and a mathematical representation of the type $X = X(T, \phi)$. Ratti et al. (1989b) presented an expression of this type and indicated that there are three major requirements for these equations if they are to be used in drying modeling. The first is to accurately represent the equilibrium data at all ranges of X. The second is to provide dependable predictions of the partial derivatives $(\partial X/\partial \phi)_T$ and $(\partial X/\partial T)_\phi$. This is because we need to write the two transport equations in terms of two variables so that, for instance, we apply the chain rule to $\partial X/\partial z$ to obtain:

$$\frac{\partial X}{\partial z} = \left(\frac{\partial X}{\partial \phi} \right)_T \frac{\partial \phi}{\partial z} + \left(\frac{\partial X}{\partial T} \right)_\phi \frac{\partial T}{\partial z} \tag{89}$$

Inaccurate values of the partial derivatives result in severe undershooting or overshooting in the numerical solution. Finally, the correlation should provide good estimates of the heat of sorption.

DRYING OF CELLULAR TISSUES

Food systems subject to drying are usually heterogeneous and complex materials. The more we understand their structure, the easier it is for us to model it and to develop predictive procedures for transport properties and equilibrium. Systems made of cellular tissues are abundant throughout the industry of dried foods: fruits, vegetables, fish, and meats are used as raw materials for dried products. The cellular structure is a well-ordered multiphase system, and as such it can be modeled. This brings up interesting opportunities.

The water–food equilibrium relationships can be predicted by modeling the different phases (cell wall, membranes, cytoplasm, vacuole) and assuming equal water chemical potential for each phase, in equilibrium with moist air. Most transport properties can be predicted on the basis of simplified models of the cellular structure. These issues have been investigated and reported by Rotstein (1987) and Crapiste et al. (1988).

SYMBOLS

A   area
a   evaporation area per unit volume
B   exergy
C   molar concentration
c   specific heat
D   diffusivity coefficient
f   fugacity or relative evaporation rate
f′ relative evaporation rate, as per Eq. (84)
G   mass flow rate of dry air
h   convective heat transfer coefficient
H   enthalpy
K   mass transfer coefficient

l   width
L   mass flow rate of bone dry material
m   mass
M   molecular weight
n   number of moles or mass flux with respect to stationary coordinates
N   molar flux with respect to stationary coordinates
p   pressure; without subscript, total pressure
Pr   dimensionless Prandtl number
q   heat flux
Q   heat flow
R   gas law constant
R_s production of entropy
S   entropy
Sc   dimensionless Schmidt number
T   absolute temperature
t   time
u   velocity
U   internal energy
V   volume
v   volume per unit mass of dry air
w   mass flow rate
X   moisture content per unit mass of bone dry material
x   rectangular coordinate
Y   absolute humidity, mass of water vapor per unit mass of dry air
y   mole fraction
z   rectangular coordinate

Subscripts

a   state of the surroundings
b   bulk properties defined on the basis of molar concentration
cr   critical
d   dry matter
D   property of moist material subject to drying
db   dry bulb
dp   dew point
e   equilibrium
eff   effective
G   property defined on the basis of partial pressure
g   dry air
H   property of moist air

l   exchange surface for interactions without mass exchange
l   liquid water
lm   logarithmic mean
loc   local property
nw   nonaqueous component
o   reference state
q   entrance or exit port for convective flows
r   reversible
s   property at saturation state
sh   shrinkage
sp   sorption property
S   entropy
v   vaporation, vapor
w   water; when there is need for phase specification, water vapor
wb   wet bulb property
ws   saturated water property
Y   property defined on the basis of absolute humidity
y   property defined on the basis of mole fraction

Superscripts

m   maximum
s   stagnant gas phase

Greek Letters

β proportionality constant
ε porosity
μ chemical potential
Φ normalized moisture content
ϕ relative humidity
ρ density
Δ property change

Other

·   time derivative
°   reference state

REFERENCES

American Society of Agricultural Engineers. 1986. *ASAE Standards 1986*, 33rd ed. ASAE, St. Joseph, MI.

Crapiste, G.H., Whitaker, S., and Rotstein, E. 1988. Drying of cellular material. I. A mass transfer theory. *Chem. Eng. Sci.* 43: 2919.

Fornell, A., Bimbenet, J.J., and Admin, I. 1980. Experimental study and modelization for air drying of vegetable products. *Lebensm.-Wiss. U.-Technol.* 14: 96.

Keey, R.B. 1972. *Drying Principles and Practice.* Pergamon Press, Oxford, England.

Keey, R.B. 1978. *Introduction to Industrial Drying Operations.* Pergamon Press, Oxford, England.

Keey, R.B. and Suzuki, M. 1974. On the characteristic drying curve. *Intl. J. Heat Mass Transfer* 17: 1455.

Loncin, M. 1976. *Genie Industriel Alimentaire-Aspects Fondamentaux.* Masson, Paris.

Perry, R.H. and Green, D. 1984. *Chemical Engineers' Handbook,* 6th ed. McGraw-Hill Book Co., New York.

Ratti, C., Crapiste, G.H., and Rotstein, E. 1989a. PSYCHR: A computer program to calculate psychrometric properties. *Drying Technol.* 7(3): 575.

Ratti, C., Crapiste, G.H., and Rotstein, E. 1989b. A new water sorption equilibrium expression for solid foods based on thermodynamic considerations. *J. Food Sci.* 54: 738.

Rotstein, E. 1987. The prediction of diffusivities and diffusion-related transport properties in the drying of cellular foods. in *Physical Properties of Foods,* Jowitt, R. (Ed.). Elsevier Applied Science Publishers Ltd., Barking, Essex, U.K.

Smith, J.M. and Van Ness, H.C. 1975. *Introduction to Chemical Engineering Thermodynamics.* McGraw-Hill Book Co., New York.

Treybal, R.E. 1980. *Mass Transfer Operations,* 3rd ed. McGraw-Hill Book Co., New York.

Van Meel, D.A. 1958. Adiabatic convection batch drying with recirculation of air. *Chem. Eng. Sci.* 9: 36.

8

Aseptic Processing of Foods

M. A. Rao

Cornell University—Geneva
Geneva, New York

INTRODUCTION

Aseptic processing offers several advantages, such as continuous operation and higher retention of nutrients, over conventional canning. An aseptic processing (AP) system must provide continuously sterilized containers that are closed or sealed in a sterile atmosphere over long periods of time.

Because of the potential for producing high-quality food products, the developments in the scientific and technological aspects of aseptic processing have been the subject of many studies. For example, Atherton (1981) reviewed the capabilities and limitations of aseptic processing systems; Toledo (1975) and Ito and Stevenson (1984) reviewed sterilization of packaging systems; Mitchell (1988) reviewed the historical developments in aseptic processing technology; and Nelson et al. (1987) and Lenges (1988) covered the principles of some of the scientific and engineering aspects of aseptic processing. The regulatory and engineering aspects of aseptic processing of fluid foods containing particulates were discussed by several authors at a symposium of the Institute of Food Technologists (Dignan et al, 1989; Heldman, 1989; Larkin, 1989; Sastry et al., 1989, Chandrana et al., 1989). Earlier studies related to AP systems of liquid foods without particulates include estimation of lethality in flow systems and their design by Deindoerfer and Humphrey (1959a,b), Simpson and Williams (1974), and Lin (1976) and reviews of sterilization by Holdsworth (1970) and of residence by Rao and Loncin (1974a,b).

In this review, the developments in transport phenomena in aseptic processing systems will be covered. An attempt is made to include valuable information in some of the early studies in heat transfer and fluid flow mentioned above.

COMPONENTS OF AN ASEPTIC PROCESSING SYSTEM

A typical AP system consists of a heating section made up of heat exchangers (plate, tubular, or scraped surface), a holding section made up of a tube, and a cooling section again made up of heat exchangers (Fig. 8.1). The Food and Drug Administration (FDA) requires, as a measure of safety, that in AP systems processing liquid foods containing particulate solids the process lethality be attained in the holding section. When the lethality accumulated in the heating and cooling sections cannot be determined with certainty, the lethality achieved in the holding tube should be equal to the process lethality. However, if one can achieve the necessary lethality without employing a holding section for liquid foods without particulates, one can achieve higher nutrient retention; such a system was quantitatively analyzed by Simpson and Williams (1974).

Studies on heat transfer in AP systems can be divided on the basis of lumped parameter analysis, distributed parameter analysis, and a combination of both as

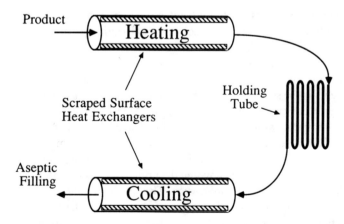

Figure 8.1 Schematic diagram of components of an aseptic processing system.

in heat transfer to solid particulates in liquids. In both the lumped and the distributed parameter approach, as shown later, knowledge of the rheological parameters will be useful. For example, rheological behavior is important in determining the length of holding tubes of fluid foods and in developing dimensionless correlations.

Rheological Behavior of Foods for Sterilization

The rheological behavior of fluid foods in relation to sterilization will be discussed in brief here and extensive reviews on the different types of rheological behavior of fluid foods, the measurement of rheological properties, and the applicability of rheological models can be found elsewhere (Holdsworth, 1971; Rao, 1986).

The simple power law model has been used extensively to describe the shear rate, $\dot{\gamma}$ (sec^{-1}), versus shear stress, σ (Pa), relationship of fluid foods:

$$\sigma = K \dot{\gamma}^{n} \tag{1}$$

Many fluid foods are shear-thinning (pseudoplastic) fluids so that the magnitudes of their flow behavior indexes are less than 1.0. Sastry (1989a) suggested that because the power law model cannot be applied at very low shear rates, the shear rate–shear stress data on liquid foods be interpreted in terms of the Ellis model:

$$\dot{\gamma} = \left[\frac{1}{\eta_0} + K_E \, \sigma^{(1/n)-1} \right] \sigma \tag{2}$$

While this model has not yet found extensive applications in the food literature, its applicability at low shear rates must be recognized. The effect of temperature on the apparent viscosity of fluid foods at a specific shear rate ($\eta_{a\dot{\gamma}}$) can be described by the Arrhenius relationship:

$$\eta = \eta_\infty e^{E_{a\eta}/RT} \tag{3}$$

In place of the apparent viscosity in the Arrhenius equation, one can employ the consistency index of the power law model of a non-Newtonian food:

$$K = K_\infty e^{E_{aK}/RT} \tag{4}$$

In nonisothermal flow, the influence of both the shear and temperature must be known. One can combine the power law and Arrhenius models to obtain the desired expressions; two such expressions that have been employed were that of Christensen and Craig (1962) (Eq. 5) and of Harper andEl-Sahrigi (1965) (Eq. 6):

$$\sigma = K_{TC} \left(\dot{\gamma} \, \exp \left[\frac{E_{ac}}{RT} \right] \right)^{\bar{n}}$$ (5)

$$\sigma = K_{TH} \, \exp \left(\frac{E_{aH}}{RT} \right) \dot{\gamma}^{\bar{n}}$$ (6)

It can be shown that $E_{aH} = n\,(E_{aC})$ so that one must be aware of the equation used to determine the activation energy.

It should be noted that few studies have been reported on the rheological properties of fluid foods at aseptic processing temperatures. The main reason for this state of affairs is that pressurized measuring systems must be employed in order to avoid boiling off of water in foods; apparently, this additional complexity adds to an already difficult task of determining shear rate–shear stress data using well-defined flow geometries such as capillary, concentric cylinder, parallel plate, and cone and plate.

STERILIZATION OF FLUIDS WITHOUT PARTICULATES

Production of aseptically processed and packaged low viscosity liquid foods such as fruit juices and milk on a commercial scale has been achieved on an impressive scale (Tillotson, 1984; David, 1989). In spite of the commercial success with certain types of liquid food products, there is considerable room for improvement in the existing processes and there is a need to develop processes based on sound scientific and engineering principles for existing and new food products.

Lumped Parameter Analysis

Deindoerfer and Humphrey (1959a) derived expressions for isothermal and nonisothermal destruction of bacterial spores, defined as $M = \ln\,(N_0/N)$ in liquid media; in the nonisothermal case, the Arrhenius expression was assumed for the temperature dependence of the first order reaction rate constant, k:

$$k = B \, \exp\,(-E_{aR}/RT)$$ (7)

Various common methods of sterilization were considered and analytical equations were derived for estimating (M) in batch and continuous sterilization for several reactor and temperature scenarios:
1. Batch sterilization
   a. Constant rate of heat flow

       Constant rate of addition to mass to be sterilized

       No change in medium mass

   b. Changing rate of heat flow

       Isothermal heat source or sink

       Nonisothermal heat source or sink

2. Continuous sterilization

   a. Constant rate of heat flow

       Constant energy loss

       Nonisothermal heat source or sink with equal and countercurrent mass flow

   b. Changing rate of heat flow

       Isothermal heat source or sink

Sterilization systems can be analyzed either in terms of M-values (Deindoerfer and Humphrey, 1959a) or in terms of G-values defined (Swartzel, 1982) as:

$$G = \log (N_0/N)/B = M/B \qquad (8)$$

It is important to recognize that the thermal properties, heat transfer coefficients, and the kinetic parameters for spore or nutrient inactivation must be available. When such data are available, the cumulative effects of inactivation in the heating, holding, and cooling sections can be summed so that the reduction in spores can be estimated for the entire sterilization process (G_P):

$$G_P = G_H + G_{H0} + G_C \qquad (9)$$

Specific expressions can be derived for G_P for different sterilization scenarios as suggested by Deindoerfer and Humphrey (1959a, b).

    For ultra-high-temperature heating processes in which the product is heated by direct steam injection and neglecting the lethality in the cooling stage, only the holding section will be important, i.e., $G_P = G_{Ho}$. To establish an equivalent heat effect by indirect UHT processing, Swartzel (1982) assumed an exponential heating regime, no holding time, and an instantaneous cooling regime, i.e., $(G_{Ho})_D = (G_H)_I$, where the additional subscripts D and I denote direct heating and indirect heating, respectively. For a given magnitude of E_a, an infinite number of time-temperature conditions would match the reduction level G. By repeating the calculations for a particular indirect system for various E_a values, a series of time-temperature lines (on semilogarithmic coordinates) of a direct system can be developed that intersect at a point called the equivalent time (t_E)-temperature (T_E) point. Once the magnitudes of t_E and T_E are known, the log reduction values of any heat labile constiuent can be calculated from the equation:

$$\log (N_0/N) = (Bt_E/2.303) \exp (-E_a/R\ T_E) \tag{10}$$

Examples of application of the equivalent point method to pasteurization of liquid whole egg and to whole milk can be found in Swartzel (1986), and its use for obtaining reaction kinetic data in flow systems can be found in Swartzel (1984) and Rhim et al. (1989).

Heat Transfer in Heat Exchangers

The heating and cooling rates in heat exchangers can be calculated from a knowledge of the temperatures of the heating and cooling media, their mass flow rates and specific heats, the heat transfer area, and the overall heat transfer coefficient. An useful review on fluid flow, heat transfer, and mixing in scraped surface heat exchangers can be found in Harrod (1986). The spatial average temperatures of the fluid food being cooled or heated were estimated by energy balances in a number of investigations (Deindoerfer and Humphrey, 1959a; Sastry, 1986; Chandarana et al., 1989; Larkin, 1989). For example, for heating in a flow system using an isothermal heat source, the temperature of the fluid is given by (Deindoerfer and Humphrey, 1959a):

$$\frac{T - T_H}{T_0 - T_H} = \exp \left(\frac{UA}{mC_p} \right) t \tag{11}$$

and the temperature of a fluid being cooled is given by (Larkin, 1989):

$$\frac{T_i - T_0}{T_i - T_{ci}} = \frac{1 - \exp [(1 - r_a) (-UA/C_p m)]}{1 - r_a \cdot \exp [(1 - r_a) (-UA/C_p m)]} \tag{12}$$

where $r_a = C_p m/C_{pc} m_c$. Many expressions for calculating the temperature of a fluid being heated or cooled can be found in texts on heat exchanger such as that of Kays and London (1964). It should be noted that few product side heat transfer coefficient data are available for non-Newtonian foods in plate, tubular, and scraped surface heat exchangers.

Distributed Parameter Approach

In the lumped parameter approach to analyzing transport phenomena in systems such as heat exchangers, one can work with mass and energy balances. However, in the distributed parameter approach, one solves the appropriate PDEs to obtain temporal and spatial distribution of velocity and temperature, and knowledge of the rheological behavior of the fluid foods is important.

Transport equations in aseptic processing systems. The fluid flow and heat transfer to fluid foods in AP systems are governed by the basic equations of continuity, motion, and energy:

$$\nabla \cdot \mathbf{V} = 0 \tag{13}$$

$$\rho \frac{\mathbf{DV}}{\mathbf{DT}} = - \nabla p - [\nabla \cdot \tau] + p \, \mathbf{g} \tag{14}$$

$$k \nabla^2 T = - (\nabla \cdot q) - (\tau : \nabla \mathbf{V}) \tag{15}$$

Equations 13 to 15 expressed in Cartesian, cylindrical, and spherical coordinates can be found in texts on transport phenomena such as that of Bird et al. (1960), and they can be solved with the appropriate boundary conditions for specific problems.

For heating and cooling of fluid foods, Eqs. (13) to (15) can be solved with the appropriate boundary conditions and transport coefficients. While analytical solutions can be derived for heat transfer in laminar flow of Newtonian fluids, solutions to many practical situations of heat transfer to Newtonian fluids in turbulent flow and to non-Newtonian fluids are not available. In addition, the lack of reliable food property data, especially at the aseptic process temperatures, makes it difficult to confide in theoretical solutions without experimental verification of the solutions. Nevertheless, analytical and numerical solutions can provide valuable information with respect to the role of important variables.

Heat transfer to non-Newtonian fluids. Heat transfer to non-Newtonian fluids in laminar tube flow of was studied (Simpson and Williams, 1974; Lin, 1976; Guariguata et al., 1979) to develop heat-cool systems with the aim of optimizing quality of fluid foods without particulates. Because the fluids were subjected to shear and temperature gradients, the rheological models included the effects of both variables. Simpson and Williams (1974) and Lin (1976) used the Arrhenius model to describe the effect of temperature; the former employed the model [Eq. (5)] of Christensen and Craig (1962) that described the combined effect of shear and temperature; whereas Guariguata et al. (1979) employed a temperature-dependent Bingham plastic model.

While Simpson and Williams (1974) and Guariguata et al. (1979) determined the desired heating-medium temperatures, Lin (1976) determined the sterilizer lengths for various heating-medium temperatures for a sterilizer with a steam-injection heating section, a holding section, and a cooling section. Because the results obtained by Lin (1976) and Guariguata et al. (1979) were similar to those obtained by Simpson and Williams (1974), a discussion of the latter's results only will be presented here.

Simpson and Williams (1974) examined heat transfer and sterilization of a power law fluid food in a sterilizer tube shown in Fig. 8.2. The equation to be solved can be derived from Eqs. (13) to (15) (Skelland, 1967) as:

$$\frac{\partial T}{\partial z} = \frac{\alpha}{v} \left[\frac{\partial^2 T}{\partial r^2} + \frac{1}{r} \frac{\partial T}{\partial r} \right] \tag{16}$$

The boundary conditions for solving Eq. (16) are:

$$T(r,0) = T_0 \tag{17}$$

$$T(r,z) = T_W \tag{18}$$

$$\frac{\partial T}{\partial r} = 0 \quad \text{at } r = 0 \tag{19}$$

The influence of temperature on viscosity was taken into consideration by the temperature coupling relationship:

$$\psi = \frac{E_a C}{R} \left(\frac{1}{T_0} - \frac{1}{T_W} \right) \tag{20}$$

When $\psi = 0$, the situation was designated as an iso-viscous case and suggested that the results of the non-Newtonian Graetz problem can be used; when ψ is large, correction factors need to be incorporated. It was pointed out that distortion of the

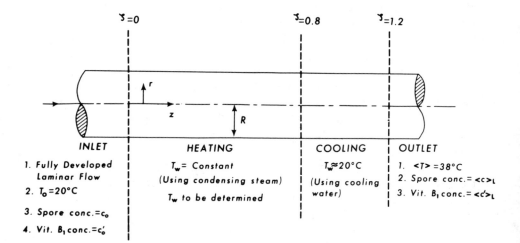

Figure 8.2 Schematic diagram of a heat-cool process sterilizer. (*Source: Simpson and Williams, 1974.*)

velocity profiles occurs at the beginning of the heating and the cooling sections, but that a correction factor would be necessary only in the cooling section. The heating section wall temperature was determined for different magnitudes of tube radii, flow behavior indexes, and of ψ.

Bacterial spores were assumed to remain on streamlines throughout the sterilization process. It was also assumed that the sterilization achieved at the center line must be equivalent to a conventional batch sterilization process carried out at 121.1°C for 2.45 min., i.e., log $(C/C_0) = -11.5$; the heating section is terminated when the center line temperature $(T_{cl}) = T_0 + 0.95 (T_W - T_0)$, and the average temperature at the end of the cooling section is 38°C. The dimensionless length of the sterilizer, ζ, defined as $\zeta = z_L \lambda a / R^2 v$, was 1.2, with the heating section being 2/3 of the length and the cooling section making up the rest. This result is valid for $0.3 \leq n \leq 1.0$ and $-4.0 \leq \psi \leq 4.0$.

Because the dimensionless length is constant, the total length increases as the square of the radius of the sterilizer tube. The use of smaller-diameter tubes results in better nutrient without compromising product sterility. The effect of neglecting the pseudoplastic nature of the fluid results in higher wall temperatures and an unnecessarily high degree of sterility as well as compromising nutrient retention. Table 8.1 shows the effect of the non-Newtonian nature of the fluid food and of the radius of the tube. If the heating-medium temperature was held constant instead of being a variable, as illustrated by Lin (1976), the cooling section length and the total sterilizer length would increase with increase in the magnitude of the flow behavior index once again emphasizing the important role of the rheological behavior of the fluid food. Guariguata et al., (1979) analyzing fluid foods following the Bingham plastic model also reported similar results; a comparison of the iso-viscous results with the nonisoviscous results showed that, for the latter, the heating section was 22% shorter and the cooling section was 18% longer.

RESIDENCE TIME DISTRIBUTION IN ASEPTIC PROCESSING SYSTEMS

In flow systems not all fluid and solid food particles remain for the same time periods, i.e., the particles have a distribution of residence times. Danckwerts (1953) proposed the concept of RTD, and the theoretical and experimental principles of residence time distribution (RTD) have been well reviewed by Levenspiel (1972) and Himmelblau and Bischoff (1968). The RTD functions and the role of RTD in continuous pasteurization systems was reviewed (Rao and Loncin, 1974a,b). Here only the necessary principles of RTD are discussed, and for additional information the above references must be consulted.

First, we note that with few exceptions RTD studies represent lumped parameter analysis. Only in a few instances, such as that of fully developed laminar flow of

Table 8.1 Effect of Tube Radius and Flow Behavior on Sterility and
Nutrient Retention

Radius/ Flow behavior index	Heating wall temperature (°C)	Probability of sterility[a]	Nutrient retention[b]
Radius (m)			
3×10^{-3}	141.0	7×10^{-13}	0.75
6×10^{-3}	134.5	1×10^{-13}	0.47
Flow behavior index, n[c]			
0.33	141.0	7×10^{-13}	0.75
1.0	146.0	2×10^{-36}	0.68

[a]Sterility estimates are based upon the temperature-dependent rate expression: $k' = 2 \times 10^{40} \exp(-3.73 \times 10^4/T) \ \mathrm{sec}^{-1}$.

[b]Nutrient retention is based upon the temperature-dependent rate expression: $k'' = 4.39 \times 10^{10} \exp(-1.18 \times 10^4/T) \ \mathrm{sec}^{-1}$.

[c]This section illustrates the effect of incorrectly assuming Newtonian behavior for a non-Newtonian fluid food which can be described by the power-law model.

fluids in tubes, one can attempt distributed parameter analysis in that the expressions for the RTD functions C and F can be derived from known velocity profiles.

Experimental RTD data on any equipment can be obtained by imposing at the inlet a pulse, step, or sinusoidal impulse and monitoring the response at the outlet. The tracer used must have physical properties such as density and viscosity similar to the test fluid food, and it should not in any manner alter the properties of the test fluid food. The sinusoidal impulse and response require more care and "sophisticated" equipment than the pulse and the step inputs. The response to a pulse impulse is called a C-curve, while the response to a step change is called a F-curve. For a closed vessel, i.e., in which material passes in and out by bulk flow (i.e., no diffusion at inlet and outlet), the C-and F-functions are related as:

$$C(\theta) = \frac{dF}{d\theta} \qquad (21)$$

One measure of the distribution of residence times (ages) of the fluid within a reactor is the E-function, defined so that $E \, d\theta$ is the fraction of material in the exit stream with age between θ and $\theta + d\theta$ (Levenspiel, 1972). It can be shown (Levenspiel, 1972) that the C- and F-functions are identical, and that for an isothermal process the ratio of the final (C) to initial (C_0) concentration of either microorganisms or nutrients can be determined from the expression:

$$\frac{C}{C_0} = \int_0^\infty \exp\left(- \frac{2.303\ t}{D(T)}\right) E(t)\ dt \tag{22}$$

where D(T) is the D-value of the microorganism or a nutrient at a temperature T. For a nonisothermal process, (C/C_0) can be determined by recognizing that the D-value is dependent on temperature and taking this dependence into consideration. At this point we note that $E(t) = [E(\theta)/t]$ is used in estimating the number of survivors or percent nutrient retention according to Eq. (22).

Interpretation of RTD data.   Experimental RTD studies are usually interpreted in terms of the the dispersion model [Eq. (23)] or the equal sized tanks in series model [Eq. (24)].

$$\frac{\partial C}{\partial \theta} = \left(\frac{D}{vL}\right) \frac{\partial^2 C}{\partial X^2} - \frac{\partial C}{\partial X} \tag{23}$$

where $X = z/L$ and $\theta = t/\bar{t} = t{\cdot}v/L$.

$$C(\theta) = \frac{N\ (N\theta)^{N-1}}{(N - 1)\ !}\ e^{-N\theta} \tag{24}$$

The former model is usually employed for flows in tubes and thus should be applicable to the interpretation of RTD in holding tubes. The theoretical RTD for laminar flow of Newtonian fluids can be derived as:

$$E\ (\theta) = \frac{1}{2\ \theta^3} \tag{25}$$

For non-Newtonian fluids that can be described by the power law model, Lin (1979) has shown that the RTD is given by the expression:

$$E\ (\theta) = \frac{d\ F(\theta)}{d\theta} = \frac{2n}{3n + 1}\ \frac{1}{\theta^3}\left(1 - \frac{n + 1}{3n + 1}\ \frac{1}{\theta}\right)^{(n-1)/(n+1)} \tag{26}$$

This expression derived by Lin (1979) appears to be correct and needs to be tested with experimental data.

There are a number of models for nonideal flows, i.e., flows that fall between the ideal conditions of a perfect mixer and plug flow. Some of the models for non-ideal flow were discussed in Levenspiel (1972) and Rao and Loncin (1974a).

Sizing Holding Tube Length

Holding tubes play an important role in thermal processing of fluid foods in HTST or aseptic flow systems. One simple but important relationship for determining the length of a holding tube (L) is based on the residence time of the fastest particle, i.e., $L = v_{max} \times$ holding time. Because the average velocity (v_{av}) of a fluid food can be readily calculated in many instances, a relationship between v_{av} and v_{max} will be useful. Such a relationship applicable for the fully developed laminar flow of fluid foods obeying the power law model is given by the relationship:

$$\frac{v_{avg}}{v_{max}} = \left(\frac{n + 1}{3n + 1} \right) \tag{27}$$

For the special case of a Newtonian fluid ($n = 1$), $v_{max} = 2 \times (v_{avg})$; this relationship provides a safe design criterion for Newtonian fluids and shear-thinning fluids in laminar and turbulent flow, but it will not provide a safe design for shear-thickening (dilatant) fluids. Even though there have been few reliable reports on dilatant behavior of fluid foods, the absence of such behavior must be confirmed by proper rheological tests in order to avoid under processing of fluid foods in holding tubes.

In practical terms, the Food and Drug Administration requires that holding tubes have a positive slope of at least 0.25 inch per foot to avoid dead spaces. Also, to conserve plant space, holding tubes are coiled. Therefore, one can expect the flow conditions in holding tubes to be far from those in fully developed laminar flow in straight tubes with the result that the maximum velocities are less than twice the average velocities. The experimental data of Sancho (1988), to be discussed later, confirm the deviations from ideal fully developed laminar flow.

Experimental RTD data.   Most of the RTD data on thermal processing flow systems found in the literature were obtained using water as the test fluid. For a plate heat exchanger, Roig et al. (1976) found that the RTD data could be fitted with five tanks-in-series model. Veerkamp et al. (1974) also found that the RTD data in a holding tube could be fitted with the tanks-in-series model.

Heppell (1985) demonstrated that milk has a broader RTD than water in an infusion-type UHT sterilizer. Therefore, it is necessary to perform RTD studies on holding tubes and heat exchangers with either fluid foods that are to be processed or with model solutions that possess physical properties similar to the fluid foods. Sancho (1988) found that the maximum velocities in a holding tube were less than those estimated from fully developed flow assumption for both Newtonian and non-Newtonian fluids. Magnitudes of the dispersion number (D/vd) shown in Fig. 8.3 as a function of the Reynolds number for the Newtonian liquids and the generalized Reynolds number for the non-Newtonian liquids followed three

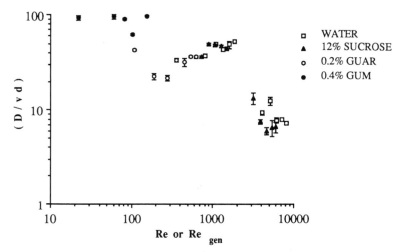

Figure 8.3   Dispersion number as a function of Reynold and generalized Reynolds number in a holding tube. (*Source: Sancho, 1988.*)

distinct trends depending on the magnitude of Re or Re gen (Sancho, 1988): (1) over the range 10–100, the disperson number was nearly constant, (2) over the range 100–2,000, the dispersion number increased with increase in Reynolds number, and (3) over the range 2,000–10,000, the dispersion number decreased with increase in Reynolds number. Because the rheological behavior of the studied non-Newtonian guar gum solutions was known, interpretation of the data in terms of the generalized Reynolds number and grouping together with the Newtonian fluid data were possible.

Helical flow.   It is known that plug flow conditions are approached in laminar helical flow, and for this reason Rao and Loncin (1974a) suggested that this type of flow could be used advantageously in continuous sterilization of foods. It is interesting to note that one continuous thermal processing system in which plug flow conditions are claimed to exist is based on helical flow (Anonymous, undated).

RTD data on liquids with particulates.   Relatively few data are available on the RTD of liquids with particulates. Taeymans et al. (1986) presented data on a scraped surface heat exchanger (SSHE) with liquids containing particulate solids. The results shown in Fig. 8.4 clearly show that the RTD of the solid particulates and the liquid carrier medium were different, and that the operating conditions of the SSHE affect the RTDs of both phases.

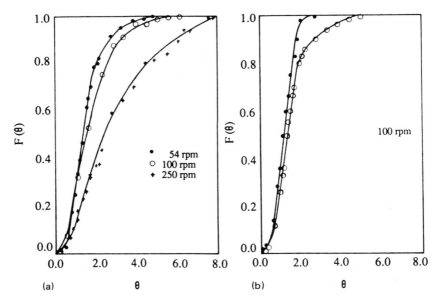

Figure 8.4   Residence time distribution in a scraped surface heat exchanger: (a) RTD of solid particulates at three RPM values of the rotor, and (b) solid (O) and liquid (●) RTDs at 100 rpm. (*Source: Taeymans et al., 1986.*)

ASEPTIC PROCESSING OF LIQUIDS CONTAINING PARTICULATES

Critical Factors in Processing Particulates

The major concern in aseptic processing of liquids containing particulates is with adequate sterility of low-acid solid foods. Thermal process evaluation for suspensions of large particles is more difficult than for liquids because while the liquid portion heats rapidly in heat exchangers, the heat transfer into the solid particles is much slower. In addition, one must take into consideration the residence time distribution and the fluid to particle heat transfer coefficient (h_{fp}). Because the fluid temperature must be known to determine the fluid to particle heat transfer, in the absence of experimentally determined values of temperature, those calculated by energy balance on the fluid and the particles assuming well-mixed conditions were employed (Sastry, 1986). Dignan et al. (1989) pointed out several important practical considerations in aseptic processing of liquids containing particulates: (1) to achieve commercial sterility, the recommended F_0 values must be higher than minimum acceptable values; (2) it is not prudent to credit any lethality achieved during cooling and all credit for lethality should be obtained exclusively in the holding tube; and (3) because energy is transferred from the carrier liquid to solid particles, the holding tube should be termed and treated as an equilibration tube.

The most critical factors in mathematical models used in computer simulation were identified (Dignan et al., 1989) as the particle size and shape, the fluid-particle heat transfer coefficient, and the RTD in the heat exchangers and holding tube determined with the actual food flowing in the system. With regards to fluid-particle heat transfer simulation, several assumptions must be justified or verified: the contribution of the heating system to lethality, justification for assumed RTD for the solid particles in the heating and holding sections, and the verification of the applicability of any empirically determined heat transfer coefficients or other thermophysical properties that influence heat transfer to the particle. The limiting conditions for application of the mathematical model with regard to the operation of the commercial system must be noted and experimental validation of any mathematical models must be demonstrated.

Other critical factors in heat-hold-cool systems were the ratio of mass of particles to mass of carrier fluid, the flow rate of fluid and particles, influence of clumping or nonuniform pumping of particles, the processing system components and its configuration, particularly that of holding tube, control of fluid temperature entering holding tube, and any other factors peculiar to the actual commercial system that may influence process delivery.

Computer Simulation of Particulate Systems

Most of the computer simulation studies have employed a lumped parameter approach to determining the temperature of the liquid phase of the food product in the heating, holding, and cooling sections, and a distributed parameter approach to determining the temperature distribution as well as the lethality in the solid. Attempts to study heat transfer into solid particles can be divided into two categories: (1) mathematical modeling of the heat transfer phenomena taking into consideration the aforementioned factors, and (2) a simplified approach based on the application of Ball's method.

One early attempt in modeling was that of de Ruyter and Brunet (1973), who studied heat transfer to spherical particles assuming that initial fluid and solid temperatures are equal, that particles are isotropic with respect to thermal and physical properties, total particle mass is not greater than 25%, the product is well mixed, the temperature of the liquid phase changes linearly (ramp function), and that the fluid and the particles are at the same temperature, i.e., an infinite magnitude of h_{fp}. Manson and Cullen (1974) analyzed heat transfer to cylindrical solids assuming a linear increase in temperature of fluid in heating section and infinite fluid-particle heat transfer coefficient, but considered residence time distribution, a constant and an exponential decrease of liquid temperature in the holding section, as well as the non-Newtonian nature of the fluid. They showed that the rheological behavior of the liquid food and the residence time distribution are important parameters. Sastry (1986) modeled particles of regular and irregular shapes, finite magnitudes of h_{fp} with Nu = 2.0 as the case when the liquid is stationary, liquid temperatures were determined using incremental energy balances over sections of the heat exchanger and the holding tube, and the liquid medium was considered to be well mixed in the radial direction, in contact with particles of uniform size that are moving with zero relative velocity.

Because the Food and Drug Administration prefers that when there is a question about the lethality accumulated in the heating section the total process lethality be accumulated in the holding section, most of the research efforts thus far have dealt with determination of the holding tube length for adequate process lethality (Sastry, 1986; Chandarana et al., 1989). Sastry (1986) showed that because of heat transfer from the liquid medium to the suspended solids, there will be a drop in the temperature of the of the medium. The precise amount of temperature drop will depend upon the number of solid particles and their heat capacities, as well as the rate of heat loss from the holding tube to the surrounding atmosphere.

The equation to be solved for the temperature distribution in a solid is:

$$\rho C_p (\partial T/\partial t) = \nabla \cdot (k \nabla T) \tag{28}$$

with the time-dependent convective boundary condition:

$$k \nabla T \cdot \mathbf{n} = h_{fp}(T_s - T_f) \tag{29}$$

It is clear that the operating variables of the heating system, the holding system, the residence time distribution, and the fluid-particle heat transfer coefficient can influence the design of an AP system and the sterility of the particulates. The effect of several variables studied by Sastry (1986) will be illustrated here in part because the variables were expressed in dimensionless terms, thus affording the opportunity to generalize the influence of the variables. Figures 8.5, 8.6 and 8.7 show the influence of dimensionless particle size, fluid to particle heat transfer coefficient, and residence time on the dimensionless holding tube length, respectively. While the general trends of the variables, i.e., increase or decrease in holding tube length, are as expected, the differences in the effects of each of the variables are very striking.

Application of Ball's method for particulates.   Ball's method continues to play an important role in thermal processing of canned foods. Therefore, its application to thermal processing of particulates in liquids in holding tubes would be of

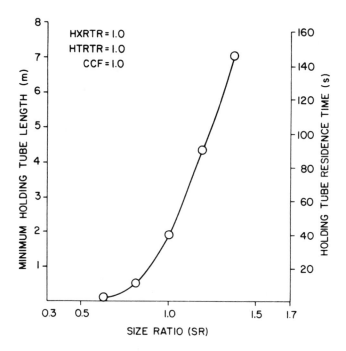

Figure 8.5 Effect of dimensionless particle size on minimum holding tube length. (*Source: Sastry, 1986.*)

considerable interest. Dail (1985) proposed that Ball's method be used for the estimation of holding time necessary to achieve sterility of solid foods. It is important to note some of the key assumptions in Dail's work: (1) the particles are assumed to be of standard shapes (cube, cylinder, or sphere); (2) magnitude of fluid-particle heat transfer coefficient is infinite; and (3) the applicability of Ball's method is contingent on long process times and convergence of the infinite series solution. Castaigne and Lacroix (1987) in applying Ball's method assumed a fluid-particle heat transfer coefficient based on the minimum value of Nu = 2.0.

In a critical examination of the application of Ball's method to aseptic processing, Larkin (1989) pointed out that: (1) because of temperature drop in the holding section (Sastry, 1986), the semilogarithmic plot for determination of the heat penetration parameters f_h and j_h will not be linear but can show considerable deviations ("tailing"); (2) if f_h is estimated from the time-temperature data early in the process, overprediction of lethalities results, whereas if f_h is estimated from data at the end of the process, underprediction of lethalities results; (3) therefore, it is important to choose a proper reference temperature from among the several

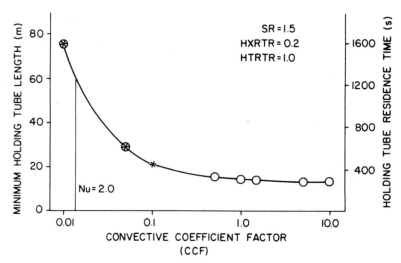

Figure 8.6 Effect of dimensionless fluid-particle heat transfer coefficient on minimum required holding tube length. (*Source: Sastry, 1986.*)

temperatures (one temperature that could be satisfactory is that of the heating medium temperature); and (4) a modified hyperbolic function should be used to fit the transition from heating to holding:

$$\frac{y^p}{a_B^p} - \frac{t_B^p}{b_B^p} = 1 \tag{30}$$

Larking (1989) showed that using a fixed value of $p = 1.7$ resulted in slightly conservative, therefore acceptable, estimates of center line lethalities in the heating and holding sections compared to calculations based on the General method. The modified Ball's method has the potential to be very useful, and additional work is necessary prior to its implementation in aseptic processing calculations.

Relative Particle-Fluid Velocity

Sastry et al. (1989) pointed out that in the holding tube a solid particle is initially at zero velocity and that it undergoes a period of acceleration during which time a high relative velocity between the particle and fluid will exist. The relative velocity data will be needed to determine heat transfer coefficients between fluid and particles employing empirical correlations such as that of Ranz and Marshall for external flow over a sphere cited in Sastry et al. (1989):

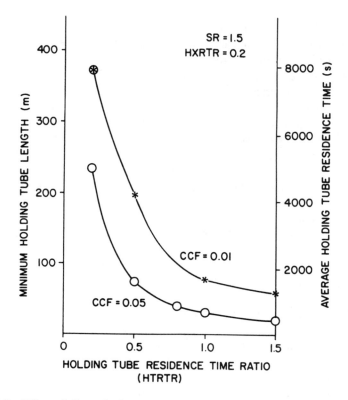

Figure 8.7 Effect of dimensionless residence time ratio on minimum required holding tube length. (*Source: Sastry, 1986.*)

$$Nu = 2.0 + 0.60 \ Re^{0.5}Pr^{0.33} \tag{31}$$

When the particle and fluid are traveling at the same velocity, the Reynolds number becomes zero, resulting in the lower limit for the Nusselt number and hence a conservative estimate for h_{fp}, i.e., the previously stated limiting value of Nu = 2.0.

The relative particle-fluid velocities are difficult to determine for practical situations where a large number of particles are present in a tube. The difficulty in calculating arises from the complex particle–particle and particle–tube wall interactions that must be taken into consideration. Recent efforts (Sastry and Zuritz, 1987; Sastry et al., 1989) have dealt with a single particle in a Newtonian fluid in turbulent flow taking into consideration the linear and angular momentum, and the various forces acting on the particle that permitted calculation of particle velocity as a function of time, which in turn was used in the Ranz and Marshall correlation. The heat transfer coefficients predicted were of the same order of magnitude as experimental data with a hollow aluminum sphere 2.39 cm in diameter, 7.17 g in mass, with a density of 1,006 kg/m^3. Magnitudes of experimental values of h$_{fp}$ ranged between 2039 and 2507 Q/m^2°C.

Merson (1989) reported that Teflon and aluminum spheres painted black and coated with liquid crystals can be used to determine fluid-particle heat transfer coefficients because in the phase transition (from crystalline to liquid) temperature range, the liquid crystals exhibit distinct colors. Further, by correlating color with temperature, surface temperatures of the spheres were measured as a function of time. This technique was applied to particles being carried through a constant temperature fluid in a tube (Stoforos et al., 1989). While liquid crystals do provide temperatures of solid surfaces, the reliability and precision of the measured temperatures in view of response times of color changes must be demonstrated so that confidence limits of the calculated heat transfer coefficients can be determined.

Batch Aseptic Processing of Particulates in a Liquid

As an illustration of the difficulties in sterilizing solid particles in a liquid medium, Hersom and Shore (1981) pointed out that the thermal process time for a safe F$_0$ of a 0.5–0.75 in carrot cube at 270°F is 3–5 minutes, while the thermal process time for sauce is 30 sec or less. Therefore, with particles greater than 0.75 inches, excessive cooking and softening at the surface can result and inactivation of lipases and peroxidases may not reach acceptable levels. A system named Jupiter, in which the solid particles and the sauce are sterilized separately, was suggested. The system has several steps for the production of aseptically processed large solids in sauce: presterilization and loading solids, heating using steam in headspace and jacket of double-cone aseptic processing vessel (DCAPV), liquor injection when mass is at 88°C; liquor maintained at 130°C; heat until desired F$_0$ obtained, cooling of vessel with sterile air compensation, draining into the sterile liquor tank, cold sterile sauce transfer and mixing in DCAPV, emptying-transfer of product to aseptic filler reservoir, and filling sterile containers. One weakness of this system is that for a product with many ingredients, the capital cost would be high due to the number of subsystems required.

Ohmic Heating

Ohmic or electroconductive heating has been developed for blanching or thawing of solids, and it can also be used for aseptic processing of fragile particulate foods because there is little mechanical agitation imposed on the particulates. Other advantages of ohmic heating are relatively uniform heating of particulate solids and minimal fouling of heat transfer surfaces (Mizrahi et al, 1975; Atherton, 1981; Murray, 1985; Lenges, 1988). In ohmic heating, the temperature of a product rises due to thermal resistance when a voltage is applied. The ohmic heater assembly is arranged so that the product is progressively heated as it flows continuously between pairs of electrodes. The electrodes and the product heater tube are insulated to prevent leakage to earth, and the heating rate is controlled by either varying the mass flow of product or the voltage input (Murray, 1985).

The rate of heat penetration to the center of a particle is dependent on the electrical conductivity and is independent of the size of the particle or its thermal diffusivity. In addition, for practical purposes, there will be no temperature gradients within particles and both the liquid and the particles are heated simultaneously, thus eliminating the need for overprocessing the liquid phase to ensure that the particulate solids are heated adequately. Even though the cost of electricity is high, because of the high efficiency of conversion of electricity to heat, the overall costs to heat a unit weight of product could be comparable to those using conventional heating methods (Murray, 1985).

The underlying principle of ohmic heating is the relationship between the heat generated (Q), on one hand, and the current (I) and the electrical resistance (R) on the other:

$$Q = I^2R \tag{32}$$

The electrical resistance is dependent on the electrical conductivity, length (L), and area of cross section (A):

$$R = L/A\sigma \tag{33}$$

Sastry (1989a) cited an unpublished document to the effect that pumpable foods with water content above 30% and containing dissolved ionic salts have the potential for ohmic heating and that nonionic materials such as fats and oils would not be suitable for this method. For a well-mixed fluid, the energy equation for ohmic heating can be written as (Sastry, 1989a):

$$\rho \, C_p(dT/dt) = I^2 \, R/V \tag{34}$$

and the rate of change of temperature in a well-mixed liquid can be written as:

$$dT/dt = J^2 \, /\sigma_T \rho \, C_p \tag{35}$$

where σ_T indicates the temperature dependence of conductivity; σ_T was found to increase with temperature.

For foods containing particulates, thermal process evaluation can be done by solving the conduction heat transfer equation [Eq. (28)] subject to the boundary condition Eq. (29). Heat generation per unit volume within a particle would depend on the location within the particle and the temperature. Because the cross-section of a particle exposed to heating current changes with depth of penetration, for rotating irregularly shaped particles, the heat generated would be dependent on time, temperature, and spatial location (Sastry, 1989a). For particles of uneven composition, the heat generation would be uneven and difficult to predict. These complexities and others such as RTD need to be considered in simulation studies. Sastry (1989b) presented preliminary results of one simulation study of heating cubic particles. The computer simulation studies must be confirmed by experimental data.

Biological Validation

Biological validation of an aseptic thermal process of particulate solids in a liquid medium would be a definitive test when there are uncertainties in one or more critical parameters. The technique of placing bacterial spores encapsulated in a carrier was employed successfully in a number of studies on lethality of canned food (Jones et al., 1980; Pflug et al., 1980; Pflug, 1982). The technique was successfully adapted for a batch aseptic process using small glass bulbs at the center of vegetable pieces (Hersom and Shore, 1981). Briefly, bacterial spores are encapsulated in a carrier and do not come in contact with food, hundreds of inoculated carriers are required, count reduction technique is used, and a calibration is generated, i.e., the number of surviving spores vs. the sterilization process equivalent time (F_T) is known. Some of the advantages of this technique are (Dignan et al., 1989): the location of all the spores in the particle is known; when a bulb is recovered, all the spores are recovered; all the spores are located near the geometric center of the particle; from the calibration procedure an accurate experimentally determined relationship of the number of survivors at an F_T value is known; and spore recovery and fit of D-model to data are not concerns.

Particles of solid food inoculated with bacterial spores is an obvious but difficult technique. Some problems and concerns with this technique are (Dignan et al., 1989): conditions affecting spore destruction during calibration and actual system must be the same, possibility of spores leaching from particles (effective N_0 must be determined), interpolation of count reduction data must be done with parity of N_0 value or a validated calibration must be done, size of inoculated particles must be controlled to be the largest size used in the manufactured product, residence times of the inoculated particles must be representative of the fastest particle residence time, location/distribution of spores in the particles must be known; if uniformly distributed in particles, an integrated sterilization value rather than a slowest-heating-zone sterilization is measured, and the z-value of the spores must be known to obtain the correct F_0 value. In spite of the aforementioned concerns, inoculation of solid foods with bacterial spores was successfully employed in a simulated aseptic process with mushrooms (Sastry et al., 1988).

SYMBOLS

a	positive distance from center of hyperbola to its vertex
A	area
b	positive distance from vertex of hyperbola to its asymptote
B	constant in Arrhenius viscosity equation; process time to end of holding section
C	concentration
C_p, C_{pc}	heat capacity, of cooling fluid
$C(t)$, $C(\theta)$	residence time distribution function for impulse input
CCF	ratio of fluid to particle heat transfer coefficient to the wall to fluid heat transfer coefficient
d	diameter of tube
D	D-value; axial dispersion coefficient
$E(t)$, $E(\theta)$	residence time distribution function for impulse input
E_{ac}, E_{ak}, $E_{a\eta}$	activation energy for viscous flow
E_{aR}	activation energy for reaction rate constant
f_h	heat penetration parameter
F	residence time distribution function for step change input
F_0	lethality at 121.1°C and for z = 10°C
G	defined as $(N/N_0)/B$
h_{fp}	fluid-particle heat transfer coefficient
HTRTR	ratio of residence time of a particle to the mean residence time for the food in the holding tube
HXRTR	ratio of residence time of a particle to the mean residence time for the food in the heat exchanger
I	electric current
j_h	heat penetration parameter
J	current density
k	thermal conductivity; first-order reaction rate constant
K	power law consistency index model
L	length
m, m_c	mass flow rate, of cooling fluid
M	log (N/N_0)
n	flow behavior index
n	unit normal vector

N	number of microbial spores or cells; number of equal size stirred tanks
p	exponent in modified hyperbolic function
q	heat flux
Q	rate of heat generation
r	radial coordinate
r_a	defined as $C_p m / V_{pc} m_c$
R	gas constant; electrical resistance; radius
t	time
t	average residence time
t_B	defined as difference between process time to end of holding section and time
t_E	equivalent time
T	temperature
T_E	equivalent temperature
U	overall heat transfer coefficient
v	velocity in axial direction
V	volume
z	axial direction

Greek Letters

α	thermal diffusivity
η	viscosity
η_a	apparent viscosity
η_0	zero shear viscosity
η_∞	viscosity at infinite temperature
σ	shear stress
$\dot{\gamma}$	shear rate
θ	dimensionless time, t/t; t V/L
ρ	density

Dimensionless Numbers

(D/ v d)	dispersion number
Nu	Nusselt number
Pr	Prandtl number
Re	Reynolds number
Re_{gen}	generalized Reynolds number
X	z/L

Mathematical Operations

D/Dt	substantial derivative = $(\partial/\partial t) + \nabla \cdot \mathbf{V}$
d/dt	derivative with respect to time
$\partial/\partial t$	partial derivative with respect to time
exp x	e^x = exponential function of x
ln x	logarithm of x to base e
$\log_{10} x$	logarithm of x to base 10
∇	del or nabla operator = $\dfrac{\partial}{\partial x} + \dfrac{\partial}{\partial y} + \dfrac{\partial}{\partial x}$

Subscripts

B	end of process time
c	cooling section, cooling fluid
ci	cooling fluid at inlet
f	of fluid
H	heating section, heating fluid
Ho	holding section
i	at inlet
O	inlet value
P	process
s	value at surface
w	value at wall

REFERENCES

Anonymous. Undated. Rotafil—a totally new heat exchanger concept designed for the food industry. Graham Corporation, Batavia, NY.

Atherton, D. 1981. Aseptic processing. A study to establish the capabilities and limitations of available machinery for aseptic processing and packaging foodstuffs, Technical Memorandum No. 270, The Campden Food Preservation Association, Chipping Campden, Gloucestershire, U.K.

Bird, R. B., Stewart, W. E., and Lightfoot, E. N. 1960. *Transport Phenomena*. John Wiley and Sons, New York.

Castaigne, F. and Lacroix, C. 1987. Calculation of residence times for the sterilization of low acid liquid foods containing particulates in continuous aseptic packaging process. Paper No. 87505 presented at the meeting of Canadian Society of Agricultural Engineers, Montreal, May 18-22.

Chandarana, D. I., Gavin, A., and Wheaton, F. W. 1989. Simulation of parameters for modeling aseptic processing of foods containing particulates. *Food Technol.* 43(3): 137.

Christiansen, E. B. and Craig, S. E. 1962. Heat transfer to pseudoplastic fluids in laminar flow. *A. I. Ch. E. J.* 8: 154.

Dail, R. 1985. Calculation of required hold time of aseptically processed low acid foods containing particulates utilizing the Ball method. *J. Food Sci.* 50: 1703.

Danckwerts, P. V. 1953. Continuous flow systems. *Chem. Eng. Sci.* 2(1): 1.

David, J. 1989. Quality assurance for UHT-sterilized and aseptically-packaged food. Aseptic Processing and Packaging Course, University of Wisconsin, Madison, WI, January 9–10.

Deindoerfer, F. H. and Humphrey, A. E. 1959a. Analytical method for calculating heat sterilization times. *Applied Microbiol.* 7: 256.

Deindoerfer, F. H. and Humphrey, A. E. 1959b. Principles in the design of continuous sterilizers. *Applied Microbiol.* 7: 264.

De Ruyter, P. W. and Brunet, R. 1973. Estimation of process conditions for continuous sterilization of foods containing particulates. *Food Technol.* 27(7): 44.

Dickerson, R. W., Scalzo, A. M., Read, R. B., and Parker, R. W. 1968. Residence time of milk products in holding tubes of high-temperature short-time pasteurizers. *J. Dairy Sci.* 51(11): 1731.

Dignan, D. M., Berry, M. R., Pflug, I. J., and Gardine, T. D. 1989. Safety considerations in establishing aseptic processes for low-acid foods containing particulates. *Food Technol.* 43(3): 118.

Guariguata, C., Barreiro, J. A., and Guariguata, G. 1979. Analysis of continuous sterilization processes for Bingham plastic fluids in laminar flow. *J. Food Sci.* 44: 905.

Harper, J. C. and El-Sahrigi, A. F. 1965. Viscometric behavior of tomato concentrates. *J. Food Sci.* 30: 470.

Harrod, M. 1986. Scraped surface heat exchangers. A literature survey of flow patterns, mixing effects, residence time distribution, heat transfer and power requirements. *J. Food Proc. Eng.* 9: 1.

Heldman, D. R. 1989. Establishing aseptic thermal processes for low-acid foods containing particulates. *Food Technol.* 43(3): 122.

Heppell, N. J. 1985. Comparison of the residence time distributions of water and milk in an experimental UHT sterilizer. *J. Food Eng.* 4: 71.

Hersom, A. C. and Shore, D. T. 1981. Aseptic processing of foods comprising sauce and solids. *Food Technol.* 35(5): 53.

Himmelblau, D. M. and Bischoff, K. B. 1968. *Process Analysis and Simulation.* John Wiley and Sons, New York.

Holdsworth, S. D. 1970. Continuous sterilization of foods. *Process Biochem.* 5(3)-57.

Holdsworth, S. D. 1971. Applicability of rheological models to the interpretation of flow and processing behavior of fluid food products. *J. Texture Studies* 2: 393.

Ito, K. A. and Stevenson, K. E. 1984. Sterilization of packaging materials using aseptic systems. *Food Technol.* 38(3): 60.

IUFoST, 1985. *Proceedings of Symposium on Aseptic Processing and Packaging of Foods.* Lund University-SIK, Lund, Sweden.

Jones, A. T., Pflug, I. J., and Blanchett, R. 1980. Effect of fill weight on the F-value delivered to two styles of green beans processed in a Sterilmatic retort. *J. Food Sci.* 45: 940.

Kays, W. M. and London, A. L. 1964. *Compact Heat Exchangers.* McGraw-Hill, New York.

Larkin, J. W. 1989. Use of a modified Ball's formula method to evaluate aseptic processing of foods containing particulates. *Food Technol.* 43(3): 124.

Levenspiel, O. 1972. *Chemical Reaction Engineering,* 2nd ed. John Wiley and Sons, New York.

Lenges, J. 1988. Aseptic processing-an overview. In *Progress in Food Preservation Processes,* (Ed.) Lenges, J. and Taeymans, D. CERIA, Brussels.

Lin, S. H. 1979. Residence time distribution of flow in continuous sterilization process. *Process Biochem.* 14(7): 23.

Lin, S. H. 1976. Continuous high-temperature/short-time sterilization of liquid foods with steam-injection heating. *Chem. Eng. Sci.* 31: 77.

Manson, J. E. and Cullen, J.F. 1974. Thermal process simulation for aseptic processing of foods containing discrete particulate matter. *J. Food Sci.* 39: 1084.

Merson, R. L. 1989. Heat penetration and bacteriological properties of food for heat sterilization process design. In *Food Properties and Computer-Aided Engineering of Food Processing Systems,* p. 93-98, (Ed.) Singh, R. P, and Medina, A. G. Kluwer Academic Publishers, Dordrecht, The Netherlands.

Mitchell, E. L. 1988. A review of aseptic processing. *Advances in Food Research* 32: 1–37.

Mizahi, S., Kopelman, I. J., and Perlman, J. 1975. Electro-conductive heating. *J. Food Technol.* 10: 281.

Murray, S. A. 1985. Thermal processing of particulate containing liquid foods. In *Proceedings of IUFoST Symposium on Aseptic Processing and Packaging of Foods,* p. 81–99. Lund University-SIK, Lund, Sweden.

Nelson, P. E., Chambers, J. V., and Rodriguez, J. H. 1987. *Principles of Aseptic Processing and Packaging.* The Food Processors Institute, Washington, DC.

Pflug, I. J. 1982. Measuring the integrated time-temperature effect of a heat sterilization process using bacterial spores. In *Food Process Engineering.* (Ed.) Schwartzburg, H. G., Lund, D., and Bomben, J. L. AIChE Symposium Series, Vol. 78, p. 68, Am. Inst. of Chem. Eng., New York.

Pflug, I. J. Smith, G., Holcomb, R., and Blanchett, R. 1980. Measuring sterilizing values in containers of food using thermocouples and biological indicator units.

J. Food Protect. 43: 119.

Rao, M. A. 1986. Rheological properties of fluid foods. In *Engineering Properties of Foods.* (Ed.), Rao, M. A., and Rizvi, S.S.H. p. 1–47. Marcel Dekker, New York.

Rao, M. A. and Loncin, M. 1974a. Residence time distribution and its role in continuous pasteurization (Part I). *Lebensm.-Wiss. u.-Technol.* 7: 5.

Rao, M. A. and Loncin, M. 1974b. Residence time distribution and its role in continuous pasteurization (Part II). *Lebensm.-Wiss. u.-Technol.* 7: 14.

Rhim, J. W., Nunes, R. V., Jones, V. A., and Swartzel, K. R. 1989a. Determination of kinetic parameters using linearly increasing temperature. *J. Food Sci.* 54: 446.

Roig, S. M., Vitali, A. A., Ortega Rodriguez, E., and Rao, M. A. 1976. Residence time distribution in the holding section of a plate heat exchanger. *Lebensm.-Wiss. u.-Technol.* 9: 255.

Sancho, M. F. 1988. A study on the residence time distribution in the holding tube of a pasteurizer. M.S. thesis, Cornell University, Ithaca.

Sastry, S. K. 1986. Mathematical evaluation of process schedules for aseptic processing of low-acid foods containing discrete particulates. *J. Food Sci.* 51: 1323.

Sastry, S. K. 1989a. Process evaluation in aseptic processsing. In *Developments in Food Preservation-5,* p. 177–206, (Ed.) Thorne, S., Elsevier Applied Science Publishers, New York.

Sastry, S. K. 1989b. A model for ohmic heating of liquid-particle mixtures. Paper presented at the 50th Annual Meeting of the Institute of Food Technologists, June 25–29, Chicago, IL.

Sastry, S. K. and Zuritz, C. A. 1987. A review of particle behavior in tube flow: applications to aseptic processing. *J. Food Process Eng.* 10: 27.

Sastry, S., Li, S. F., Patel, P., Konanayakam, M., Bafina, P., Doores, S., and Beelman, R. B. 1988. A bioindicator for verification of thermal processes for particulate foods. *J. Food Sci.* 53: 1528.

Sastry, S. K., Heskitt, B. F., and Blaisdell, J. L. 1989. Experimental and modeling studies on convective heat transfer at the particle-liquid interface in aseptic processing systems. *Food Technol.* 43(3): 132.

Scalzo, A. M., Dickerson, R. W., Read, R. B., and Parker, R. W. 1969. Residence times of egg products in holding tubes of egg pasteurizers. *Food Technol.* 23(5): 80.

Simpson, S. G. and Williams, M. C. 1974. An analysis of high temperature/short time sterilization during laminar flow. *J. Food Sci.* 39: 1047.

Singh, R. K. 1987. Residence time distribution in aseptic processing. In *Principles of Aseptic Processing and Packaging,* (Ed.) Nelson, P., Chambers, J., and Rodriguez, J. P. 17–27. The Food Processors Institute, Washington, DC.

Skelland, A. H. P. 1967. *Non-Newtonian Flow and Heat Transfer.* John Wiley and Sons, New York.

Stoforos, N. G., Park, K. H, and Merson, R. L. 1989. Heat transfer in particulate foods. Paper presented at the 50th Annual Meeting of the Institute of Food Technologists, June 25–29, Chicago, IL.

Swartzel, K. R. 1982. Arrhenius kinetics as applied to product constituent losses in ultra high temperature processing. *J. Food Sci.* 47: 1886.

Swartzel, K. R. 1984. A continuous flow procedure for reaction kinetic data generation. *J. Food Sci.* 49: 803.

Swartzel, K. R. 1986. Equivalent-point method for thermal evaluation of continuous flow systems. *J. Agric. Food Chem.* 34: 396.

Taeymans, D., Roelans, E. and Lenges, J. 1986. Residence time distribution in a horizontal SSHE used for UHT processing of liquids containing solids. *Proc. 4th International Congress on Engineering & Food,* p. 247–258, (Ed.) Le Maguer, M., and Jelen, P. P. Elsevier Applied Science Publishers, New York.

Tillotson, J. E. 1984. Aseptic packaging of fruit juices. *Food Technol.* 38(3): 63.

Toledo, R. Y. 1975. Chemical sterilants for aseptic processing. *Food Technol.* 29(5): 102.

Veerkamp, C. H., Romijn, A. J. M., and Pol, J. C. 1974. Influence of varying residence time distribution on inactivation of microorganisms during pasteurization of egg products. *Lebensm.-Wiss. u.-Technol.* 7: 306.

9

Encapsulation and Controlled Release of Food Components

Marcus Karel

Rutgers University
New Brunswick, New Jersey

INTRODUCTION

The reasons for desiring controlled release in foods are summarized in Table 9.1. These needs are obvious to food technologists and nutritionists. However, it may be valuable to review the major current and potential future needs in food processing.

The current status of applications was reviewed recently (Dziezak, 1988). This report and the brief summaries listing commercial products which were appended to the report indicate the following major applications:

1. **Acidulants**, in particular citric and ascorbic acid used as dough conditioners, taste adjuncts, and pH-lowering agents: The pH-lowering may be desired at some stage of the process either to change characteristics of the product (meat processing) or to assist in a crosslinking process (as in calcium alginate gel formation).
2. **Flavors**. The major need for encapsulation is to retard or to eliminate

Table 9.1   Functions of Controlled Release in Foods

A substance in formulated food is released *upon consumption* but prevented from diffusion during the series of operations in food processing (e.g., flavors, nutrients).

A substance is *released* in *a specific processing step*, but *protected* in *preceding steps* (e.g., leavening agents, crosslinking agents).

Preservatives needed in given portion of food (e.g., surface) are prevented from diffusing to another portion (to avoid dilution).

volatilization, and to minimize or to eliminate contact with atmospheric oxygen. Another function of controlled release systems, which is not discussed much in literature, but which offers potential advantage in engineered foods, is the potential for changing the composition of the flavor mixture being released with time, thus compensating for differential rates of loss of the components of the original food flavor.

3. **Leavening agents** are widely encapsulated to be released under optimum baking conditions.

4. **Salts**, including sodium chloride, are encapsulated to avoid undesirable changes in physical properties of the food product (changes in flowability, solubility, water sorption) prior to consumption. They may also be encapsulated to avoid contact with oxidizable lipids, some of which have been reported to undergo catalytic peroxidation in presence of table salt.

5. **Some minerals**, in particular iron, are encapsulated for the same reasons as the salts. Vitamins are primarily encapsulated to avoid loss due to reactions with atmospheric components or with food components (De Man, 1986).

A review of the "Product Update" in the same issue of Food Technology in which the encapsulation review appeared indicates that almost all encapsulated ingredients being advertised fall into the above categories. They include: citric and fumaric acids encapsulated in fat (Durkee), spices and flavors (Felton; Hansen's; McCormick, Bush Boake Allen; Fritzche Dodge Olcott), and vitamins (Desmo).

Some of the less well known or discussed applications of the controlled release are the following:

1. Release of color-, flavor-, and possible texture-modifying agents timed and synchronized with microwave cooking. This is an area of very active interest in a number of food companies, flavor companies, and in some research and academic institutions.

An interesting invention granted a U.S. patent (#4413069) to J. W. Marshall claims in situ release of various reactive materials from microcapsules sensitive to microwave heating because they incorporate agents heating preferentially. The applications cited here are in nonfood applications, but the principle seems to offer great potential in microwave heating of prepared foods, *if* materials of food- grade can be substituted for the nonedible materials.

Claims for specific nonthermal effects of microwaves (2450 MHz) on drug release from liposomes have been made by Liburdy and Magin (1985). This effect was oxygen sensitive and seemed to be free- radical mediated. Liburdy et al. (1986) also reported enhanced release from liposomes due to static magnetic fields. These results remain to be confirmed under conditions relevant to food applications.

2. Controlled release of enzymes, either during ripening, storage, or after consumption. An area of increasing interest is the release of enzymes at a predetermined rate or at a predetermined time or upon a predetermined signal. Some of the pioneering work in this area has been done at Wisconsin Alumni Foundation (Magee and Olson, 1981; U.S. Patent 4310554; Braun et al. 1982; Braun and Olson, 1986). In this work the emphasis has been on fat-encapsulated proteases or cell-free extracts of bacterial cells to be added to cheese milk prior to curd formation. Premature enzyme loss is thereby avoided. Subsequent work in this field (Law and King, 1985; Piard et al., 1986; Koide and Karel, 1987) has been directed to the utilization of liposomes for controlled release of proteolytic enzymes. The major application is the accelerated and controlled ripening of cheddar cheese, which is of major interest to various users (Lawrence et al., 1987.)

Additional controlled-release applications for enzymes in the food industry have not been reported. However, this field offers much promise for imaginative applications. In the area of aroma formation lipases, lipoxidases, and isomerases seem of particular interest, in addition to the proteases discussed above. In the area of taste modification and texture control, amylases and proteases may be important.

There has been considerable interest in incorporation of selected enzymes (e.g., Z8-galactosidase) in the diet in an encapsulated form surviving the passage through the stomach and being released at the pH of intestines. This application is aimed at people with lactose intolerance (Houts, 1988). Various "enteric" encapsulating systems are available (Deasy, 1984), but systems suitable for incorporation in foods have been very limited (Sada et al., 1988; Matsuzaki et al., 1989).

3. Protection of sensitive ingredients during a processing operation. An example in this area is the sensitivity of aspartame. This sweetner is being used in a great variety of food applications, and the number of uses is growing daily. It is also sensitive to heat, but the rate constant for heat inactivation is strongly dependent on concentration and on pH. By providing an internal

environment with a combination of high concentration (low water content) and optimum pH, it may be possible to allow retention of a high degree of sweetness through such transient heating stages as pasteurization and baking.

ENCAPSULATION OF FOOD ADDITIVES IN HYDROPHILIC MATRICES

The principal methods of encapsulation and controlled release are shown in Tables 9.2 and 9.3. Of these, the most commonly used in foods is the encapsulation in hydrophilic matrices. The underlying theoretical principle in these systems is selective diffusion in amorphous glasses. These glasses are formed of sugars, sugar–polymer mixtures, and from other edible mixtures capable of forming glasses relatively impermeable to organic compounds and to "fixed" gases including oxygen and carbon dioxide. Starch hydrolysis products (maltodextrins) are particularly useful. Many industrial products including freeze-dried coffee depend in their success on the properties of these impermeable glasses (Flink, 1975). The retention of flavors within the matrix and the protection against atmospheric oxygen remains excellent, provided the temperature remains below the transition temperature for the matrix. Since these transition temperatures are strongly dependent on plasticizer content, and in the case of sugars and many food polymers the plasticizer is water, protection against water must be provided either by external packaging, or by double encapsulation, that is, a hydrophilic matrix encapsulating a lipophilic flavor mixture is in turn encapsulated in a lipophilic matrix, usually a fat with a known melting point.

The principle of "selective diffusion" developed by Thijssen and Rulkens (1968) is best illustrated by the dependence of the diffusion coefficients of water and of organic compounds on water content and on temperature. Figures 9.1 and 9.2 show this dependence. The data can be interpreted as being due to the sensitivity of glass transition temperatures to water content. This dependence has been investigated extensively by many investigators. Most recently it was reviewed by Levine and Slade (1986). If we assume that the diffusion coefficient for the encapsulated substance is very low below the glass transition temperature (Tg), but increases substantially above Tg, then it becomes obvious that reduction of water content and/or temperature rapidly to a level below Tg will minimize losses. The dependence of Tg or of collapse temperature (Tc) on water content is shown for maltodextrin in Fig. 9.3. Similar and very extensive data have been reported by Levine and Slade (1986) for starch hydrolysis products.

The rate of release above the glass transition temperature depends on the value of diffusion coefficient D for the encapsulated substance. This value depends on temperature *and* moisture content, and may be correlated by Eq. (1) (Williams et al., 1955):

Table 9.2 Principles of Encapsulation of Compounds Used in Food Formulation

1. Entrapment in amorphous (glassy) matrices achieved by rapid cooling or by drying.
2. Encapsulation in fat-based matrices.
3. Encapsulation in crosslinked or coacervated polymers.
4. Entrapment within sugar crystals.
5. Strong physical adsorption.
6. Chemisorbed compounds.
7. Incorporation in liposomes.

Table 9.3 Methods of Release Used in Foods

Commonly used:
   For hydrophilic encapsultants: temperature and moisture control. For fat capsules: thermal release.
Other release methods:
   pH control
Addition of surfactants
Enzymatic release (e.g., proteolysis)
Ultrasonics
Grinding
Photo-release
Control of ionic strength
Magnetic methods

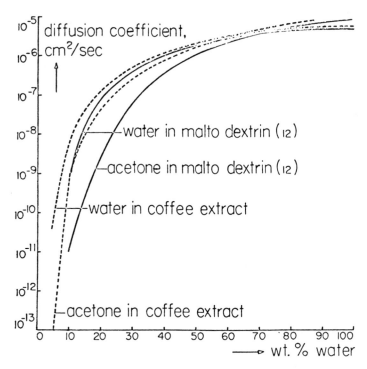

Figure 9.1 Basis of "selective diffusion" theory of Thijssen: Dependence of diffusion coefficients on water content.

$$\log \frac{D}{Dg} = \frac{C_1(T - Tg)}{C_2 + (T - Tg)} \tag{1}$$

This relation, however, assumes that the degree of mobility corresponding to the new temperature-moisture condition above Tg has already been attained. In reality, however, the glass transitions (or collapse) take time, and the rate at attainment of the new state is in itself dependent on temperature and moisture, probably by the same WLF relationship:

$$\frac{D - D_e}{D_i - D_e} = e^{-t/\tau} \tag{2}$$

where τ = relaxation time

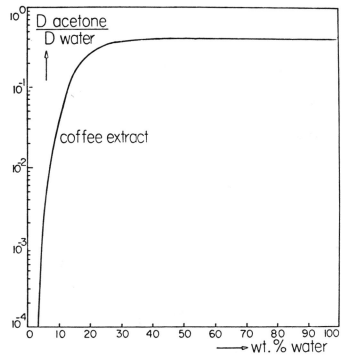

Figure 9.2 Ratio of diffusion coefficients for water and for acetone in maltodextrin as a function of water content.

$$\log \tau - \log \tau_{TG} = \frac{-C_1(T - T_G)}{C_2 + (T - T_G)} \tag{3}$$

As a result, release kinetics can be tailored to be rapid by exceeding Tg conditions (by raising water content or temperature) substantially or to be muchslower by maintaining the encapsulation system close to the relevant $T_G$value. This time course of release has been discussed by Karel and Flink (1983) and Karel (1985a), and a mechanism specifically relating rate of release to the progress of collapse in a "web" of encapsulating matrix was proposed by Omatete and King (1978).

The technology of these encapsulations is well known industrially, and is based on preparation of a suspension or solution of the flavor and of the encapsulating material and subsequent dehydration through spray drying, freeze drying (Karel, 1985b) or solvent dehydration (Zilberboim et al., 1986).

The effectiveness of encapsulation depends strongly on the concentration of the

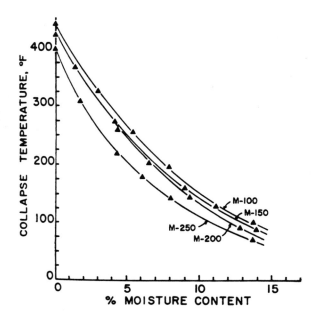

Figure 9.3 Dependence of "collapse temperature" for maltodextrins on water content. (M-100 to M-250 are different maltodextrin fractions.)

encapsulating material in the solution from which it is dried. Very extensive studies on the dependence of efficiency of encapsulation on the concentration of solute and on conditions of dehydration, cooling, or desolventizing have been reviewed by Karel and Flink (1983), King et al. (1984), and Toei (1986). In spray drying, upon droplet formation, rapid evaporation from the surface produces a surface layer in which the *selective diffusion* mechanism comes into play. In order to be most effective, the spray-drying process should have the following features:

1. Short residence time prior to droplet formation,
2. Rapid evaporation once droplets are formed,
3. High degree of selectivity of the relatively "dry" surface layers,
4. Absence of liquid circulation within the droplets,
5. Avoidance of surface "cracks."

In freeze drying, upon water crystallization, the nonfrozen solution is viscous and the diffusion of flavors is retarded. Upon beginning of freeze drying, the surface of this solution becomes an amorphous solid in which selective diffusion comes into play.

The success of freeze drying as an encapsulation method depends on the following critical factors:

1. Formation of thick layers between pores, which are created by ice crystals,
2. Selectivity of the amorphous solid,
3. Absence of structure "collapse," which implies mobility (possibly transient) allowing flavor diffusion,
4. Absence of "cracks."

Encapsulation of food additives has been severely limited by the requirement that the ingredients be edible and approved by the FDA. In addition the food industry is currently emphasizing the use of "natural" rather than synthetic ingredients. The formulations are therefore based on maltodextrins or starch hydrolysis products, on sugars, including sucrose, maltose, and lactose, on polysaccharides derived from plants either terrestrial (cellulose and derivatives, pectin, gum arabic, starch and derivatives, gum acacia) or marine (alginates, agar, carrageenan), or from microorganisms (xanthah). The use of some other polysaccharides (chitosan) is at present only a potential, rather than reality. There are a number of synthetic polymers (Langer, 1980; Langer and Peppas, 1983; Deasy, 1984) but their use in food additive encapsulation has been very limited. An attempt by Dynapol to introduce synthetic polymers into the domain of food additives through their uses as "leashing" agents (Furia and Bellanca, 1977; Havinghorst, 1978) (i.e., preventing in vivo absorption of additives functional in vitro) failed, probably primarily because of regulatory and consumer considerations.

Given these facts, which severely limit the possibility for engineering appropriate polymeric systems through synthetic chemistry, the emphasis in the industry and in research institutes working on encapsulation has been on technology of matrix formation (dispersion of encapsulated material; method of cooling, drying, or desolventizing the encapsulant; postencapsulation treatments), and on formulation of appropriate mixtures of approved materials to give desired properties.

In as far as general theory or even sets of guiding principles are concerned, these appear to be well developed only in the fields of spray drying and freeze drying (Kerkhof and Thijssen, 1977; Karel and Flink, 1983; King et al., 1984; Toei, 1986; Karel and Langer, 1988; King, 1988). In as far as industrial knowledge of spray drying, fluidized bed evaporation and cooling, solvent dehydration, and similar techniques are concerned, these are available in a number of companies conducting research, development, and production on a contract basis. Some of the important groups are Southwest Research Institute in San Antonio, Washington University Technology Associates, and Balchem of State Hill, New York. In-house activity is of course present in many food and ingredient companies as well.

ENCAPSULATION OF FLAVOR, COLOR, NUTRIENTS, ETC. IN FAT CAPSULES

This technology is fairly well established, and recent developments have been entirely in the area of new applications (especially for new processing or food preparation technologies) and improvements in equipment. The most advanced technologies exist in specialized ingredient companies and in contract companies (Andres, 1975; Drieziak, 1988).

ENCAPSULATION BY EXTRUSION

Most of the encapsulation technologies have focused on the use of fluidized bed driers or desolventizers (Wurster Process) and of spray driers for formation of hydrophilic glasses encapsulating flavors. Variants including dehydration of the matrix in solvents (Zilberboim et al., 1986). However, the advances in extrusion have led to considerable activity in the use of extruders for microencapsulation. One of the oldest processes in this respect is that used in Oslo, Norway by a company (Collet Co.), which was using sucrose with very low water content as a feed to an extruder which melts it. Droplets of oil containing vitamin A are injected at extruder exit. Sucrose glass is formed, entrapping the drop and protecting it from atmospheric oxygen.

Griffith Co. in Chicago (Dr. Ralph Sair) developed a process in the 1970s based on encapsulation of perfumes and of flavors in extruded soy protein or other protein matrices.

The most recent review on the subject of extrusion as a means of encapsulating flavors is that of Risch and Reineccius (1988), who reviewed early work by Swisher (U.S. patents 2,809,895 and 3,041,180) on extrusion of flavors in low moisture corn syrup into hot immiscible liquids and then cooled, washed with solvent, and then dried. This is a variant of methods to bring starch hydrolyzates below Tg. Emulsifiers were necessary to disperse flavor in concentrated corn syrup. Improvement included plasticizing corn syrup with glycerol and direct extrusion into cold isopropanol, which served simultaneously three functions (dehydration, cooling, and removal of encapsulated flavor). Subsequent industrial improvements by National Starch and by American Maize and others dealt mainly with improved formulations (using mixtures of maltodextrins, sucrose, derivatized starches, lactose, and various emulsifiers). Process improvements (primarily temperature, mixing, and extrusion conditions) were patented recently (U.S. 4,610,890) (Mutka and Nelson, 1988).

Very extensive modifications of extrusion conditions for encapsulation were also undertaken by Southwest Research Institute in San Antonio. They also have capabilities in spray chilling and spray cooling (Food Technology, 1988). Many additional methods for preparing capsules from glass-forming or solidified fat materials have been tried. Fairly stable but relatively course capsules with relatively low encapsulation efficiency may be achieved by simply mixing with stirring in the material to be encapsulated with the melt, cooling, granulating, sieving, and washing the excess off the surfaces (e.g., Koishi et al., 1984). We routinely prepare capsules with 1% loading using hydrogenated soybean oil by this simple method. A comparison of microencapsulation processes in terms of capsule characteristics was prepared by Robert Sparks of Washington University at St. Louis, whose laboratory specializes in coating solid or high viscosity cores with wax-polymer melts to give walls of 3–200 microns.

USE OF LIPOSOMES

The use of liposomes as means for delivery of food additives is very recent. The most commonly used liposome systems are based on lecithin or lecithin plus cholesterol. The most relevant studies are those conducted in France by Picard (1986), by the English group of Kirby and Law (1986), and at MIT (Koide and Karel, 1987; Matsuzaki et al., 1989).

At the present time we have work underway on the use of liposomes for delivery of enzymes (Z8-galactosidase) (Matsuzaki et al., 1989) and of *small* molecules, in particular xylose, glucose, and amino acids. The purpose is to allow the controlled release of flavor- or color-generating reactants upon delivery of a stimulating signal.

Koide and Karel (1987) showed that pH change, temperature change, infusion of surface active agents or of ions as well as mechanical energy input may serve as such stimuli.

We have been successful in retaining sugars as well as peptides in liposomes prepared by the dehydration-rehydration method (Kirby and Gregoriadis, 1984a). We tried with some success the freeze-thaw method (Mayer et al., 1985) and had particularly good success using the reverse-phase evaporation method (Szoka and Papahadjopoulos, 1978). Unfortunately, the reverse-phase method requires the use of organic solvents.

Reverse-phase evaporation is used most often and has been shown to form liposomes with stability and release properties superior to other methods. In reverse-phase evaporation an organic solvent phase and an aqueous phase containing the compound to be encapsulated are emulsified with the organic solvent as the continuous phase. Subsequently, the organic solvent is evaporated under reduced pressure and large unilamellar vesicles are formed. Although technically satisfactory for the purpose of forming the liposomes, this procedure suffers from several disadvantages: (1) organic solvents are used, which may denature the material to be encapsulated, (2) a lack of control during solvent evaporation due to foaming can lead to loss of both lipid and material to be encapsulated, and (3) solvent residues can render the liposomes unsatisfactory for many potential applications such as in foods.

We have recently made the discovery that supercritical fluids can be used as solvents to prepare liposome by a variant of the reverse-phase method. Work is currently underway to determine whether this procedure can be used for encapsulation.

COACERVATION, CROSSLINKED CAPSULES, AND INTERFACIAL POLYMERIZATION

Coacervation or phase separation systems were, of course, among the very earliest capsule-forming systems. The variety of capsules that could be produced by this general method, the principles of which are described very well by Deasy (1984), is very large, and in food applications limited primarily by suitability of various polymers because of need for FDA approval or because of flavor and taste effects. The crosslinking of the wall formed by the coacervate is also often necessary to get good capsule stability, and the choice of crosslinking agents acceptable for human food is severely limited. The use of formaldehyde-crosslinked capsules for animal feeding, however, has been considered.

One such procedure for stabilizing polyunsaturated fats for feeding to ruminants uses polyunsaturated fats encapsulated in formaldehyde-crosslinked protein. The work was pioneered by Australians (Scott, 1975) and tested in animals in a number of institutions (Edmondson et al., 1974; Ashes et al., 1984; Finlayson and Armstrong, 1986). Formaldehyde-crosslinked capsules are acceptable for animal feeding in Canada.

Most synthetic polymeric systems are limited in their use to drugs (Rooda et al., 1986). The major opportunities in the field of coacervate capsules lie in utilization of natural colloids. A number of possibilities exist especially in the area of marine biopolymers (Renn, 1984) in microbial or plant polysaccharides and proteins. There is at present, however, no good theoretical basis for selection of appropriate systems, and reports of successful encapsulation are often based on empirical discoveries (e.g., Shomer, 1988; Chilvers et al., 1988).

The above discussion is based primarily on aqueous coacervation systems. A number of coacervation-phase separation systems involving nonaqueous manufacturing vehicles have also been used (Deasy, 1984). However, even if the solvents used can be completely removed, leaving no residues, the polymer systems investigated are often not suitable for food applications. Cellulose esters are the major polymers studied that are applicable to foods.

The same problems arise with interfacial polymerization. The groups of Professor Langer at MIT and of Dr. Jorge Heller at Stanford Research Institute in

California have extensive work on polyesters, polyamino compounds, and polyanhydrides, but the work is directed to drug release. One of the advantages of interfacial polymerization-derived systems is the ability to "build in" sensitivity to specific release signals, such as pH (Bala and Vasudevan, 1982).

USE OF COMPLEXES, OF STRONG PHYSICAL ADSORPTION, OF CHEMISORPTION AND OF CRYSTAL INCLUSION

In addition to encapsulation in fats, in crosslinked polymers, and in manomeric or polymeric glasses below Tg, attention has been given to various modes of complexing or of binding. These have usually the advantage of being more *selective* than encapsulation, but conversely suffer from a relatively low volumetric load capacity.

Complexes

Recently condiderable attention has been given to the use of cyclodextrins, which are produced by the action of *Bacillus macerans* amylase on starch. Three types (alpha, beta, and gamma, composed respectively of 6, 7, and 8 Z7-D-glucopyranose units) are formed (Takeo and Kuge, 1972). The use of Z8-cyclodextrins in biotechnology has been promoted by the Hungarians who manufacture the compounds (Szenti and Szejtli, 1986). A publication called *Cyclodextrin News* is published in England (FDS Publ. P.O.B. 41, Trowbridge, Wiltshire, England BA148UE).

Selective removal of bitter compounds from grapefruit juice was studied by Wagner et al., 1988, and the use of complexes for stabilization and for controlled release of drugs have been reported (Matsuyama et al., 1987; Gandhi and Karara, 1988; Celebi and Nagai, 1988). Food applications have been explored by Szenti and Szejtli (1986) and Reineccius and Risch (1986). An excellent review of the complexes was prepared by Saenger in 1980. The potential use is very attractive, especially in combination with encapsulation in glasses after an initial entrapment in cyclodextrin, but as of now the use of cyclodextrins is not approved in the United States. Formation of inclusion complexes of flavor models with starch was studied by Osman-Ismail and Solms (1973), . . . by Japanese workers in Kyoto (Kuge and Takeo, 1968; Takeo et al., 1973), and Tokyo (Suzuki and Hasue, 1976).

"Co-Crystallization"

A related subject is the inclusion of compounds within sugar *crystals*. Chen et al. (1988) presented a paper at the 1988 IFT meeting claiming "co-crystallization" of various flavor mixtures ranging from diacetyl to essential oils in a mixture with sucrose. They claim good stability of "entrapped flavor" at a loading of about 7%. The photomicrographs show "flavor" solids located in interstices between crystals. The stability of such a system and reproducibility of results are, in my opinion, likely to depend on partial pressure of water in the system, since the "flavor" is accessible to water vapor, and the sucrose will have only minor water-sorbing potential as long as it remains crystalline. General Foods obtained patents in 1976 and 1977 (U.S. 3970765 and 4031255) on "fixing acetaldehyde in sucrose." These patents claim the acetaldehyde "fixed" within the individual crystals and therefore could be expected to give formulations of greater stability. The "loading," however, was limited to less than 0.5% by weight.

Adsorption (Physical and Possibly Chemisorption)

The sorption of flavor compounds and other organic compounds by food polymers has been the subject of numerous studies. In terms of producing binding adequate for stabilization of volatile and/or soluble compounds in food storage, the field is much more limited. Most of the work has been done by Maier in Germany (Maier, 1970), Solms in Switzerland, and their disciples, as well as Japanese, including the Kyoto group mentioned previously and Miyagi and Tanaka (1973).

Maier et al. studied binding of flavors to lipids (Maier and Kessler, 1977; Maier and Balcke, 1979), protein (Maier, 1974), sugars and salts (Maier, 1969), and starch (Maier and Bauer, 1972; Maier and Hartmann, 1977; Maier et al., 1987).

This work provides much useful information on specific interactions, but it is difficult to apply to complex food systems. Even in purified polymeric systems in which both surface adsorption and sorption within a glass can occur simultaneously, it is difficult to discern the contributions of each mechanism. In any case, the contribution of specific adsorption to encapsulation of flavor compounds is usually less significant than entrapment (Chirife and Karel, 1973).

The group of Professor Solms in Switzerland (E.T.H.) studied the binding of model flavor compounds with food components. (Solms et al., 1973; Beyeler et al., 1974; Beyeler and Solms, 1974). The major conclusions of their work are consistent with a physical adsorption model with rather weak binding forces.

REFERENCES

Andres, C. 1976. *Food Processing* 37(5):72.
Ashes, J.R., Cook, L.J., and Sidhu, G.S., 1984. *Lipids* 19:159.
Bala, K. and Vasudevan, P. 1982. *J. Pharmac. Sci.* 71:960.
Beyeler, M., Solms, J., Wiegand, U.W., and Weder, H.G., 1974. *L.W.T.* 7(6) 313.

Beyeler, M. and Solms, J., 1974. *L.W.T.* 7(4):217.

Braun, S.D., Olson, N.F., and Lindsay, R.C., 1982. *J. Food Sci.* 47:1803.

Braun, S.D., and Olson, N.F., 1986. *J. Dairy Sci.* 69:102.

Celebi, N. and Nagai, T., 1988. *Drug Dev. Ind. Pharm.* 14:63.

Chen, A.C., Veiga, M.F., and Rizzuto, A.B., 1988. Presented at IFT.

Chilvers, G.R., Gunning, A.P., and Morris, V.J., 1988. *Carboh. Polymers* 8:55.

Chirife, J., and Karel, M., 1973. *J. Food Sci.* 38:768.

Deasy, P.B., 1984. *Microencapsulation and related drug processes,* Marcel Dekker, Inc., New York.

DeMann, J.M., DeMann, L., and Wygerde, T., 1986. *Milchwissegch.* 41(8) 468.

Dziezak, J., 1988. Special Report of Microencapsulation and Encapsulated Ingredients. *Food Technol.* 42(4):135.

Edmondson, L.F., Yoncoskie, R.A., Rainey, N.H., Douglas, F.W., and Bitman, J., 1974. *JAOCS* 51:72.

Finlayson, H.J., and Armstrong, D.G., 1986. *J. Sci. Food Agric.* 37:742.

Flink, J.M., 1975. In: *Freeze Drying and Advanced Food Technology*, Goldblith, S.A., Rey, L., and Rothmayr, W., (Eds.). Academic Press, New York, pp. 309–332.

Furia, T.E., and Bellanca, N., 1977. *JAOCS* 54:239.

Gandhi, R.B., and Karara, A.H., 1988. *Drug Dev. Ind. Pharm.* 14:657.

Havinghorst, C.H., 1978. *Food Engineering,* June, 1974, p. 46.

Houts, S.S., 1988. *Food Technol.* 42(3):110.

Karel, M., 1985a. In: *Properties of Water in Foods.* Simatos, D. and Multon, J. (Eds.). Nijhoff Publ. Dordrecht, Holland. p. 153.

Karel, M., 1985b. *Int. J. Food Tech. Food Process Eng.* (ZFL) 36:138.

Karel, M., and Flink, J., 1983. *Advances in Drying II.* Hemisphere Press, New York., p. 103.

Karel, M., and Langer, R., 1988. In: *Flavor Encapsulation.* Risch, S.J., and Reineccius, G.A., (Eds.) ACS Symp. Series 370., p. 177.

Kerkhof, P.J.A.M., and Thijssen, H.A.C., 1977. *AICHE Symp. Series* 163:33.

King, B., 1978. Interaction of Flavor Compounds in Model Food Systems. Ph.D thesis, E.T.H. Zurich.

King, C.J., 1988. In: *Preconcentration and Drying of Food Materials.* Bruin, S., (Ed.). Elsevier, Amsterdam, p. 147.

King, C.J., Kieckbusch, T.G., and Greenwald, C.G., 1984. *Advances in Drying.* Mujumdar, A.S., (Ed.). Vol. 3, p. 71. Hemisphere, New York.

Kirby, C.J., Brooker, B.E., and Law, B.A., 1987. Int. *J. Food Sci. Technol.*22:355.

Kirby, C., and Gregoriadis, G., 1984a. Dehydration-Rehydration Vesicles: A simple method for high yield drug entrapment in liposomes. *Bio/Technology*, November: 979–984.

Kirby, C., and Gregoriadis, G., 1984b. In: *Liposome Technology*, Vol 1., pp. 19–27. Gregoriadis, G., (Ed.). CRC Press, Inc., Boca Raton.

Kirby, C.J., and Law, B.A., 1986. In: *Biotechnology in the Food Industry*, p. 17. Online Publications, Pinner, UK.

Koide, K., and Karel, M., 1987. Int. *J. Food Sci. Technol.* 22:207.

Koishi, M., Ishizaka, T., and Nakajima, T., 1984. *Appl. Biochem. 2nd Biotech.* 10:259.

Kuge, T., and Takeo, K., 1968. *Agr. Biol. Chem.* 32:753.

Langer, R., 1980. *Chem. Eng. Commun.* 6:1.

Langer, R., and Peppas, N.A., 1983. *J. Macromol. Sci. Rev. Macromol. Chem.* 23:61.

Law, B.A., and King, J.S., 1985. Use of liposomes for proteinase addition to cheddar cheese. *J. Dairy Res.* 52:183–188.

Lawrence, R.C., Creamer, L.K., and Gilles, J., 1987. *J. Dairy Sci.* 70:1748.

Levine, H., and Slade, L., 1986. *Carbohyd. Polymers* 6:213.

Liburdy, R.P., Tenforde, T.S., and Magin, R.L., 1986. *Radiation Res.* 108:102.

Liburdy, R.P., and Magin, R.L., 1985. *Radiation Res.* 103:266.

Magee, E.L., Olson, N.F., and Lindsay, R.C., 1981. *J. Dairy Sci.* 64:616.

Magee, E.L., and Olson, N.F., 1981. *J. Dairy Sci.* 64:600.

Maier, H.G., 1969. *J. Chromatogr.* 45:57.

Maier, H.G., 1970a. *Angew. Chemie.* (Intern.) 9:917.

Maier, H.G., 1970b. *Z. Lebensm. Untersuch. Forsch.* 144:1.

Maier, H.G., 1971. *Z. Lebensm. Untersuch Forsch.* 145:213.

Maier, H.G., 1972. *Z. Lebensm. Untersuch Forsch.* 149:65.

Maier, H.G., 1973. *Z. Lenensm. Untersuch. Forsch.* 151:384.

Maier, H.G., 1974. *Deutsche Lebensm. Rdschau.* 70(10):349.

Maier, H.G., and Balcke, C., 1979. *Fette-Seifen* 81:441.

Maier, H.G., and Bauer, A., 1972. *Starch* 24:101.

Maier, H.G., and Harman, R.U., 1977. *Z. Lebensm. Untersuch. Forsch.* 163:251.

Maier, H.G., and Kessler, K., 1977. *Fette-Seifen* 79:241.

Maier, H.G., Moritz, K., and Rummler, U., 1987. *Starch* 39:126.

Matsuyama, K., El-Gizawy, S., and Perrin, J.H., 1987. *Drug Dev. Ind. Pharm.* 13:2687.

Matsuzaki, M., McCafferty, F., and Karel, M., 1989. *Int. J. Food Technology.* 24:451.

Mayer, L.D., Hope, M.J., Cullis, P.R., and Janoff, A.J., 1985. Solute distributions and trappings efficiencies observed in freeze-thawed multilamellar vesicles. *Biochim. Biophys. Acta.* 817:193–196.

Miyagi, Z., and Tanaka, K., 1973. *Kolloid Z.* 251:739.

Moskowitz, G.J., and Noelck, S.S., 1987. Enzyme-modified cheese technology. *J. Dairy Sci.* 70:1761.

Mutka, J.R., and Nelson, D.B., 1988. *Food Technol.* 42(4):154.

Omatete, O.O., and King, C.J., 1978. *J. Fd Technol.* 13:265.

Osman-Ismail, F., and Solms, J., 1973. *L-W-T.* 6(4):147.

Picard, J.C., El Soda, M., Alkhalaf, W., Rosseau, A., Desmazeaud, M., Vassal, L., and Gripon, J.C., 1986. *Biotechnology Letters* 8(4):241–246.

Reineccius, G.A., and Risch, S.J., 1988.

Risch, S.J., and Reineccius, G.A., 1988. (Eds.). *Flavor Encapsulation.* ASC Symp. Series No. 370. ASC, Washington, D.C.

Renn, D.W., 1984. Marine algae and their role in biotechnology. In: *Biotechnology in Marine Science*, Colwell, R.R., Sinskey, A.J., and Pariser, E.R., (Eds.). John Wiley and Sons, New York, p. 191.

Roorda, W.E., Bodde', H.E., DeBoer, A.G., and Junginger, H.E., 1986. *Pharmac. Weekblad* 8:165.

Sada, E., Katoh, S., Terashima, M., and Tsukiyama, K.-I., 1988. *Biotechnol. and Bioeng.* 32:826.

Saenger, W., 1980. *Angew. Chemie.* 19:344.

Scott, T.W., 1975. *Cereal Foods World* 20(2):72.

Shomer, I., 1988. *J. Sci. Food Agric.* 42:55.

Solms, J., Osman-Ismail, F., and Beeler, M., 1973. *Can. Inst. Food Sci. Tech. J.* 6(1):A10.

Szenti, L., and Szejtli, J., 1986. *J. Food Sci.* 51:1024.

Szoka, F., and Papahadjopoulos, D., 1978. *Proc. Natl. Acad. Sci. USA* 75:494–498.

Suzuki, H., and Hasue, K., 1976. *J. Jap. Soc. Starch Sci.* 23:131.

Takeo, K., and Kuge, T., 1972. *Starch* 24:281.

Takeo, K., Tokumura, A., and Kuge, T., 1973. *Starch* 25:357.

Thijssen, H.A.C., and Rulkens, W.H., 1968. *DeIngenieur* 80:45.

Tirrell, D.A., Donaruma, L.G., and Turek, A.B., 1985. Macromolecules as drugs and carriers for biologically active molecules. *Annals N.Y. Acad. Sci.* 446.

Toei, R., 1986. *In Drying '86.* Mujumdar, A.S., (Ed.). Vol. 2, p. 880. Hemisphere, New York.

Torres, J.A., and Karel, M., 1985. *J. Food Process. Preserv.* 9:107.

Wagner, Jr., C.J., Wilson, III, C.W., and Shaw, P.E., 1988. *J. Food Sci.* 53:516.

Williams, M.L., Landell, R.F., and Ferry, J.D., 1955. *J.A.C.S.* 77:3701.

Zilberboim R., Kopelman, I.J., and Talmon, Y., 1986. *J. Food Sci.* 51:1301.

10

Extrusion of Foods

Judson M. Harper

Colorado State University
Fort Collins, Colorado

INTRODUCTION

Have you wondered how all the variety of shaped and flavored cereals and snack foods are manufactured? Many of these and other convenience food products are produced on extruders, which both cook and/or form raw food ingredients. It all started in 1935 when screw extruders were developed to continuously form viscous macaroni dough into a variety of shapes and replace the batch rolling/sheeting and cutting processes previously used. The cooking extruder is an extension of the macaroni extruder and is capable of heating food ingredients up to 200°C under pressure, thereby both cooking and texturizing the ingredients (Harper, 1978).

The food extruder has seen increased utility because it has many characteristics that manufacturers can apply to both expand and enhance their product lines. These features include:

- Low cost — Labor requirements and processing space per unit of production are less than for other cooking/forming systems.

- Product characteristics — A variety of shapes, textures, colors, and appearances can be produced.
- Energy efficient — Cooking is performed at low moistures so less drying is required.
- New foods — Most carbohydrate and protein ingredients can be processed to increase product variety.
- No effluents — Few solids are lost, which reduces the BOD of effluent streams from the processing plant.
- Control of thermal/mechanical environment to change food ingredients.

New designs of extrusion equipment have increased the range of their application in food processing. Cooking extrusion has been described as a high temperature–short time (HTST) process for food ingredients without detrimental side effects while destroying undesirable enzymes, microorganisms, and heat-labile substances occurring in foods that cause spoilage, lessen storage stability, or reduce nutritional value (Linko et al., 1981).

The most common applications of cooking extruders are the production of expanded snack foods, ready-to-eat breakfast cereals, soup and beverage bases, weaning foods, pregelantinized starches, textured vegetable proteins, precooked composite flours, and confection products. More recently, the food extruder has been used as a chemical reactor to produce modified starch, cook and hydrolyse starch with acid or enzymes, restructure and upgrade meat by products, and roast cocoa and coffee (Linko, 1983).

The purpose of this paper is to describe the extruder operation applied to the cooking and shaping of food ingredients. Next, the operational characteristics of the significant types of food extruders are compared. The transformation of starch and protein into restructured and texturized convenience foods by the extruder is described. Finally, newer applications of the extruder are enumerated.

OPERATING PRINCIPLES OF FOOD EXTRUDERS

Extrusion Process

The verb "extrude" describes a process of shaping by forcing softened or plasticized materials through dies or holes by pressure. A food extruder is a device that facilitates the shaping and restructuring process for food ingredients. This paper will begin with single-screw extruders, which are those commonly used for this purpose and an example of which is shown in Fig. 10.1.

The screw is the heart of the extrusion process, and its design and speed of rotation greatly influence the extrusion operation. The movement of the feed ingredients through the extruder occurs because of the pitched flights along the

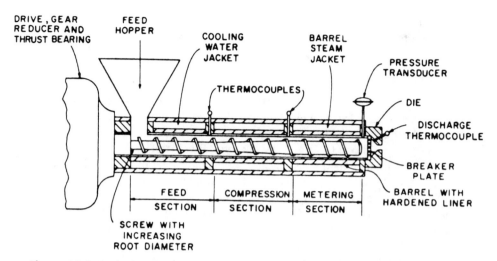

Figure 10.1 A single-screw extruder of the type used to cook and restructure food ingredients.

longitudinal axis of the screw. As the screw turns, the flights push the food ingredients forward when the material adjacent to the barrel wall does not slip and turn with the screw. To reduce slip, the barrel wall is often grooved to increase friction.

The screw performs three functions: conveying, working and heating, and mixing. The feed (conveying) section is characterized by deep flights which accept moistened granular feed materials and move them down the length of the screw. As the feed materials move along the screw, they encounter greater restriction or compression, causing them to completely fill the channel or the space existing between the screw flights.

The transition or compression section follows, having decreasing flight heights or other means to reduce the conveying capacity of the screw and cause compression. The food materials circulate in the screw channel as they move forward and are worked into a continuous mass. The energy necessary to make the viscous materials flow is supplied by a large drive motor turning the screw. Shearing during the flow process dissipates the mechanical energy, increases ingredient temperatures, and transforms the ingredients into a plasticized mass.

The later sections of the screw are designed to mix the plasticized ingredients. Shallower flights increase shear and improve mixing to assure uniformity of the high temperature/pressure dough.

When the flow path in the channel of the screw is continuously open, the extruder is not a positive displacement pump, and the conveying capacity is reduced when the pressure at the discharge increases. The negative pressure flow has the effect of reducing the net output from the extruder.

Extruder Operation

Raw ingredients used in cooking extruders consist of ground cereals or mill feed fractions, starch, and vegetable protein meal. Minor ingredients include fat, sugar, salt, acid or base, emulsifiers, color, and flavor. Water is added to achieve moisture contents ranging from 10 to 40%, with a typical value of 25%.

A combination of energy inputs causes the temperature of the feed ingredients to rise quickly as they pass through the extruder. Of all these sources, the mechanical energy addition by the turning screw is by far the most significant and normally accounts for 50–100% of the total energy addition. Heat transferred from the heated or cooled barrel wall, and/or from steam injected and mixed directly into the food within the extruder, is less important.

In cooking extrusion, the temperature of the food ingredients increases rapidly to 150–200°C after they are compressed and worked. The highest temperatures exist <20 sec, otherwise burning and off-flavors would occur. The high pressure within the extruder creates an environment similar to that of a pressure cooker and prevents flashing of steam until the pressure is released when the product emerges from the die.

At high temperature spontaneous puffing results once the dough exits the die and the pressure is released. The puffing is largely due to the escaping steam from the heated water in the product. Loss of moisture and heat in the product accompanies puffing, causing the product's temperature to drop rapidly to about 60°C.

Expanded extrudates have an open cellular structure consisting of individual cells surrounded by a membrane of starch or protein. The size of these cells controls product texture and density, two important product characteristics. These extrudates are normally cut at the face of the die with a rotating knife blade (Fig. 10.2) and then dried in a hot air oven to 2–12% depending on the product's requirements. Finished products can be coated with colors, flavor, oil, and/or sugar.

Control of finished product attributes is largely a function of the moisture content of the extruded ingredients, discharge temperature, the extent of ingredient damage occurring during the flow process, and the die configuration.

FOOD EXTRUDER TYPES

Single-Screw Extruders

The single-screw extruder is used either to pressurize food dough for cold forming through a die with essentially no puffing or for cooking and forming with extensive

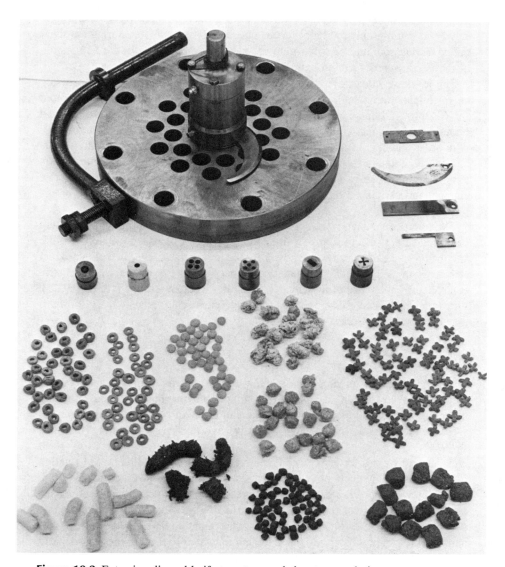

Figure 10.2 Extrusion die and knife to cut expanded restructured pieces.

heating and product puffing. In the former case, the screw has deep flights and turns slowly so a minimum amount of mechanical energy is dissipated during the shear process. Macaroni and chewing gum production are examples of the application of the single-screw forming extruder.

Single-screw cooking extruders have compressive screws with decreasing channel depth turning at high speeds to increase shear and mechanical energy input for heating. To increase capacity and efficiency, it is common to preheat ingredients in a preconditioner by adding steam before they enter the extruder.

In order to work, mix, and pressurize ingredients, the channel of the screw must become completely filled. Increasing pressure behind the die causes a reduction in net throughput since a reverse pressure flow occurs down the continuous screw channel and partially counteracts the forward flow imparted by the flights on the turning screw.

The operational characteristics of the extruder are dependent upon the pressure behind the die, slip at the barrel wall (controlled largely by barrel wall temperature and the presence of barrel wall grooves), the length of the screw filled with product (which is related to feed rate and screw speed), and the dough characteristics controlled by the types of ingredients, moisture, and temperature (Harper, 1978). Interactions between these variables limit the operational range and flexibility of the single-screw extruder.

Twin-Screw Extruders

The operating characteristics of twin-screw extruders differ from their single-screw analogs and have been responsible for increasing the popularity of extrusion processing and expanding its range of application. Twin-screw extruders have two screws which lie side by side and normally have intermeshing flights (Dziezak, 1988).

Twin-screw extruders can be characterized by the relative direction of rotation of the screws. A typical co-rotating twin-screw profile is illustrated in Fig. 10.3. Initial screw sections are designed to convey granular ingredients. The enhanced conveying capability of the twin-screw extruder allows these machines to handle sticky or otherwise difficult ingredients better than single-screw extruders. The parallel screws are self-wiping when the intermeshing flights on one screw sweep the bottom of the channel of the adjacent screw.

To increase mechanical energy dissipation during conveying, kneading disks are employed. An example of two-lobed kneading elements is shown in Fig. 10.4. Food materials pass from one disk to another, which provides a mechanical mixing action. Heat transfer through the barrel walls and the dissipation of mechanical energy are enhanced when these sections are completely filled by food ingredients. Filling occurs before any restrictions in the flow channel. Reverse pitch screw elements or dies are common types of restrictors (Harper, 1986a).

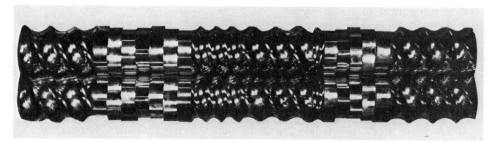

Figure 10.3. Twin-screw profile showing alternate conveying and mixing (kneading) sections.

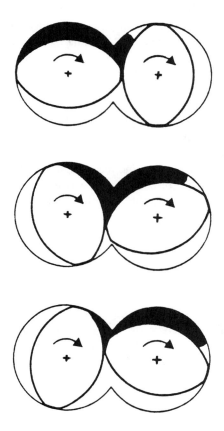

Figure 10.4 Kneading elements on a twin-screw extruder showing dough transfer.

Comparison of Extruder Types

Table 10.1 compares single- and twin-screw extruder characteristics.

To analyze costs, it is necessary to consider the cost of the extrusion equipment alone and in conjunction with the entire system. Twin-screw extruders cost more due to a relatively complex screw design, greater sophistication of the drive, and more complexity associated with barrel design and heat transfer jackets. This added equipment cost can be offset by the twin-screw extruder's capability of processing at lower moistures so that less finished product drying equipment is required.

Preconditioning ingredients with steam and water is widely used to provide about one-half of the heat necessary for cooking/processing in single-screw extrusion. The remainder comes from the mechanical energy input. In twin-screw extrusion, more heat is derived from mechanical energy input with the remainder from heat transferred through the barrel wall jackets.

The twin-screw extruder offers some advantages because of the characteristics of the screw. First, the ingredient-conveying capability exceeds that of the single-screw extruder, making it possible to handle a wide variety of ingredients that can cause feeding problems with single-screw machines. Twin-screw extruders are less prone to feed interruption and related surging because the screws' self-wiping feature.

Mechanically, twin-screw extruders are more complex and have greater constraints on operating ranges than single-screw extruders. Their gear drives have torque, pressure, and thrust limitations, which can limit product moistures and back pressures at dies (Harper, 1986b).

A single-screw extruder is relatively ineffective in transferring heat from jackets surrounding the barrel because of poor mixing which occurs within the channel. Twin-screw extruders have considerably more heat transfer capability because of improved mechanical mixing, which has expanded their application to heating and cooling viscous pastes, solutions, and slurries.

RESTRUCTURING BIOPOLYMERS

Extrusion involves cooking carbohydrate/protein ingredients, shaping, texturizing, puffing, or expanding. Extrusion differs from other processes in that these processing steps occur at relatively low moistures and under conditions of high temperture and shear for short periods of time. Under these conditions, the transformation of raw starch and protein to create restructured foods is rather remarkable.

Starch

Starch is the predominant ingredient in extruded snack and ready-to-eat (RTE) cereals. In most instances, degermed cereal grits are used as an extrusion feed

Table 10.1 Comparison of Single- and Twin-Screw Extruders

Item	Single-Screw	Co-rotating Twin-Screw
1. Relative capital cost/unit capacity		
• Extruder	1.0	1.5–2.0
• System	1.0	0.9–1.3
2. Relative maintenance	1.0	1.0–2.0
3. Energy		
• With preconditioner	Half from steam	Not used
• Without preconditioner	Mechanical energy	Mix of mechanical energy and heat exchange with barrel
4. Screw		
• Conveying angle	»10°	»30°
• Positive displacement	No	No
• Self-cleaning/wiping	No	Yes
• Variable flight height	Yes	No
• Screw length/diameter	4-25	10-25
• Mixing	Poor	Good
5. Drive		
• Relative torque/pressure	5.0	1.0
• Gear reducer	Simple	Complex
6. Heat transfer	Poor	Better in filled sections
7. Operations		
• Moisture	12–35	6 to very high
• Ingredients	Flowing grandular materials	Wide range
• Flexibility	Narrow operating	Greater operating

material rather than purified starch so that some protein, fat, and fiber also occur in the feed mixture.

Starch is a very large biopolymer of the simple sugar glucose. Normally the repeating glucose sugar units are attached end-to-end to form chains of amylose starch. When some branching occurs and side chains exist, the starch molecule is called amylopetin and has a structure similar to a thickly branched bush. Most cereal starches contain 25–30% amylose, and the remainder is amylopectin. In the native form, starch molecules are crystalline and exist as spheroid granules. During high-moisture cooking, the starch in these granules expands, absorbing and holding large quantities of water, and forms thick pastes.

Under the low-moisture conditions used for extrusion, opening up or gelatiniz-ing the starch granule depends on a combination of heat and mechanical shear. The starch granules are sheared as they pass through the extruder, and this mechanical action exposes the interior of the crystalline granules (Mercier, 1980). The mechanical working and flow of the viscous ingredients in the extruder requires significant mechanical energy input to turn the screw, which is dissipated as heat, causing a temperature rise. With shear, some of the larger starch molecules are broken into shorter chains (dexturization) and exhibit greater solubility in water (Gomez and Aguilera, 1984). The result of the gelatinization and dexturization is a plasticized dough or melt which can be extruded through a die for forming and which puffs when the high temperature internal moisture changes to steam after the die.

Extruded product characteristics are largely affected by the extent of starch transformation during the extrusion cooking process (Dosady, 1986). Products with a harder, glassier starch matrix and larger pore size are characteristic of low shear cooking under higher-moisture conditions. Starch damage is reduced when less mechanical energy is applied to the product and more heat is added by steam injection or heat transfer through the barrel wall. Products which are soft and easily hydratable in water, tend to stick to the teeth when consumed, have smaller pore sizes with thinner cell walls, and are characteristic of cooking in a high shear environment where considerable mechanical energy is dissipated.

Recent advances in extrusion allow much greater control of the shear environment during the cooking process to produce the desired product density, mouthfeel, and appearance (Zuilichem et al., 1980). Control of the shear/cooking environment is a result of variation of screw design, length/diameter ratio of the barrel, screw speed, source of heat used to increase product temperature, die design, and moisture of the food ingredients.

Protein

Defatted soy protein can also be restructured in an extruder to create a layered meat-like structure. Textured soy protein (TSP) is manufactured from defatted soy flour containing approximately 50% protein. Soy protein that has received only moderate heat treatment, having a nitrogen solubility index (NSI) of 50–70, is the preferred ingredient because it exhibits many of the characteristics of the native protein.

Texturization involves a restructuring of protein molecules into a layered crosslinked mass which is resistant to disruption upon further heating and/or processing. The soy protein is moistened to 33–45% and heated and sheared in the initial turns of the extrusion screw. This causes the native globular proteins to unfold with interruption of the chemical bonds that hold their native tertiary structure.

The large unfolded protein molecules align themselves in the direction of flow in the screw channels and void space behind the die. Flow is necessary to achieve alignment of the unfolded molecules (Harper, 1986b) and, with increased temperature to 140–160°C, chemical crosslinking reoccurs to give the layered, fibrous structure which simulates meat. Stanley (1986) describes the relatively linear nature of the unfolded protein molecules which allow substantial directional alignment to occur, creating flat surfaces when the product is dissected longitudinally. The open cellular cross-sectional structure of TSP and the smooth surface created when the sample is pulled apart longitudinally are shown in the scanning electron micrographs in Fig. 10.5. These layered restructured products can be rehydrated with about three times their weight of water for use as a simulated meat or meat extender.

NEWER APPLICATIONS

The direct expansion of cereal/starch ingredients to manufacture a variety of ready-to-eat snack and breakfast cereal products is the major application of food

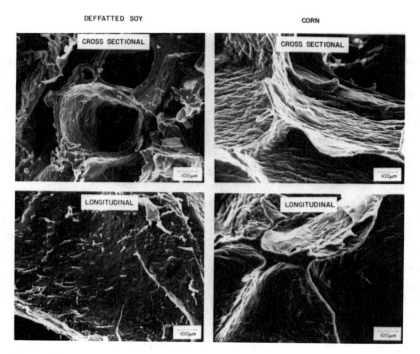

Figure 10.5 Restructured soy protein and corn grits.

extruders to the production of human foods (Hauck, 1980). To expand the variety of products, twin-screw extruders having extra-long, 40:1 length/screw diameter barrels are being used to perform the sequential operations of cooking, cooling, and forming with or without puffing. An example of such a process is shown in Fig. 10.6. This multiple-step extrusion process is being used to produce precooked macaroni, noodles, or rice for use as side dishes. The initial section of the extruder screw is designed to become completely filled with ingredients so that kneading disks can rapidly heat ingredients by dissipation of large quantities of mechanical energy. Rapid cooling is accomplished by venting steam at an intermediate screw section. Heat and moisture are removed from the product in the form of steam, which is vented. The cooled dough is pressurized by the final screw sections and extruded into precise shapes with a die at the discharge. These pellets can be subsequently puffed by hot air, frying in oil, or micro-wave heating.

The low pressure intermediate vent section can also be used to add color/flavor to the precooked dough before its final extrusion (Miller, 1988). If these materials are added to ingredients before cooking, they often deteriorate and lose much of their intended characteristics in the finished product.

To add variety to extruded products, co-extrusion is being used to make products having multiple texture characteristics. Using two extruders and a special die similar to that shown in Fig. 10.7, filled products are manufactured. Using filled tubes or U-shaped pieces, the filling can be either savory or sweet but must have the same water activity as the expanded cereal-base tube. This is accomplished using oil-based emulsions or low-moisture gels which have been plasticized with glycerol and/or other high molecular weight alcohols.

High-moisture extrusion is being used to restructure deboned muscle meat or reform low value fish into higher value products such as simulated scallops and crab legs. Various gums and/or salts are used in combination with the heat generated in the process to achieve the desired end result.

Heating sugar and starch to make confections is an increasingly important application for extruders. The application has advantages over the more conventional batch cooking process because the starch can be cooked at lower moisture, and flavor and color can be added and mixed into the product just before product forming.

"Conching" describes a process in which cocoa is finely ground and mixed with fat to produce a coating material. Extrusion will perform the time-consuming steps required by the conventional roller grinding process.

New and novel applications of extrusion continue to be developed and applied. They are replacing older, less efficient processes and/or are using new and unconventional ingredients to produce an expanded array of convenience food products.

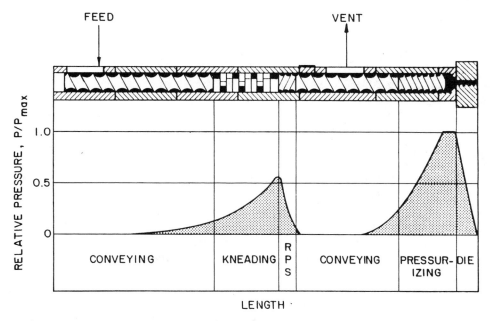

Figure 10.6 A twin-screw extruder configured to perform sequential cooking, cooling, and forming processes.

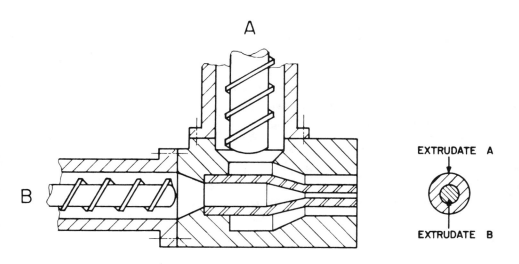

Figure 10.7 Co-extrusion die.

REFERENCES

Dosady, L. L. 1986. Review of recent studies on the mechanism of starch extrusion. In *Food Engineering and Process Applications. Unit Operation.* Vol. 2. (Ed.) LeMaguer, M. and Jelen, P., pp. 143–156. Elsevier Applied Science Publishers, New York.

Dziezak, J. D. 1988. Single- and twin-screw extruders in food processing. *Food Technol.* 43(4):163–174.

Gomez, M.H., and Aguilera, J. M. 1984. A physiochemical model for extrusion of corn starch. *J. Food Sci.* 49:40–44.

Hauch, B. W. 1980. Marketing opportunities for extrusion cooked products. *Cereal Foods World* 25:594–595.

Harper, J. M. 1978. Extrusion processing of food. *Food Technol.* 32(7):67–72.

Harper, J. M. 1986a. Extrusion texturization of foods. *Food Technol.* 40(3):70–76.

Harper, J. M. 1986b. Processing characteristics of food extruders. In *Food Engineering and Process Applications. Unit Operations.* Vol. 2, (Ed.) LeMaguer, M. and Jelen, P., pp. 101–114. Elsevier Applied Science Publishers, New York.

Linko, P. 1983. Recent trends inextrusion cooking applications. In *Research in Food Science and Nutrition; Food Science and Technology: Present Status and Future Direction.* Vol. 5, (Ed.) McLoughlin, J. V. and McKenna, B. M., p. 363–374. Bolle Press, Dublin, Ireland.

Linko, P., Colonna, P., and Mercier, C. 1981. High-temperature, short-time extrusion cooking. In *Advances in Cereal Science and Technology,* Vol. 4, (Ed.) Pomeranz, Y., pp. 145–235. American Association of Cereal Chemists, St. Paul, MN.

Mercier, C. 1980. Structure and digestibility alterations of cereal starches by twin-screw extrusion-cooking. In *Food Process Engineering. Food Processing Systems,* Vol. 1, (Ed.), Linko, P., Malkki, Y., Olkku, J., and Larinkar, J., pp. 795–807. Applied Science Publishers, Ltd., Essex, U.K.

Miller, R. C. 1988. Continuous cooking of breakfast cereals. *Cereal Foods World* 33:284–291.

Stanley, D. W. 1986. Chemical and structural determinants of texture of fabricated foods. *Food Technol.* 40(3):65–69.

van Zuilichem, D. J., Bruin, S., Janssen, L. P. B. M., and Stolp, W. 1980. Single screw extrusion of starch- and protein-rich materials. In *Food Process Engineering. Food Processing Systems,* Vol. 1, (Ed.) Linko, P., Malkki,Y., Olkku, J. and Larinkari, J., pp. 745–756. Applied Science Publishers, Ltd., Essex, U.K.

11

Developments in Food Freezing

R. Paul Singh and Jatal D. Mannapperuma

University of California at Davis
Davis, California

INTRODUCTION

Use of the freezing technique for long-term preservation of foods is now a well-established process in the food industry. While freezing offers several advantages when compared with other methods of food preservation, there are several design and operating characteristics that must be understood and carefully controlled to optimize the performance of a freezing system. The behavior of foods at subzero temperatures is influenced by the separation of ice from the system, the low temperature itself, and the crystallization or precipitation of other components, such as salts. Thermodynamic, kinetic, and diffusional properties of food systems at subzero temperatures are markedly influenced by the low temperatures. In addition, as the fraction of the frozen water increases, some of the physical properties of the system change dramatically. All these factors influence the design and operation of a freezing system. In this paper, some recent developments in mathematical approaches to describe food freezing will be presented. First, a brief overview of the ice crystallization process is given to describe the physical changes

309

that take place when freezing is initiated. The concluding section will provide a discussion of the commonly used food-freezing systems.

ICE CRYSTALLIZATION PROCESS

The freezing process in water, as shown in Fig. 11.1, involves heat removal with a consequent decrease in temperature until the freezing point is reached. The liquid phase generally solidifies at a temperature lower than the melting point of the corresponding solid phase, therefore a certain amount of supercooling (or undercooling) is necessary. Next, the latent heat is removed accompanied with phase change at a constant temperature. After the removal of latent heat, the temperature again decreases as additional heat is removed from the ice.

In case of a food product or any aqueous solution, the initial removal of sensible heat results in a decrease in temperature. Compared to the freezing curve for water, the initial freezing point for the food is depressed due to the presence of solutes. A portion of the water is crystallized during initial freezing, while the remaining solution becomes more concentrated, thus further reducing the freezing point of the unfrozen component of the solution. This phenomenon of decreasing temperature with increasing solution concentration continues until the eutectic

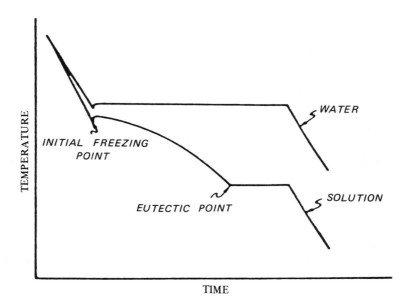

Figure 11.1 Freezing diagram of water and a binary solution.

point of the solute is reached (Fig. 11.1). Each solute in the solution has an eutectic point. At temperatures lower than the eutectic point, both ice and solute crystallize. In binary solutions, crystallization and heat removal occur at constant temperature. After the crystallization process is complete, additional heat removal results in a further decrease in temperature.

In food freezing, the ice crystallization process is of considerable interest. The number and size of ice crystals influence the product quality. Ice crystallization occurs in two steps, namely (1) nucleation, and (2) crystal growth.

Nucleation

There are two types of nucleation: homogeneous nucleation and heterogeneous nucleation. Homogeneous nucleation occurs in highly purified water as nuclei form due to random clustering of water molecules. In heterogeneous nucleation, small foreign particles present in the solution provide sites for the ice embryo to form and grow. In addition, mechanical disturbance and local variations in solute concentration also encourage nucleus formation. In food products, due to the presence of solutes, heterogeneous nucleation is most often observed. Reid (1983) has shown that food polymers (CMC, guar, carrageenan) increase the heterogeneous nucleation rate of water by up to 30 times. Charoenrein and Reid (1988) have studied the behavior of heterogeneous and homogeneous nucleation of ice in water and in sucrose solution by using differential scanning calorimeter. They induced heterogeneous nucleation by inseminating the sample with a bacterial ice nucleant, *Pseudomonas syringae*. They found that the maximum effective homogeneous nucleation rate of water sample was around 233.3 K, confirming earlier results by Rasmussen and Mackenzie (1972). The maximum effective heterogeneous nucleation rate in water was around 263.6 K, which is the active temperature for the majority population of *P. syringae*.

Crystal Growth

Crystal growth occurs after the nucleus has formed. For the nucleus to grow, it must be larger than a critical size. The rate of crystal growth is influenced by the rate of diffusion of water molecules from the unfrozen solution to the crystal surface, the rate of heat removal, and the temperature. The rate of crystal growth increases as the temperature decreases. At extremely low temperatures the viscosity becomes very high, which tends to retard the rate of crystal growth; however, the viscosity influence is overcome by the effect of lower temperatures that favors increased crystal growth.

The nucleation process has a direct influence on the size of ice crystals. If the rate of heat removal is slow and the product temperature is close to 0°C, then very few nuclei will form and they will grow to a large size. However, if the rate of heat removal is fast and the temperature of supercooling is relatively low, then numerous nuclei will form and their size will be small. Smaller ice crystals in food freezing are generally preferred especially for foods prone to increased drip loss and impairment of other quality attributes due to the formation of larger ice crystals.

THERMODYNAMICS OF FOOD FREEZING

Foods that require low temperature as a means of preservation of quality generally contain a large amount of water, usually in the range of 60–95%. Water present in the food acts as a solvent for various organic and inorganic compounds, such as sugars and salts. When the temperature of a typical food is gradually lowered, ice crystals begin to separate from the solution at a temperature slightly below the freezing point of pure water. This temperature is referred to as the initial freezing point.

Separation of ice crystals increases the concentration of the solutes in the solution and depresses the freezing point further. This phenomenon continues almost throughout the freezing process (Staph and Woolrich, 1951). Therefore, ice and water fractions in the frozen food depend on the temperature. Since the physical and thermal properties of ice and water are quite different, the properties of the frozen food become temperature dependent.

Thermodynamic analysis of freezing of dilute solutions is used extensively to predict initial freezing point and the composition of the frozen food in terms of ice and water fractions. The predicted compositions are then used to predict thermophysical properties of frozen foods.

Considering that frozen food is an equilibrium mixture of water, ice, and solutes, both Raoult's law of dilute solutions (Eq. 1) and the Claussius–Clapeyron relation for the coexisting phases (Eq. 2) are applicable.

$$p_v = X_w p_{v0} \tag{1}$$

$$\frac{dp_v}{dT} = \frac{\lambda p_v}{RT^2} \tag{2}$$

Equations (1) and (2) are then combined to yield the total differential Eq. (3).

$$\frac{dX_w}{X_w} = \frac{\lambda}{RT^2} dT \tag{3}$$

Equation (3) is integrated between the freezing point of pure water T_p (where X_w is 1) and any temperature T below T_p where water, ice and solutes remain in equilibrium.

$$X_w = \exp\left[\frac{\lambda}{R} \left\{\frac{1}{T_P} - \frac{1}{T}\right\}\right] \tag{4}$$

The mass fractions of the water and solutes in the frozen food are substituted for molar fraction of water in Eq. (4) and rearranged to obtain an explicit expression for T.

$$\frac{1}{T} = \frac{1}{T_P} - \frac{R}{\lambda} \ln\left[\frac{\dfrac{Y_w}{M_w}}{\dfrac{Y_w}{M_w} + \sum \dfrac{Y_j}{M_j}}\right] \tag{5}$$

At the initial freezing point of the food the mass fractions of the water and solutes are the same as those in the unfrozen food. This condition can be used in Eq. (5) to obtain an explicit expression for the initial freezing point temperature, T_Z.

$$\frac{1}{T_z} = \frac{1}{T_p} - \frac{R}{\lambda} \ln\left[\frac{\dfrac{Y_{wz}}{M_w}}{\dfrac{Y_{wz}}{M_w} + \sum \dfrac{Y_j}{M_j}}\right] \tag{6}$$

Equation (6) is used to calculate the initial freezing point when mass fractions of all the soluble solids are known (Choi and Okos, 1984). This equation is further simplified when T_z is close to T_p to obtain a linear equation for the depression of freezing point (Fennema et al., 1973; Heldman and Singh, 1981).

$$T_P - T_Z = \frac{RT_P^2}{\lambda} \sum \frac{Y_j}{M_j} \tag{7}$$

When the initial freezing point is known, the soluble solids term in Eqs. (6) or (7) can be evaluated. When the soluble solids term is evaluated using initial freezing point or is known by some other means, then it can be used in Eq. (5) to calculate the mass fraction of unfrozen water in the food Y_w at any temperature below the initial freezing point. On the other hand, the soluble term can be eliminated from Eqs. (5) and (6) to obtain an explicit expression for Y_w as:

$$Y_w = Y_{wz} \frac{F_z - F_P}{F - F_P} \qquad (8)$$

where

$$F = F\{T\} = \exp \left[\frac{\lambda}{RT} \right] \qquad (9)$$

Since the sum of ice and water fractions remain constant during the freezing process, ice fraction also can be calculated.

$$Y_I = Y_{WZ} - Y_W \qquad (10)$$

Heldman (1974) used this procedure to predict unfrozen water fraction in raspberries, asparagus, tomato pulp, apple juice concentrate, orange juice, cod fish, and lean beef. The predicted results were compared with the experimental observations compiled by Dickerson (1969). The comparison for all but cod fish and lean beef were reported to be satisfactory. The deviation in these two cases was attributed to the presence of significant amounts of unfreezable water. The effect of unfreezable water on the mass fraction of total unfrozen water can be accounted for by modification of Eq. (8).

$$Y_W = (Y_{WZ} - Y_A) \frac{F_Z - F_P}{F - F_P} + Y_A \qquad (11)$$

Once this correction was made, satisfactory agreement between predicted and experimental unfrozen water contents was obtained for foods containing significant amounts of unfreezable water (Heldman, 1974). Figure 11.2 illustrates the comparison of experimental and predicted unfrozen water fractions for raspberries where the unfreezable water is negligibly small. The effect of unfrozen water content on the predictions is demonstrated by the comparison shown in Fig. 11.3 where predictions by Eqs. (8) and (11) are shown together with experimental values for lean beef. A method equivalent to Eq. (11) has been used by Schwartzberg (1976) and Chen (1985) to predict unfrozen water content during freezing of foods.

The latent heat of fusion of ice decreases roughly 27% as temperature changes from 0°C to −40°C (Fig. 11.4). Mannapperuma and Singh (1989) allowed for this variation by assuming that the latent heat λ is a linear function of temperature

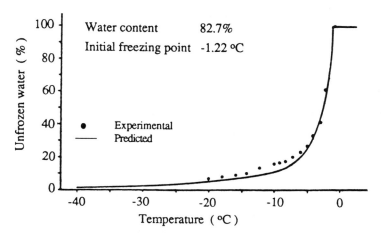

Figure 11.2   Unfrozen water fraction as a function of temperature for raspberries. (*From Heldman, 1974.*)

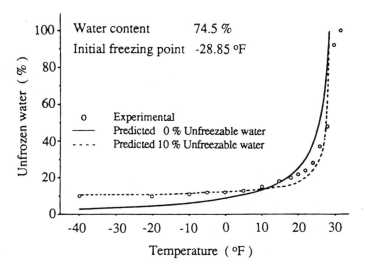

Figure 11.3   Effect of unfreezable water on prediction of unfrozen water fraction of lean beef (*From Heldman, 1974.*)

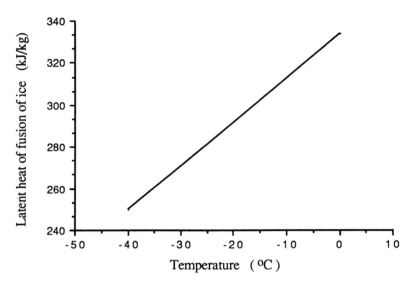

Figure 11.4 Variation of latent heat of fusion of water with temperature.

(Eq.12) and deriving a more complete scheme (Eq. 13 and 14) for the prediction of composition of frozen foods.

$$\lambda = \lambda_0 + \lambda_1 T \tag{12}$$

$$Y_W = (Y_{WZ} - Y_A) \frac{F'_Z - F'_Z}{F' - F'_P} + Y_A \tag{13}$$

where

$$F' = F'\{T\} = T^{-(\lambda_1/R)} \exp\left[\frac{\lambda_0}{RT}\right] \tag{14}$$

This expression is expected to perform better than Eqs. (11) and (9) when T_Z is much lower than T_p and result in more accurate predictions particularly when used to simulate enthalpy of frozen foods.

PROPERTIES OF FOODS

The properties of food are of prime importance in estimating freezing and thawing times and refrigeration loads. These properties include density, enthalpy, specific heat, and thermal conductivity. Published literature dealing with properties of foods can be broadly divided into experimental investigations and prediction models (Holdsworth, 1987).

The experimental investigations usually report the properties measured at known temperatures and water contents. Experimental determination of all the required food properties is a difficult task. It is also of limited use since properties of the same food vary with variations of parameters such as water content, fat content, and temperature. These reasonings have prompted many researchers to develop mathematical expressions to predict properties of a food using composition and properties of its constituents.

Experimental Investigations of Properties of Foods

The experimental methods used in the investigation of food properties show a wide variation. The early investigators of unfrozen water, enthalpy, and specific heat used mixing calorimeters while recent workers have used differential scanning calorimeters. Steady state methods, such as guarded plate and concentric cylinder apparatus, were used in the early studies on thermal conductivity while most of the recent studies resort to the use of unsteady state probe methods. Murakami and Okos (1989) contains a review of a large number of methods used in the experimental determination of thermal conductivity of foods.

Riedel (1951) presented results of calorimetric investigations on fruits, vegetables, and fruit juices. He observed that the enthalpy of all juices depended only on the dry matter content in spite of the fact that different chemical compounds such as glucose and sucrose constituted the dry matter. He used this unique behavior of juices to present a universal chart for the enthalpy of fruit and vegetable juices as a function of temperature and the dry matter fraction (Fig. 11.5). He proposed a formula to extend the application of this chart to determine the enthalpy of whole vegetables and fruits. He also presented similar charts for meats, fish, and egg components. A summary of some of Riedel's results was presented by Dickerson (1969).

Staph and Woolrich (1951) published a review of a 20-year research project on calorimetric investigations of a variety of fruits, vegetables, and fish. They found that the depression of freezing point of water in the foods was caused by the dissolved matter. They proposed the now widely accepted hypothesis that the separation of ice, as food freezes, increases the concentration of solutes and depresses the freezing point further. They also introduced the concept of apparent specific heat to account for the distribution of phase change enthalpy over the extended freezing temperature range.

Other early experimental studies of properties of foods include thermal conductivity of a variety of frozen vegetables, fruits, egg components, and fish by

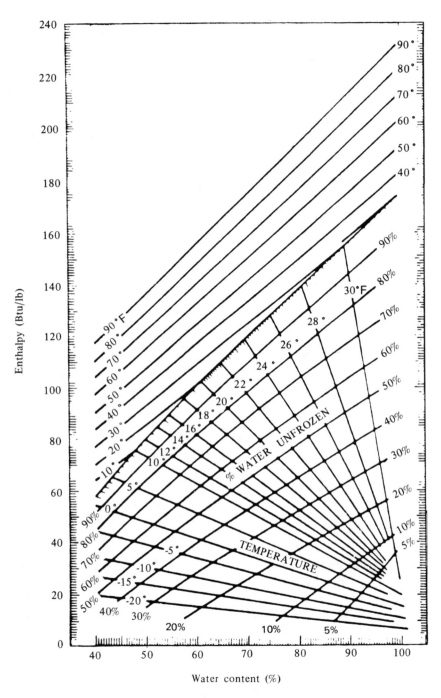

Figure 11.5 Enthalpy of fruit and vegetable juices. (*From Dickerson, 1969.*)

Smith et al. (1952), enthalpy, specific heat, and thermal conductivity of cod fish by Jason and Long (1955), and thermal conductivity of a number of meats, fats, and gelatin gels by Lentz (1961).

Ramaswamy and Tung (1981) reported an experimental study of properties of apples over a wide temperature range. The properties included density, specific heat, thermal conductivity, and thermal diffusivity. ASHRAE (1985) and Polly et al. (1980) contain listings of experimentally obtained values of food properties. These two references and Holdsworth (1987) contain listings of publications dealing with food properties.

Prediction of Food Properties

An extensive collection of property correlations for a variety of foods is listed by Miles et al. (1983). They also proposed a computer model to predict the food properties using properties of the food components. In this model the depression of freezing point is assumed to be proportional to the mass fraction of solutes in the food. This assumption is used to calculate the mass of ice as a function of temperature.

Heldman (1982) and Heldman and Singh (1981) used the compositions predicted by the freezing point depression method to predict density, thermal conductivity, and apparent specific heat of foods. Choi and Okos (1984) used the same method to predict the ice and water fractions in a frozen food. These predictions were then used in constitutive relations to determine density, thermal conductivity, and specific heat of foods. They listed regression equations for the density, thermal conductivity, thermal diffusivity, and specific heat of proteins, carbohydrates, fats, fibers, and ash as functions of temperature, to be used in these constitutive relationships.

Prediction of Specific Heat and Enthalpy

Specific heat and enthalpy are thermodynamic properties that can be derived by summing up the component properties in gravimetric proportion. The specific heat of foods in the temperature range where phase change does not take place is modeled as the summation of the products of mass fraction and specific heat of components (Eq. 15).

$$c_U = \Sigma c_i Y_i \qquad (15)$$

In the temperature range where phase change takes place, the apparent specific heat must account for the latent heat of the phase change. The common approach has been to express enthalpy of the frozen food, which includes both latent and sensible effects, as a function of temperature and to differentiate it with respect to temperature to obtain an expression for the apparent specific heat.

Schwartzberg (1976), Miles et al. (1983), Chen (1985), and Mannapperuma and Singh (1989) are some of the references that contain such models. All these models use freezing point depression to estimate the ice and water fractions. They differ from one another only by the nature of approximations used.

Schwartzberg (1976) derived Eqs. (16) and (17) for apparent specific heat and enthalpy of the frozen food below the initial freezing point.

$$c_F = c_U + (Y_A - Y_{WZ})(c_W - c_I) + (1 - Y_{WZ}) \frac{M_W}{M_G}$$

$$\left[\frac{RT_P^2}{(T - T_p)^2} - 0.8 (c_W - c_I) \right] \tag{16}$$

$$H_F = (T - T_D) \left[c_U + (Y_A - Y_{WZ})(c_W - c_I) + (1 - Y_{WZ}) \frac{M_W}{M_G} \right.$$

$$\left. \left[\frac{RT_P^2}{(T_P - T)(T_P - T_D)} - 0.8 (c_W - c_I) \right] \right] \tag{17}$$

Rearranged forms of Eq. (16) were used for the determination of unfreezable water fraction Y_A and effective molecular weight of soluble solids M_G using calorimetric data. This procedure was used to predict apparent specific heat of a number of fish and meat products with good accuracy. Figure 11.6 is an illustration of the apparent specific heat of cod fish predicted by Eq. (16) with experimental data reported by Riedel (1956).

Chen (1985) proposed Eqs. (18) to (21) for the determination of specific heat and enthalpy of foods below and above the initial freezing point.

$$\frac{c_F}{J} = 0.37 + 0.3 \, Y_G + Y_G \, \frac{RM_W T_P^2}{M_G(T - T_p)^2} \tag{18}$$

$$\frac{c_U}{J} = 1 - 0.55 \, Y_G - 0.15 \, Y_G^3 \tag{19}$$

$$\frac{H_F}{J} = (T - T_D) \left[0.37 + 0.3 Y_G + Y_G \, \frac{RM_W T_P^2}{M_G(T - T_P)(T_D - T_P)} \right] \tag{20}$$

$$\frac{H_U}{J} = H_Z + (T - T_Z)(1 - 0.55 Y_G - 0.15 Y_G^3) \tag{21}$$

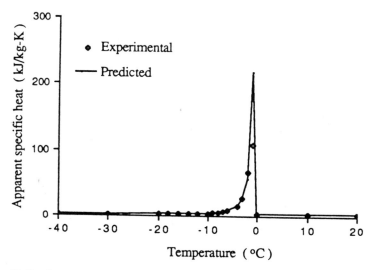

Figure 11.6 Apparent specific heat of cod fish.

These equations were used to predict specific heat and enthalpy of a number of food products. The predictions were compared with experimental data and with predictions by a number of other methods. Table 11.1 illustrates a comparison presented by Chen (1985) for the enthalpy and apparent specific heat of cod fish.

Mannapperuma and Singh (1989) used the unfrozen water fractions as expressed by Eq. (13) and specific heat of ice expressed by Eq. (22) to model apparent specific heat and enthalpy of frozen foods by Eqs. (23) and (24). They proposed Eqs. (25) and (26) for the unfrozen foods. A comparison of enthalpy of cod fish predicted by this method with experimental values (Riedel 1956) is shown in Fig. 11.7.

$$c_I = c_{I0} + c_{I1}T \tag{22}$$

$$c_F = (1 - Y_{WZ})c_B + Y_{WZ}(c_{I0} + c_{I1}T)$$

$$+ (Y_{WZ} - Y_A)\left[\frac{F'[\lambda_0 + \lambda_1 T]^2}{RT^2[F' - F_F']} + \lambda_1 \right] \frac{F_Z' - F_P'}{F' - F_P'} + Y_A\lambda_1 \tag{23}$$

Table 11.1 Comparison of Predictions of Enthalpy and Apparent Specific Heat

Temperature (°C)	Apparent specific heat (kJ/kg-K)			Enthalpy (kJ/kg)		
	Experimental[a]	Predicted[b]	Predicted[c]	Experimental[a]	Predicted[b]	Predicted[c]
-40	1.8	2.3	1.9	0	0	0
-30	2.0	2.4	2.0	19.2	23.3	19.6
-20	2.5	2.7	2.3	42.1	48.5	41.6
-18	2.7	2.8	2.5	47.5	54.1	45.9
-16	2.9	3.0	2.6	53.2	60.0	51.0
-14	3.2	3.2	2.9	59.3	66.3	56.6
-12	3.6	3.6	3.3	66.0	73.3	62.8
-10	4.1	4.3	4.0	73.6	81.3	69.7
-9	4.6	4.8	5.1	78.0	85.5	74.4
-8	5.3	5.5	5.3	82.9	91.3	79.3
-7	6.2	6.5	6.4	88.6	97.5	85.1
-6	7.7	8.1	8.0	95.5	105.1	92.3
-5	6.2	10.8	10.8	104.3	114.8	101.6
-4	15.3	15.6	15.9	116.7	129.1	114.6
-3	26.8	26.1	26.9	136.4	149.3	135.2
-2	67.4	55.9	58.2	176.4	189.3	174.6
-1	108.6	217.4	227.5	302.4	304.3	289.3
0	4.1	3.8	3.8	330.2		326.3
10	3.7	3.8	3.8	366.9		363.8
20	3.7	3.8	3.8	403.8		401.3

[a]From Reidel, 1956.
[b]From Schwartzberg, 1976.
[c]From Chen, 1985.

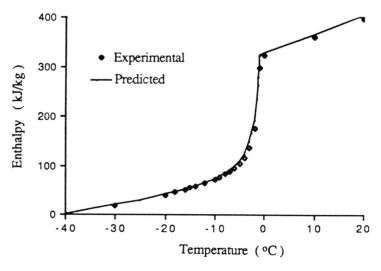

Figure 11.7   Enthalpy of cod fish.

$$H_F = (1 - Y_{WZ})c_B(T - T_D) + Y_{WZ}[c_{I0}(T - T_D) + \tfrac{1}{2}c_{I1}(T^2 - T_D^2)]$$

$$+ \left[(Y_{WZ} - Y_A)\frac{F_Z' - F_P'}{F' - F_P'} + Y_A\right](\lambda_0 + \lambda_1 T)$$

$$- \left[(Y_{WZ} - Y_A)\frac{F_Z' - F_P'}{F_D' - F_P'} + Y_A\right](\lambda_0 + \lambda_1 T_D) \qquad (24)$$

$$c_U = c_B Y_B + c_W Y_W \qquad (25)$$

$$H_U = H_{FZ} + (c_W Y_W + c_B Y_B)(T - T_Z) \qquad (26)$$

Prediction of Density and Thermal Conductivity

Modeling of density and thermal conductivity of composite materials require more information than mass fractions and component properties. In case of density, added knowledge of porosity and the assumption of conservation of both mass and volume in the process of mixing the components to constitute the food are required. Equation (27) represents this model. The difference in densities between ice and water causes the density of food to undergo a rapid change near the freezing point.

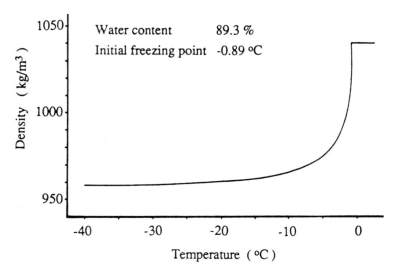

Figure 11.8   Density of strawberries.(*From Heldman, 1982.*)

Figure 11.8 depicts the variation of density with temperature predicted by this model for strawberries (Heldman, 1982).

$$\frac{1}{\rho} = \frac{1}{1 - \varepsilon} \; \Sigma \; \frac{Y_i}{\rho_i} \tag{27}$$

Thermal conductivity of composite materials such as foods depends not only on the properties of its components but also on its structure. A common food material such as meat has different thermal conductivities parallel to its fibers than perpendicular to the fibers. The effect of porosity on thermal conductivity is also not as simple compared to its effect on density. Freezing process can affect porosity significantly and directional growth of ice crystals can affect the thermal conductivity. Therefore, modeling of thermal conductivity is far more difficult than modeling of other thermal properties.

The Maxwell-Euken model (Jason and Long, 1955; Lentz, 1961) is an adaptation of an electrical conductivity model for this purpose. This model [Eqs. (28) and (29)] treats the food composite as a mixture of a continuous phase (subscript c) and a dispersed phase (subscript d).

$$k = k_c \left[\frac{(3 - \Psi) - 2\Psi V_d}{(3 - \Psi) + \Psi V_d} \right] \tag{28}$$

where

$$\Psi = 1 - \frac{k_d}{k_c} \tag{29}$$

When this model is used to model unfrozen foods, the food solids are treated as the dispersed phase and water is treated as the continuous phase. In frozen foods the choice of phases is not this clear. Jason and Long (1955) used a two-stage model. Ice dispersed in water is treated in the first stage, food solids dispersed in ice water mixture treated in the second stage. Figure 11.9 shows the comparison of thermal conductivity predicted by this method for cod fish muscle compared with experimental data.

Kopelman (1966) considered the different structural features of foods and proposed three separate models for homogeneous, fibrous, and layered food systems (Fig. 11.10). He derived Eqs. (31) to (34) to express thermal conductivity of the three systems. These models have the advantage of recognizing the

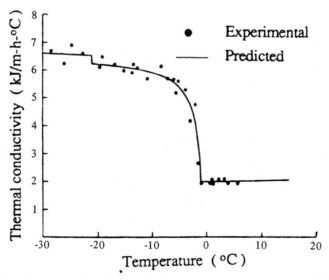

Figure 11.9 Thermal conductivity of cod fish. (*From Jason and Long, 1955.*)

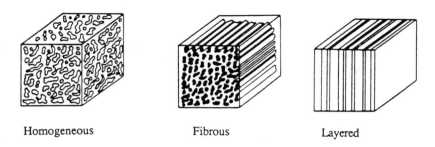

| Homogeneous | Fibrous | Layered |

Figure 11.10 Structural models of food. (*From Kopelman, 1966.*)

difference between the thermal conductivity parallel to the fibers and that perpendicular to the fibers. However, it poses the same problems as the Maxwell-Euken model when more than two distinguishable phases are present.

Homogeneous system

$$k = k_c \left[\frac{1 - \Psi V_d^{2/3}}{1 - \Psi V_d^{2/3}(1 - V_d^{1/3})} \right] \qquad (30)$$

Fibrous system

$$k_{||} = k_c \left[1 - \Psi V_d \right] \qquad (31)$$

$$k_\perp = k_c \left[\frac{\Psi - V_d^{1/2}}{\Psi - V_d^{1/2}(1 - V_d^{1/2})} \right] \qquad (32)$$

Layered system

$$k_{||} = k_c \left[1 - \Psi V_d \right] \qquad (33)$$

$$k_\perp = k_c \left[\frac{\Psi - 1}{\Psi - 1 + \Psi V_d} \right] \qquad (34)$$

Choi and Okos (1984) used a simple model where thermal conductivities of food components are multiplied by volume fractions and added to obtain the thermal conductivity of the food composite (Eq. 35). They compared predictions of this model with the experimental data for a number of food products (Fig. 11.11). Miles et al. (1983) and Mannapperuma and Singh (1989) have also used this model. This model is simple and can accommodate any number of components without further complications.

$$k = \rho \sum k_i \frac{Y_i}{\rho_i} \qquad (35)$$

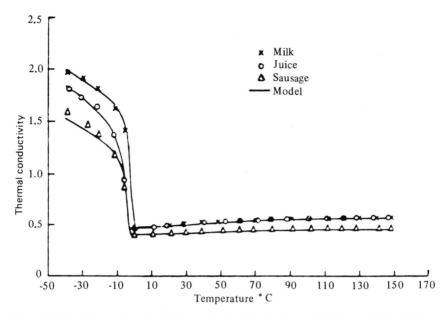

Figure 11.11 Thermal conductivity predictions by simple model. (*From Choi and Okos, 1984.*)

Porosity of the food affects its thermal conductivity considerably. Miles et al. (1983) and Mannapperuma and Singh (1989) corrected the density values for the food porosity using Eq. (27) and used the corrected density in place of ρ in Eq. (35) to calculate thermal conductivity.

Murakami and Okos (1989) presented a detailed study on thermal conductivity modeling of porous foods. They evaluated nine models and selected Keey's (1972) model where the series and parallel thermal conductivities are coupled using a distribution factor ϕ as described by Eq. (37).

$$\frac{1}{k} = \frac{1 - \phi}{k_c(1 - V_d) + k_d V_d} + \phi\left[\frac{1 - V_d}{k_c} + \frac{V_d}{k_d}\right] \tag{36}$$

where

$$\phi = \sum_{n=0}^{3} [B_{0n}\zeta_w^n + B_{1n}\zeta_w^n(\varepsilon - 0.4) + B_{2n}(\zeta_w - 0.25)^n(\varepsilon - 0.4)^2$$

$$+ B_{3n}(\zeta_w - 0.1627)^n(\varepsilon - 0.4)^3] \tag{37}$$

The coefficients B_{0n}, B_{1n}, B_{2n}, and B_{3n} (12 in all) were obtained by regression analysis of a large collection of experimental data on thermal conductivity of porous foods. Figure 11.12 graphically illustrates the decrease in distribution factor with the increase of porosity and moisture content of the food.

Choi and Okos (1984) reported mathematical models for the variation of density, specific heat, thermal conductivity, and thermal diffusivity of proteins, fat, carbohydrates, fiber, and ash with temperature over the range −40 to 150°C. These models can be used with the food composition data published by Watt and Merrill (1975) to predict the properties of a wide variety of foods within this temperature

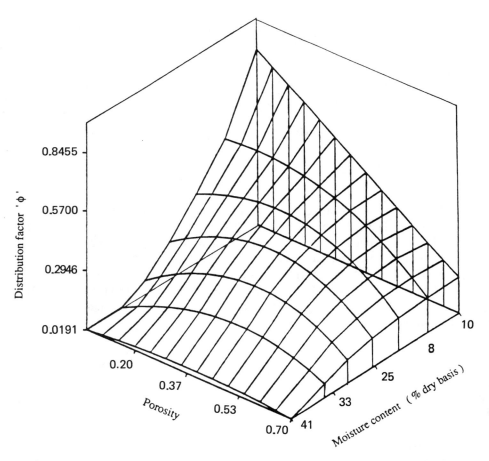

Figure 11.12   Response surface of distribution factor φ (*From Murakami and Okos, 1988.*)

range with a good degree of accuracy. A computer program package developed by Mannapperuma and Singh (1989) to predict thermophysical properties of foods uses this as one of the approaches.

TRANSPORT PHENOMENA DURING FREEZING OF FOODS

Freezing of foods requires the removal of large quantities of heat in relatively short periods of time. Therefore, proper application of principles of heat transfer plays an important role in calculations involving food freezing. Evaporative weight loss during freezing, freezer burn during frozen storage, and surface dripping during thawing involve combined heat and mass transfer processes that require careful investigation.

Heat Transfer Phenomena

Heat transfer by conduction takes place inside the food during initial stages of sensible heat removal prior to freezing. This heat transfer process is described by Fourier's law of heat conduction:

$$q = -k \frac{dT}{dx} \tag{38}$$

The removal of sensible heat during unsteady state heat transfer processes is accounted for by a heat balance performed over a small element:

$$\rho c \frac{\partial T}{\partial t} = \text{div} [k \text{ grad} [T]] \tag{39}$$

The exact form of the vector operators in Eq. (39) depends on the coordinate system used. Simplified forms result when the number of dimensions is less than the maximum allowed by reason of axisymmetry and/or infinite dimensions. The six simplified cases generally referred to as regular geometries are infinite slab, infinite cylinder, sphere, finite cylinder, infinite rectangular rod, and rectangular parallelepiped. Analytical solutions of Eq. (39) are available for these regular geometries under a variety of conditions. Most food items are approximated by one of these shapes to enable use of standard solution procedures.

The freezing process begins when the temperature starts to drop below the initial freezing point. Freezing is accompanied by the liberation of the latent heat of phase change. This stage is usually described as heat conduction with phase change. The incorporation of phase change process in the heat conduction process becomes a difficult mathematical problem. Many approaches to this problem are reviewed later. When most of the phase change process is complete, the situation again approaches a heat conduction problem.

Heat removed from the food at the surface is carried away by a medium employed in the freezer. The nature of the heat removal process at the surface of the food is usually described by a boundary condition. The type of the boundary condition depends on the type of the freezer employed in the freezing process.

In case of the plate freezer the freezing medium is a metal plate cooled by an internal flow of a refrigerant. This situation is usually described as a constant temperature type boundary condition:

$$T_s = T_a$$

In freezing of foods by direct contact with a fluid medium such as in air blast, fluidized bed, and liquid immersion freezing techniques, the heat removal at the surface is by convective heat transfer:

$$-k \ \frac{dT}{dx} = h(T_a - T_s) \tag{41}$$

The convective heat transfer coefficient h in Eq. (41) is not a property of the food. It is determined by properties of the fluid, velocity of the fluid, and the nature of the fluid flow pattern around the food. Geometrical shape of the food and its size influence the flow pattern, hence the value of h.

Cleland and Earle (1976) reported on new methods for the experimental evaluation of heat transfer coefficients. Arce and Sweat (1980) conducted a survey of heat transfer coefficients. Arce and Sweat (1980) conducted a survey of heat transfer coefficients encountered in food refrigeration processes and reported the methods used in their evaluation. The range of heat transfer coefficients in this collection vary from 10 to 67 W/m^2K for freezing in air and from 420 to 440 W/m^2K for freezing in brine.

Convective type boundary conditions are used to account for the resistance of the packaging material in plate freezing and in fluid medium freezing by a properly defined overall heat transfer coefficient.

$$\frac{1}{h} = \frac{1}{h_{film}} + \frac{d_{package}}{k_{package}} \tag{42}$$

Holdsworth (1987) listed d/k ratio for a number of packaging materials which can be incorporated in Eq. (42). Magnitude of the inverse of this ratio (equivalent of a surface heat transfer coefficient) varied from the of 7.7 W/m²K for 5 mm corrugated cardboard to 714 W/m²K for 76 μm paper.

Radiation heat transfer takes place between the food and the surfaces of the freezer seen by the food. In the typical temperature range encountered in freezing practice this component of heat transfer is quite small. However, it can be modeled by an equation of the form (Hayakawa et al., 1983b):

$$-k \frac{dT}{dx} = e\sigma S(T_a^4 - T_s^4) \tag{43}$$

A more complicated heat removal process takes place during cryogenic freezing of foods using freezants that boil off at the surface of the food. Freezing by immersion in liquid nitrogen and in Refrigerant-12 are examples of this boiling boundary condition. Heat removal rate in boiling is dependent on the surface temperature of the food. Figures 11.13 and 11.14 illustrate this dependence for liquid nitrogen and for Refrigerant-12, respectively. This type of surface heat removal rate can be modeled as surface temperature dependent heat flux type

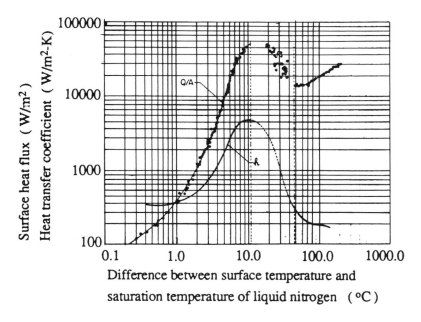

Figure 11.13 Boiling curves for liquid nitrogen at atmospheric pressure. (*From Flynn et al., 1961.*)

boundary condition (Eq. 44) or as a surface temperature dependent convective type boundary condition (Eq. 45).

$$-k \, \frac{dT}{dx} = q_s\{T_s\} \tag{44}$$

$$-k \, \frac{dT}{dx} = h\{T_s\}(T_a - T_s) \tag{45}$$

The boiling curve for liquid nitrogen (Fig. 11.13) indicates that the surface heat transfer coefficient varies from 200 W/m^2K in the film boiling region to about 5000 W/m^2K at the critical heat flux. In case of Refrigerant-12, this range is from about 200 W/m^2K to about 8000 W/m^2K. Evans (1989) modeled immersion freezing of foods in Refrigerant-12 using surface temperature–dependent convective type boundary condition based on the boiling curve shown in Fig. 11.14.

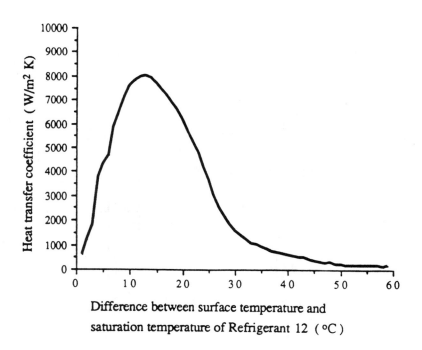

Figure 11.14   Boiling curve for Refrigerant-12 at atmospheric pressure. (*From Evans, 1989.*)

Average values of surface heat transfer coefficients for freezing of foods in liquid nitrogen and in Refrigerant-12 are also found in the literature. Dinglinger (1969) estimated the average heat transfer coefficient in spray freezing with liquid nitrogen to be about 175 W/m^2K. Thompson et al. (1984) reported heat transfer coefficients of 1190 W/m^2K for spray freezing and 1700 W/m^2K for immersion freezing of corn cobs in Refrigerant-12.

Mass Transfer Phenomena

When cold air is used as the freezing medium for foods, a gradient of water vapor pressure exists from the surface of the food to the medium. This gradient causes water to evaporate from the surface of the food, resulting in evaporation loss. Hayakawa et al. (1983b) incorporated heat removal due to evaporation in a boundary condition of the form:

$$-k \ \frac{dT}{dx} = g\lambda(\mu_a - \mu_s) \tag{46}$$

The rate of evaporation loss at the surface is dependent primarily on the relative humidity, temperature, and velocity of the cooling air. Rapid freezing in fluidized beds and by use of cryogens reduces weight loss by crust freezing of the surface layers. Table 11.2 is a comparison of weight loss when four freezing methods are used with four different food products as presented by Astrom and Londahl (1969). This information ranks the four freezing methods in the order of decreasing weight loss as, air blast, liquid nitrogen, fluidized bed, and Refrigerant-12. The unexpectedly high weight loss in liquid nitrogen freezing was attributed to the use of nitrogen vapor blast prior to immersion for better thermal efficiency.

Moreno (1979) reported on experimental studies of moisture loss during freezing of a variety of foods and a number of mathematical models. The symposium on weight loss (International Institute of Refrigeration, 1970) and the review of Norwig and Thompson (1984) contain valuable information on weight loss during freezing.

Freezer burn is the surface desiccation of foods during freezing and frozen storage due to sublimation of ice crystals on the food surface. Kaess and Weidemann (1969) observed that rapid freezing increased the freezer burn during subsequent frozen storage in comparison to slow freezing. A model to simulate the thawing process incorporating the surface mass transport aspects is proposed by Mannapperuma and Singh (1988b). This source also reviews the mass transfer aspects during thawing of foods.

Table 11.2   Weight Loss during Freezing

	Fluidized bed or multilevel belt freezer	Liquid freon freezant	Liquid nitrogen freezant	Conventional air blast freezer
A. Strawberries, variety 1				
1. Freezing time (min)	14	3	5	900
2. Final temperature (°C)	−23	−25	−28	−20
3. Weight loss (%)	1.5	0.0	1.4	2.7
A. Mushrooms				
1. Freezing time (min)	22	3	5	180
2. Final temperature (°C)	−20	−30	−26	−20
3. Weight loss (%)	1.2	0.0	1.9	2.5
A. Plaice Fillets				
1. Freezing time (min)	18	4	8	180
2. Final temperature (°C)	−20	−30	−50	−20
3. Weight loss (%)	0.8	0.1	1.4	1.3
A. Beef				
1. Freezing time (min)	35	4	8	180
2. Final temperature (°C)	−20	−28	−50	−20
3. Weight loss (%)	0.7	0.1	1.4	1.3

Source: Astrom and Londahl (1969).

PREDICTION OF FREEZING TIMES

The freezing time of a food is influenced by the heat transfer process within the food and by the nature of heat transfer between the food and the cooling medium. Shape, size, and thermal properties of the food control the heat transfer within the food while the effect of the cooling medium is treated as a boundary condition. Versatility of a freezing time prediction method may be judged by a number of factors. Its applicability to different geometrical shapes, a variety of different boundary conditions, convenience of its use, and accuracy of its predictions are the leading factors that should be considered.

The accuracy of prediction of freezing times should be measured primarily in comparison with actual experimental data. The experimental data available for this purpose is small compared to the number of publications on prediction methods. Cleland (1977, 1985), de Michelis and Calvelo (1983), Lescano (1973), Hayakawa et al. (1983a), and Hung and Thompson (1983) contain the bulk of the experimental data used to compare the accuracy of prediction methods. Cleland and Earle (1984) discussed criteria for the assessment of freezing time prediction methods. Errors in thermal conductivity, specific heat or enthalpy, surface heat transfer coefficient, and poor experimental techniques are observed to affect the comparisons. They strongly recommended comparison of predictions with a proven prediction method in addition to experimental data.

The freezing process in solid foods is treated as a heat conduction process with phase change. The transition of the food from unfrozen to frozen state takes place over a range of temperature. However, a majority of the published methods are based on the simplifying assumption that the phase change takes place isothermally. Such methods constitute a broad class, which can be further subdivided into exact analytical and approximate analytical solutions. The published methods that allow for the gradual phase change over a temperature range are based on apparent specific heat formulation and enthalpy formulation. Freezing time prediction methods are reviewed in the following four sections based on these subdivisions.

The reviews in this presentation are intended to introduce the basic concepts and key references of the four methods. Holdsworth (1987), Cleland (1985, 1977), Cleland and Earle (1984), Ramaswamy and Tung (1984), and Bakal and Hayakawa (1973) are some of the other recent reviews of freezing time prediction methods.

Exact Solution for Isothermal Phase Change: Neumann Solution

This problem represents the freezing of a pool of pure liquid when the top surface is maintained at a subfreezing temperature. Cowell (1967), Bakal and Hayakawa (1973), and Charm (1978) have discussed the use of this solution for the estimation of freezing times of foods. This method is reviewed as an exercise in appreciation of the complexity of the problem.

Heat conduction in unfrozen and frozen regions are governed by two partial differential equations [Eqs. (47) and (48)]. The two regions are assumed to have different thermal properties. Figures 11.15 shows the temperature profiles at an arbitrary time t.

$$\frac{\partial T_1}{\partial t} = \alpha_1 \frac{\partial_2 T_1}{\partial x^2} \tag{47}$$

$$\frac{\partial T_2}{\partial t} = \alpha_2 \frac{\partial_2 T_2}{\partial x^2} \tag{48}$$

The phase change takes place at the freezing point temperature, T_F. Four initial

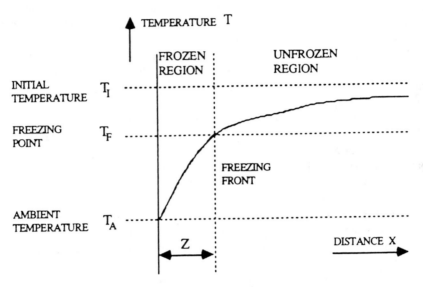

Figure 11.15   Freezing of a semi-infinite space—Stefan's problem.

and boundary conditions [Eqs. (49) to (52)] are written to specify fixed temperature at the boundaries and at the freezing front.

$$T_1 = T_F \qquad \text{at } x = Z(t) \tag{49}$$

$$T_2 = T_F \qquad \text{at } x = Z(t) \tag{50}$$

$$T_1 = T_A \qquad \text{at } x = 0 \tag{51}$$

$$T_2 = T_{IN} \qquad \text{at } t = 0 \tag{52}$$

The location of the freezing front, Z, varies with time, which adds a new variable. The additional equation required to compensate is obtained by writing a heat balance for the movement of the freezing front.

$$k_1 \frac{\partial T_1}{\partial x} - k_2 \frac{\partial T_2}{\partial x} = \lambda \rho_2 \frac{dZ}{dt} \tag{53}$$

The solution of this problem is commonly referred to as Stefan's solution

(Luikov, 1968) or as Neumann's solution (Charm, 1978). Equations (54) to (56) present this solution as given by Luikov (1968).

$$T_1 = T_a + (T_F - T_a) \; \frac{\text{erf } (x/\sqrt{4\alpha_1 t})}{\text{erf } (\beta/\sqrt{4\alpha_1})} \tag{54}$$

$$T_2 = T_{IN} - (T_{IN} - T_F) \; \frac{\text{erfc } (x/\sqrt{4\alpha_2 t})}{\text{erfc } (\beta/\sqrt{4\alpha_2})} \tag{55}$$

Equations (54) and (55) are substituted in the moving boundary condition Eq. (53) to obtain the transcendental Eq. (56) for the parameter β. This equation is usually solved by trial and error or by a graphical procedure.

$$\frac{k_1(T_F - T_a)}{\sqrt{\alpha_1}} \; \frac{\exp (-\beta^2/4\alpha_1)}{\text{erf } (\beta/\sqrt{4\alpha_1})} + \frac{k_2(T_{IN} - T_F)}{\sqrt{\alpha_2}} \; \frac{\exp (-\beta^2/4\alpha_2)}{\text{erfc } (\beta/\sqrt{4\alpha_2})}$$

$$= \tfrac{1}{2} \, \rho_2 Y_W \lambda \beta \sqrt{\pi} \tag{56}$$

Charm (1987) predicted freezing times using this method under fixed temperature boundary conditions and using the same method with some modifications under convective type boundary conditions. Cleland (1977) used the method modified by Charm (1978) to predict the freezing times of Tylose slabs under convective boundary conditions and compared the results with a large numbers of experimental data. He reported that this prediction method resulted in a range of errors from −18% to 630% with a mean of 160% and concluded that these modifications were very misleading.

This solution illustrates the complex nature of the freezing process even under highly idealized situations and serves as a warning about the problems that lie ahead.

Approximate Solution for Isothermal Phase Change: Plank's Equation

The method used by Plank (1913, 1941) is based on two simplifying assumptions. First, it assumes that the unfrozen region is initially at the freezing point, which forces this region to always remain at the freezing point, the only change being the movement of the freezing front. Second, it neglects the heat capacity of the frozen region, which results in a linear temperature profile in the unfrozen region. Convective type heat transfer from the medium to the food is the boundary condition used. Figure 11.16 illustrates the assumed temperature profiles in case of an infinite slab.

Equations (57), (58), and (59) describe heat transfer from the medium to the food

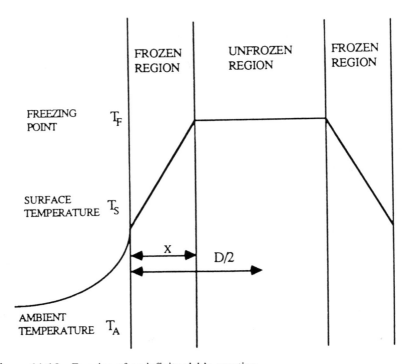

Figure 11.16 Freezing of an infinite slab's equation.

surface, conductive heat transfer through the frozen zone, and liberation of latent heat at the moving freezing front:

$$q = -hA \ \ (T_a - T_s) \tag{57}$$

$$q = -kA \ \frac{(T_s - T_F)}{Z} \tag{58}$$

$$q = A\lambda\rho \ \frac{dZ}{dt} \tag{59}$$

These three equations are combined to eliminate q and T_s:

$$\frac{dt}{dZ} = \frac{\lambda\rho}{(T_F - T_a)} \left[\frac{1}{h} + \frac{Z}{k} \right] \tag{60}$$

Equation (60) is integrated to obtain the time taken by the freezing front to move to the center of the slab.

$$ t = \frac{\lambda \rho}{(T_F - T_a)} \left[\frac{D}{2h} + \frac{D^2}{8k} \right] \tag{61} $$

The same treatment can be applied to an infinite cylinder and a sphere to obtain similar equations. Equations for these different shapes are combined into the following:

$$ t = \frac{\lambda \rho}{(T_F - T_a)} \left[\frac{Pd}{h} + \frac{Rd^2}{k} \right] \tag{62} $$

where P and R are parameters decided by the geometry. The specific values are: infinite slab, $\frac{1}{2}$ and $\frac{1}{8}$; infinite cylinder, $\frac{1}{4}$ and $\frac{1}{16}$; sphere $\frac{1}{6}$ and $\frac{1}{24}$.

The equation was applied to brick-shaped geometries by use of a chart (Ede, 1949; Charm, 1978; Heldman and Singh, 1981) which gives P and R as functions of ratios of major dimensions.

The assumptions used in the derivation of these equations exclude the sensible heat of the food above and below the freezing point altogether. Therefore, its application to cases with significant sensible heat removal has to result in underestimation of freezing times. In spite of this shortcoming, this method had been used to predict freezing times with reasonable accuracy mainly because of its simplicity (Ede, 1949).

Modifications to Plank's equation to improve the prediction accuracy by including the effects of sensible heats above and below the freezing point were reported by Nagaoka et al. (1955), Levy (1958), Cowell (1967), IIR (1972), Cleland (1977), Mascheroni and Calvelo (1982), Hung and Thompson (1983), de Machelis and Calvelo (1983), and Pham (1984). Cleland and Earle (1984) presented a comparison of predictions by three of these modified methods (Hung and Thompson, 1983; de Machelis and Calvelo, 1983; Pham, 1984), a numerical method (Cleland and Earle, 1976), and a simple formula (Cleland and Earle, 1982) with results of 275 freezing experiments. The experiments selected were from Hung and Thompson (1983), de Machelis and Calvelo (1983), and Cleland (1977). These experiments included four food materials and Tylose.

Table 11.3 is a summary of this comparison. This table indicates that the method proposed by Pham (1984) performed better than other methods based on modifications to Plank's equation. The method of Pham (1984) divides the freezing process into three stages: precooling, freezing, and subcooling. The total process time is obtained by summing up the time for these three stages. Equations (63) to (70) summarize this procedure.

Table 11.3 Comparison of Freezing Time Prediction Methods

Source	Finite differences		Cleland and Earle (1982)		Methods Pham (1984)		Hung and Thompson (1983)		de Michelis and Calvelo (1983)	
	Mean	S.D.	Mean	S.D.	Mean	S.D.	Mean	S.D.	Mean	S.D.
Hung and Thompson (1983)										
23 Tylose slabs	-1.3	8.7	-12.7	6.0	-3.8	7.7	1.1	2.9	-4.3	14.1
9 potato slabs	-3.9	4.9	-14.5	4.2	-5.1	3.5	2.9	4.0	-4.2	8.4
9 carp slabs	3.1	12.9	-12.6	9.0	-2.4	11.1	2.8	4.1	6.1	16.9
9 ground beef slabs	6.6	16.6	-11.8	12.7	-1.8	13.9	2.5	4.8	10.1	18.9
9 lean beef slabs	2.8	11.6	-14.7	6.4	-6.0	8.9	0.8	4.0	7.9	14.7
Cleland (1977)										
43 Tylose slabs	0.0	5.3	1.6	2.8	3.9	4.4	23.1	11.5	-4.1	6.2
6 potato slabs	-0.5	5.1	-0.5	1.8	3.7	2.3	21.9	6.5	-2.6	3.5
6 lean beef slabs	4.8	4.7	2.1	4.0	4.1	4.6	23.8	12.6	5.6	4.4
30 Tylose cylinders	-1.8	5.2	-1.0	3.6	0.8	5.1	15.4	9.7	-6.5	8.2
30 Tylose spheres	-0.3	3.3	1.6	5.2	2.3	5.1	15.8	9.6	-0.5	10.3
72 Tylose bricks	-3.8	5.8	-0.9	5.7	-1.3	4.7	12.5	10.5	-26.1	9.6
de Michelis and Calvelo (1983)										
5 lean beef slabs	-0.1	7.4	0.9	1.9	-0.9	4.9	1.8	9.2	-2.2	10.5
24 lean beef bricks, rods, finite cylinders	8.6	6.3	6.1	10.3	4.1	6.6	11.8	8.3	-11.3	10.8

Source: *Cleland (1985)*.

$$t = \frac{V}{hA} \left[\frac{c_U(T_{IN} - T_{FA})}{\Delta T_{m1}} \left\{ 1 + \frac{Bi_1}{6} \right\} + \frac{\Delta H}{\Delta T_{m2}} \left\{ 1 + \frac{Bi_2}{4} \right\} \right.$$

$$\left. + \frac{c_F(T_{FA} - T_{EA})}{\Delta T_{m3}} \left\{ 1 + \frac{Bi_3}{6} \right\} \right] \tag{63}$$

where

$$\Delta T_{ml} = \frac{[T_{IN} - T_a] - [T_{FA} - T_a]}{\ln \left[\dfrac{T_{IN} - T_a}{T_{FA} - T_a} \right]} \tag{64}$$

$$\Delta T_{m2} = T_{FA} - T_A \tag{65}$$

$$\Delta T_{m3} = \frac{[T_{FA} - T_a] - [T_{EA} - T_a]}{\ln \left[\dfrac{T_{FA} - T_a}{T_{EA} - T_a} \right]} \tag{66}$$

$$T_{EA} = T_E - \frac{[T_E - T_a]}{\left[2 + \dfrac{4}{Bi_2} \right]} \tag{67}$$

$$Bi_1 = \frac{1}{2} \left[\frac{hD}{k_F} + \frac{hD}{k_U} \right] \tag{68}$$

$$Bi_2 = \frac{hD}{k_F} \tag{69}$$

$$Bi_3 = \frac{hD}{k_F} \tag{70}$$

Average freezing temperature T_{FA} in Eqs. (63) to (66) is defined by an integral equation to allow for the gradual phase change process. For Tylose this temperature is 1.5°C below the initial freezing temperature. Pham (1984) presents a numerical example to illustrate the use of this method.

Approximate analytical methods that can predict freezing times with reasonable accuracy such as the four methods compared by Cleland and Earle (1984) are too complicated to be termed simple prediction methods. This is to be expected due to the complex nature of the freezing problem.

Gradual Phase Change: Apparent Specific Heat Formation

In the region below the initial freezing point, heat is stored both in latent and sensible forms. Apparent specific heat is defined as the sum of these two components. The sensible heat component includes the heat required to raise the temperature of ice, water and solids in the food. Apparent specific heat, C, defined in this manner is a strong function of temperature but it is equally applicable in both frozen and unfrozen regions. Therefore, both of these regions can be represented by one governing equation:

$$C(T) = \frac{\partial T}{\partial t} = \text{div } (k\{T\} \cdot \text{grad } T) \tag{71}$$

This governing equation is nonlinear due to the temperature dependence of apparent specific heat and thermal conductivity. Analytical solutions are available only when these temperature dependencies take very simple forms. Therefore numerical solution procedures based on finite difference and finite element methods are sought.

Boundary conditions at the surfaces encountered in practice can be of fixed temperature, fixed heat flux, or convective type. All three of these boundary conditions can be treated in this method.

$$\text{Fixed temperature type} \qquad T = T_s \tag{72}$$

$$\text{Fixed heat flux type} \qquad -k \, \frac{\partial T}{\partial x} = q_s \tag{73}$$

$$\text{Convective type} \qquad -k \, \frac{\partial T}{\partial x} = h(T_a - T_s) \tag{74}$$

The condition of symmetry at the center can be used as the other boundary condition (Eq. 75). Uniform temperature throughout the food is used as the initial condition (Eq. 76).

$$\frac{\partial T}{\partial x} = 0 \tag{75}$$

$$T = T_{IN} \tag{76}$$

Cleland (1977) studied a number of finite difference methods and selected Lee's (1966) three-time level finite difference scheme over the others such as explicit and Crank-Nicholson schemes. Lee's scheme results in a linear system of difference equations which can be solved using a standard elimination procedure.

Bonacina and Comini (1971) used this method to simulate the freezing process of a slab of Tylose under fixed temperature boundary conditions. The apparent specific heat curve was modified by replacing the sharp peak with a flat top having the same area underneath to prevent peak jumping.

Cleland and Earle (1977) extended the use of this method to freezing under convective boundary conditions. This application required modifications to the nodal configuration and the updating scheme to reduce oscillations. Cleland (1977) used this method to predict the freezing times of slabs, cylinders, spheres, and parallelepipeds of Tylose, minced beef, and mashed potatoes within 9% of the experimental values with 95% confidence. Cleland and Earle (1984) compared predictions by this method with predictions by modified forms of Plank's equations. Summary of this comparison presented in Table 11.3 indicate its superiority to other methods. An added advantage of this method is that it provides a complete temperature history of the process at a number of nodes instead of a "freezing time." Figure 11.17 is an illustration of a predicted history compared to an experimental record.

A finite element method based on apparent specific heat formulation was used to simulate freezing of two-dimensional foods under fixed temperature, fixed heat flux, and convective type boundary conditions by Rebellato et al. (1978) and

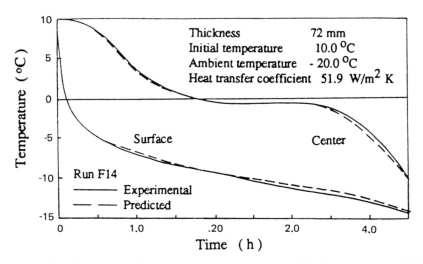

Figure 11.17 Temperature history during freezing of a tylose slab. (*From Cleland, 1977.*)

Purwadaria and Heldman (1982). Hayakawa et al. (1983b) used a similar method under combined convection, radiation, and evaporation type boundary conditions. Cleland (1985) also used a finite element method based on apparent specific heat formulation under convective boundary conditions.

Gradual Phase Change: Enthalpy Formulation

Formulation of the heat conduction with phase change problem using enthalpy as the primary dependent variable and temperature as the secondary dependent variable also allows one governing equation to be applicable for both phases:

$$\frac{\partial H}{\partial t} = \text{div } (k\{H\} \cdot \text{grad } T\{H\}) \tag{77}$$

Boundary conditions and initial conditions expressed by Eqs. (72) to (76) for apparent specific heat can also be used with enthalpy formulation. The temperature dependence of enthalpy is required to relate primary and secondary dependent variables.

The temperature dependence of enthalpy exhibits only a weak discontinuity at the initial freezing point compared to the temperature dependence of apparent specific heat, which has a strong discontinuity. This feature minimizes the undesirable phenomena such as "peak jumping" and "stable oscillations" in the numerical schemes based on the enthalpy formulation.

An explicit numerical method based on the enthalpy formulation was used to predict freezing and thawing times of foods of one-dimensional geometries under convective boundary conditions by Mannapperuma and Singh (1988). They developed a concept of area and volume factors to enable use of the same updating scheme in all three coordinate systems. A scheme to calculate the largest time increment that ensures stability was incorporated to improve the computational speed. A computer program using this method with a graphic output of the temperature profiles was made available for educational use. Figure 11.18 illustrates a graphic output of the program.

The application of the method was extended to two-and three-dimensional geometries and to fixed temperature and fixed heat flux type boundary conditions by the same authors (Mannapperuma and Singh, 1989). Accuracy of prediction was assessed by comparison with a large collection of experimental data. Table 11.14 is a summary of this comparison. A computer program package utilizing this prediction method is made available by the authors. Evans (1989) used an extended version of this program to predict freezing under boiling boundary conditions.

Pham (1985) proposed a three-time level finite difference scheme based on enthalpy method, which is unconditionally stable. Pham (1986) reported the use

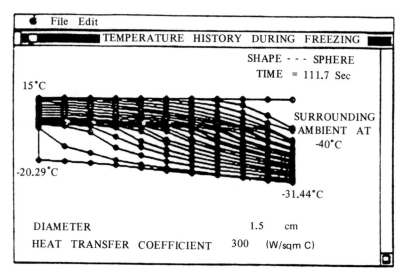

Figure 11.18 Temperature profiles during freezing of a tylose sphere. (*From Mannapperuma and Singh, 1988a.*)

of a lumped capacitance finite element scheme with explicit and three-time level enthalpy formulation. These methods reported to be superior to distributed capacitance schemes, appear to have a great deal of potential for practical application.

DEVELOPMENTS IN THE FOOD-FREEZING EQUIPMENT

There are three major types of freezing equipment used in the food industry, namely air blast, indirect contact, and direct-contact freezers. Further classifications of each of these freezing systems (as shown in Fig. 11.19) and their salient features are discussed in this section.

Air Blast Freezing

Air blast freezers use cold air (−18°C to −40°C) as the heat transfer medium. The commonly used systems are batch or stationary freezing tunnels, and in-line freezers that include straight-belt freezers, large-capacity carton freezers, and spiral belt freezers. In the early days of the freezing industry, the freezing process was

Table 11.4 Comparison of Freezing and Thawing Times Predicted by Enthalpy Formulation

Source	Food material	Geometry	Process[a]	No. of runs	Percentage error of prediction			
					Average	S.D.	Max	Min
Cleland (1977)	Tylose	Infinite slab	F	43	0.1	5.7	8.3	-13.9
		Infinite cylinder	F	30	-4.6	4.4	1.6	-14.0
		Sphere	F	30	-2.3	3.7	6.0	-10.0
		Brick	F	72	-0.8	4.6	9.9	-9.3
Cleland (1985)	Tylose	Infinite slab	T	35	-3.9	3.7	4.5	-9.3
		Infinite cylinder	T	34	-4.3	2.4	0.9	-8.3
		Sphere	T	35	-2.4	5.7	7.3	-11.5
		Brick	T	68	-3.2	3.4	4.2	-9.0
Cleland (1977)	Minced lean beef	Infinite slab	F	6	3.8	5.6	10.0	-3.9
	Mashed potato	Infinite slab	F	6	0.7	3.2	6.3	-2.2
Cleland (1985)	Minced lean beef	Infinite slab	T	6	-5.7	5.8	1.0	-13.6
		Brick	T	4	2.0	7.8	12.7	-5.2

Hung and Thompson (1983)	Tylose	Infinite slab	F	23	-3.0	8.6	20.6	-18.6
	Lean beef	Infinite slab	F	9	2.8	11.8	22.1	-10.9
	Mashed potato	Infinite slab	F	9	-3.2	4.4	2.1	-9.4
	Ground beef	Infinite slab	F	9	1.3	13.8	18.6	-24.3
	Carp fish	Infinite slab	F	9	-1.5	13.2	18.5	-23.6
de Michelis and Calvelo (1983)	Lean beef	Infinite slab	F	5	-7.6	7.3	4.5	-15.0
	Lean beef	Infinite rod	F	3	-10.1	4.4	-6.3	-14.9
	Lean beef	Finite cylinder	F	4	7.1	8.6	18.1	-0.5
	Lean beef	Brick	F	17	3.6	5.9	12.4	-9.5
Hayakawa et al. (1983a)	Lean beef	Infinite rod	F	6	11.2	8.1	24.6	1.7
	Tylose	Finite cylinder	F	6	-5.6	9.1	6.6	-15.3
Lescano (1973)	Cod fish	Infinite slab	F	8	-2.9	8.5	14.4	-13.3
Wilson (1986)	Peas	Sphere	F	4	7.2	18.9	31.7	-14.1

[a]F—freezing; T—thawing.
Source: Mannapperuma and Singh (1989).

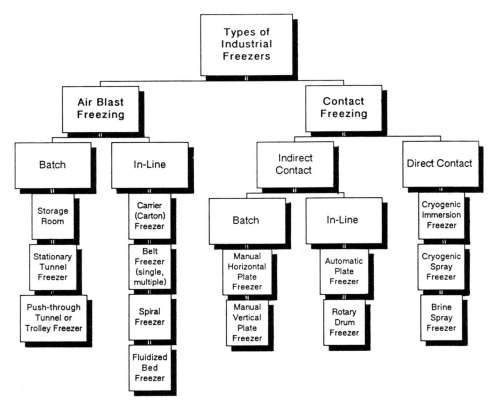

Figure 11.19 Classification of food freezing systems.

accomplished simply by placing packaged foods in rooms with extra-refrigeration capacity. Initial improvements included freezing batchwise on racks or trolleys. Further development of in-line continuous freezing systems allowed integration of the freezing process into the production line. The in-line continuous concept has proved beneficial both from economic and food quality points of view (Ricci, 1988).

When a storage room is used for freezing, one can expect a very slow rate of freezing. Heat transfer is retarded since low air velocities result in a thick boundary layer around the product being frozen. This practice can result in adversely affecting the food quality. The slow rate of freezing encourages formation of large ice crystals; the heat added to the room from the unfrozen product may warm the already frozen and adjacently stored product. In addition, the vapor pressure gradient can lead to aroma migration between the unfrozen and the frozen product leading to undesirable flavor changes.

Increasing air velocities around the product reduces the thickness of the boundary layer. This improves heat transfer. The stationary tunnel is the simplest type of freezing equipment. Trays with food are arranged on racks within the tunnel and they are exposed to high velocity air. The labor requirements for this type of batch system are high. A push-through tunnel system that uses moving carts with racks helps in decreasing the labor costs.

The in-line continuous freezing systems originated with a mesh belt conveyor used in a blast room. Although initially continuous product flow was achieved, the distribution of air flow was nonuniform, resulting in poor heat transfer. Further developments included using vertical air flow through the product, thus creating a good contact between the product and the air. In some freezing systems, a two-belt design is used. The first belt allows precooling and the formation of a crust on the product surface before the product is transferred to the second belt for completion of the freezing process. It is necessary that a uniform product spread be achieved on the belt, otherwise "air channeling" can lead to poor heat transfer.

In air blast freezers, floor space can be reduced by stacking belts above each other, thus forming single feed/single discharge multipass systems (usually three passes) or multiple, single-pass systems (multiple infeeds and discharges) stacked one atop the other (Ricci, 1988). Another belt freezing system uses a spiral belt design. The spiral belt freezer involves an endless belt stacked cylindrically one tier on top of the other. This type of freezer is particularly suited for products that must be handled carefully and those that have a relatively long freezing time. The spiral design permits a compact freezer for such products. Recent developments in spiral freezers involve a self-stacking, self-enclosing belt leading to a more compact design and a better control of air flow.

When vertical air velocities in the range of 2–10 m/sec are used in the belt freezer, the product is fluidized. The fluidization principle has been effectively used in the individual quick freezing (IQF) systems since 1960s. The IQF systems are now one of the most popular methods of freezing particulate foods such as berries and diced vegetables. The air velocity used in the fluidized bed is a function of the particulate shape, dimensions, and weight. Most of the published studies on fluidized beds are for small nonfood particles (diameter less than 2 mm). Design equations such as Egrun's equation (Kunii and Levenspiel, 1969), while suitable to calculate minimum fluidization velocity, do not give realistic results for food particulates with diameters ranging from 5 to 25 mm.

A graphic description for a range of air velocities for fluidization was presented by Persson (in Kramer and Wani, 1967). He suggested that stable fluidization was obtained for Froude numbers between 65 and 170 as shown in Fig. 11.20. In a recent study on fluidized bed freezing of foods, Khairullah (1989) showed that the Nusselt number increases linearly with the Reynolds number in a fixed bed freezer

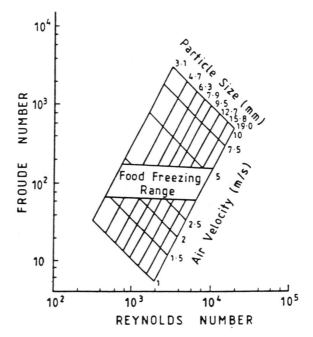

Figure 11.20   Range of food freezing in fluidized bed freezers. (*From Holdsworth, 1987.*)

until the fluidization velocity is obtained. In the fluidized range of air velocities, the Nusselt number remained unchanged.

In a theoretical study, Reynoso and Calvelo (1985) compared the production rates of a fluidized bed freezer with fixed bed freezer of the same dimensions. For green peas, they found that the fluidized bed freezer had a 20–30% higher production capacity than the fixed bed freezer.

Air blast freezers must be defrosted at regular intervals, often every 4 to 8 hours. A recent development uses high velocity jets of dry air to mechanically blast the frost from the fins. This improvement increases the operating time of the freezer between defrosting (Ricci, 1988).

Indirect-Contact Freezing

In an indirect-contact freezer, the food product is separated from the refrigerant with a metal plate. The heat transfer occurs mainly due to conduction. This type of freezing system is also classified as batch and in-line as shown in Fig. 11.19.

Plate freezers are the most common type of contact freezers. A metal plate separates the product from the refrigerant. The food product in a slab shape is pressed between the plates; a good contact assures rapid rate of heat transfer. Normally the product thickness is kept to a maximum of 6 cm, since larger thicknesses result in poor heat transfer. Void spaces in a packaged food can lead to reduced heat transfer. Both horizontal and vertical plate freezers are available, and operate manually. Continuous plate freezers are also available that offer higher production capacity for a compact size. The plate freezers are energy efficient since no fans are required. Package bulging is avoided since the metal plates provide pressure during the freezing process. Recent developments include a specialized contact freezer called PELLoFREEZE, where pumpable products such as purees, soups, and sauces are frozen into portion-controlled individual quick frozen pellets (Ricci, 1988).

Direct-Contact Freezers

Liquid nitrogen and liquid carbon dioxide are the most commonly used cryogenic liquids. The freezers are usually straight tunnels; however, other designs such as spiral conveyors, liquid immersion, and batch-type cabinets are also available. A straight-through single-belt, in-line tunnel is the most common type of cryogenic freezer. Liquid nitrogen is introduced at the exit end and sprayed directly on the product. The vapors are used to precool and initially freeze the product. While the low temperature of the refrigerant (liquid nitrogen boils at $-196°C$) allows very fast freezing, this method can also lead to cracking of the product due to internal stresses within the product. The use of liquid nitrogen is an expensive proposition, since 0.5–1 kg of nitrogen is used per kg of product. However, the initial capital cost of such equipment is low, and for small production rates the cryogenic system may be justified on an economic basis. Liquid freon (R-12) has also been used in the past; however, it is now being phased out due to the long-term effects of fluorocarbons on the ozone layer. Other commercial immersion and spray-freezing systems include the use of propylene glycol and calcium chloride.

CONCLUSIONS

While the knowlege of food properties and heat transfer occurring during freezing is essential for the design of food freezers, there is a considerable lack of reliable information. A quantitative understanding of the ice crystallization process is necessary to achieve desired quality characteristics in the frozen product. In the future, computer-aided simulations based on a mathematical description of the freezing process are expected to play an important role in the design and operation of food-freezing systems.

NOMENCLATURE

A	area (m^2)
B	regression coefficient
Bi	Biot number
c	specific heat (J/kg K)
C	apparent specific heat (J/m^3K)
D	diameter or major dimension (m)
e	emissivity
F	function defined by Eq. (9)
F'	function defined by Eq. (13)
g	convective mass transfer coefficient (kg/m^2)
H	specific enthalpy (J/kg)
H	specific enthalpy (J/m^3)
h	heat transfer coefficient (W/m^2K)
J	a conversion factor (4184 J/kcal)
k	thermal conductivity (W/m K)
M	molecular weight
p	pressure (Pa)
q	heat flux (W/m^2)
R	ideal gas constant for water vapor (J/kg K)
S	shape factor
T	temperature (K)
V	volume fraction
X	mole fraction
Z	location of the freezing front (m)
x	distance measured along x axis (m)
Y	mass fraction
α	thermal diffusivity (m^2/s)
β	a parameter in Eq. (56)
ε	porosity
λ	specific enthalpy of solid to liquid phase change of water (J/kg)
ρ	density (kg/m^3)
μ	specific humidity (kg water vapor/kg dry air)
ϕ	distribution factor defined by Eq. (37)
Ψ	a parameter defined by Eq. (29)
σ	Stefan-Boltzmann constant (W/m^2K^4)
ζ	moisture content (fraction-dry basis)

Suffixes

A   of unfreezable water
a   of ambient medium
B   of the food solids
c   of the continuous phase
D   at the datum temperature
d   of the dispersed phase
E   at the final temperature at the center
EA   at the average final temperature
F   of the frozen food
FA   at the average freezing temperature
G   of the soluble food solids
I   of ice
IN   initially
i   of the ith component of the food
j   of the jth component of soluble solids
m   mean
o   at the center at the end of freezing
p   at the freezing point of pure water
s   at the surface
U   of the unfrozen food
v   of water vapor
vo   of water vapor over pure water
W   of water in the food
WZ   of water in the food at the initial freezing point of the food
Z   at the initial freezing point of the food
⊥   perpendicular to fibers
| | parallel to fibers

Symbol { } is used to denote a function of the enclosed parameters. Symbols ( ) and [ ] are used to denote groups with or without implied multiplication. Σdenotes summation of the suffixed parameters within over the full range of the suffix.

REFERENCES

Arce, J. A., and Sweat, V. E., 1980. Survey of published heat transfer coefficients encountered in food processes. *ASHRAE Transactions,* 86(2): 235.

Astrom, S., and Londahl, G. 1969. Air blast freezing versus ultra rapid freezing. A comparison of freezing results with some various vegetables and prepared foods. *Bull. Int. Inst. Refrig.,* Annexe 6: 121.

Bakal, A., and Hayakawa, K. I. 1973. Heat transfer in freezing and thawing of foods. *Adv. Food Res.* 20: 218.

Bonacina, C., and Comini, G. 1971. On a numerical method for the solution of the unsteady state heat conduction equation with temperature dependent parameters. *Proceedings of the XIIIth International Congress of Refrigeration* 2: 329.

Charm, S. E., 1978. *Fundamentals of Food Engineering,* Ch. 6. AVI Publishing Co., Westport, CT.

Charoenien, S., and Reid, D. S. 1988. The use of DSC to study the kinetics of heterogeneous and homogeneous nucleation of ice in aqueous systems. Paper presented to the Annual Meeting of North American Thermal Analysis Society Meeting, Florida, Oct.

Chen, C. S. 1985. Thermodynamic analysis of freezing and thawing of foods: Enthalpy and apparent specific heat. *J. Food Sci.* 50(4): 1158.

Choi, Y., and Okos, M. R. 1984. Effect of temperature and composition on the thermal properties of foods. In *Food Engineering and Process Applications.* 1: 93–101. Elsevier Applied Science Publishers, New York.

Cleland, A. C. 1977. Heat transfer during freezing of foods and predictions of freezing times, Ph.D. thesis, Massey University, New Zealand.

Cleland, A. C., and Earle, R. L. 1976. A new method for the prediction of surface heat transfer coefficients in freezing. *Bull. Int. Inst. Refrig.* Annexe 1: 361.

Cleland, A. C., and Earle, R. L. 1977. The third kind of boundary condition in numerical freezing calculations. *Int. J. Heat & Mass Transfer* 20(10): 1029.

Cleland, A. C., and Earle, R. L. 1984. Assessment of freezing time prediction methods. *J. Food Sci.* 49(4): 1034.

Cleland, A. C., and Earle, R. L. 1982. Freezing times predictions for foods-a simplified procedure. *Int. J. Refrig.* 5: 134.

Cleland, D. J. 1985. Prediction of freezing and thawing times for foods. Ph.D. thesis, Massey University, New Zealand.

Cowell, N. D. 1967. The calculation of food freezing times. *Proceedings of the XIIth International Congress of Refrigeration* 11: 667.

de Michelis, A., and Calvelo, A. 1983. Freezing time predictions for brick and cylindrical shaped foods. *J. Food Sci.* 48(3): 909.

Dickerson, R. W., Jr. 1969. Thermal properties of food. In *The Freezing Preservation of Foods,* 4th ed., Vol. 2: 27.(Ed). Tressler, D. K., Van Arsdel, W. B., and Copley, M. J. AVI Publishing Co., Westport, CT.

Dinglinger, G. 1969. Problem of heat transfer when spray-freezing with liquid nitrogen. *Bull. Int. Inst. Refrig.* Annexe 6: 141.

Ede, A. J. 1949. The calculation of freezing and thawing of foodstuffs. *Modern Refrigeration* 52: 52.

Evans, K. E. 1989. Freezing of foods under boiling boundary conditions, M.S. thesis, University of California, Davis.

Fennema, O. R., Powrie, W. D., and Marth, E. H. 1973. *Low Temperature Preservation of Foods*, p. 127. Marcel Dekker, Inc., New York.

Flynn, T. M., Draper, J. W., and Roos, J. J. 1961. The nucleate and film boiling curve of liquid nitrogen at one atmosphere. *Adv. Cryogenic Engr.* 7: 539.

Hayakawa, K., Nonino, C., and Succar, J. 1983a. Two dimensional heat conduction in food undergoing freezing: Predicting freezing time of rectangular or finitely cylindrical food. *Journal of Food Science,* 48(6): 1841.

Hayakawa, K., Nonino, C., Succar, J., Comini, G., and Del Giudice, S. 1983b. Two dimensional heat conduction in food undergoing freezing: Development of computerized model. *J. Food Sci.* 48(6): 1849.

Heldman, D. R. 1974. Predicting the relationship between unfrozen water fraction and temperature during food freezing using freezing point depression. *Trans. of the ASAE* 17(1): 63.

Heldman, D. R. 1982. Food properties during freezing. *Food Technol.* 36(2): 92.

Heldman, D. R., and Singh, R. P. 1981. *Food Process Engineering,* 2nd ed., Ch. 3 and 4. AVI Publishing Co., Westport, CT.

Holdsworth, S. D. 1987. Physical and engineering aspects of food freezing. In *Developments in Food Preservation.* Vol. 4: 153. (Ed.) Thorne, S. Elsevier Applied Science.

Hung, Y. C., and Thompson, D. R. 1983. Freezing time prediction for slab shape foodstuffs by an improved analytical method. *J. Food Sci.* 48(2): 555.

International Institute of Refrigeration. 1970. Symposium on Weight Losses in Foodstuffs. *Bull. Int. Inst. Refrig.,* Annexe 3.

International Institute of Refrigeration. 1972. *Recommendations for the Processing and Handling of Frozen Foods,* 2nd ed., p. 24. Paris.

Jason, A. C., and Long, R. A. K. 1955. The specific heat and thermal conductivity of fish muscle. *IXth International Congress of Refrigeration.* 2: 160.

Kaess, G., and Weidemann, J. F. 1969. Freezer burn of animal tissue. 7. Temperature influence on development of freezer burn in liver and muscle tissue. *J. Food Sci.* 34: 394.

Keey, R. B. 1972. *Drying Principles and Practice.* Pergamon Press, New York.

Khairullah, A. 1988. Optimization of fixed and fluidized bed freezing process. M.S. thesis, University of California, Davis, CA.

Kopelman, I. J. 1966. Transient heat transfer and thermal properties in food systems. Ph.D. thesis, Michigan State University, East Lansing, MI.

Kramer, A., and Wani, K. 1967. *Proc. XII International Congress Refrigeration,* Vol. 12: 677.

Kunii, D., and Levenspiel, O. 1969. *Fluidization Engineering.* John Wiley Inc., New York.

Lees, M. 1966. A linear three level difference scheme for quasilinear parabolic equations. *Mathematics of Computation.* 20: 516.

Lentz, C. P. 1961. Thermal conductivity of meats, fats, gelatin, gel and ice. *Food Technology*, 15(5): 243.

Lescano, C. E. 1973. Predicting freezing curves in codfish fillets using the ideal binary solution assumption. M.S. thesis, Michigan State University, East Lansing, MI.

Levy, F. L. 1958. Calculating the freezing time of fish in air blast freezers. *J. Refrigeration* (March/April): 55.

Luikov, A. V. 1968. *Analytical Heat Diffusion Theory.* Ch. 11. Academic Press, New York.

Mannapperuma, J. D., and Singh, R. P. 1988a. Prediction of freezing and thawing times of foods using a numerical method based on enthalpy formulation. *J. Food Sci.* 53(2): 626.

Mannapperuma, J. D. and Singh, R. P. 1988b. Thawing of foods in humid air. *International Journal of Refrigeration*, 11(3): 113.

Mannapperuma, J. D., and Singh, R. P. 1989. A computer-aided method for the prediction of properties and freezing/thawing times of foods. *J. Food Engr.* (In press).

Mascheroni, R. H., and Calvelo, A. 1982. A simplified model for freezing time calculation in foods. *J. Food Sci.* 47(4): 1201.

Miles, C. A., van Beek, G., and Veerkamp, C. H. 1983. Calculation of thermophysical properties of foods. In *Physical Properties of Foods,* (Ed.) Jowitt, R., Escher, F., Hallstrom, B., Meffert, H. F. Th., Spiess, W. E. L. and Vos, G. Applied Science Publishers, London, and New York.

Moreno, J. 1979. A posteriori determination of the mean storage time as a random function of the natural weight losses during cold storage of perishable products. *XVth International Congress of Refrigeration* 3: 1123.

Murakami, E. G., and Okos, M. R. 1988. Measurement and prediction of thermal properties of foods. In *Food Properties and Computer-Aided Engineering of Food Processing Systems*, (Ed.) Singh, R. P., and Medina, A. Kluewer Publishing Co., Amsterdam, The Netherlands.

Nagoaka, J., Takagi, S., and Hotani, S. 1955. Experiments on the freezing of fish in an air blast freezer. *Proceedings of the IXth International Congress of Refrigeration* 4: 105.

Norwig, J. F. and Thompson, D. R. 1984. Review of dehydration during freezing. *Transactions of the ASAE* 27(5): 1619.

Pham, Q. T. 1984. Extension to Planck's equation for predicting freezing times of foodstuffs of simple shapes. *International J. Refrigeration* 7(6): 377.

Pham, Q. T. 1985. A fast unconditionally stable finite difference scheme for conduction heat transfer with phase change. *International J. Heat & Mass Transfer* 28: 2079.

Pham, Q. T. 1986. The use of lumped capacitance in the finite element solution of heat conduction problems with phase change. *International J. Heat & Mass Transfer* 29: 285.

Plank, R. 1913. *Z. ges. Kalte-Ind.* Bd. 20. Quoted in Ede, A. J. 1949. The calculation of freezing and thawing of foodstuffs. *Modern Refrigeration* 52: 52.

Plank, R. 1941. *Z. ges. Kalte-Ind.* Beih. Reihe 3, H. 10. Quoted in Ede, A. J. 1949. The calculation of freezing and thawing of foodstuffs. *Modern Refrigeration* 52: 52.

Polley, S. L., Snyder, O. P., and Kotnour, P. 1980. A compilation of thermal properties of foods. *Food Technol.* 34(11): 76.

Purwadaria, H. K., and Heldmann, D. R. 1982. A finite element model for the prediction of freezing rates in food products with anomalous shapes. *Transactions of the ASAE* 25: 827.

Ramaswamy, H. S., and Tung, M. A. 1981. Thermophysical properties of apples in relation to freezing. *J. Food Sci.* 46(3): 724.

Ramaswamy, H. S., and Tung, M. A. 1984. A review on predicting freezing times of foods. *J. Food Process Engr.* 7(3): 169.

Rasmussen, D. H., and Mackenzie, A. P. 1972. In Jellinek, H. H. J. *Water Structure in Water-Polymer Interface* pp. 131–140.

Rebellato, L., Del Giudice, S., and Comini, G. 1978. Finite element analysis of freezing processes in foodstuffs. *J. Food Sci.* 43(1): 239.

Reid, D. S. 1983. Fundamental physicochemical aspects of freezing. *Food Technology*, 37(4): 110.

Reynoso, R. O. and Calvelo, A. 1985. Comparison between fixed and fluidized bed continuous pea freezers. *International Journal of Refrigeration* 8(2): 109.

Ricci, M. J. 1988. Food freezing systems and equipment. Paper presented at the American Institute of Baking, Batter and Breading Technology Seminar, Manhattan, Kansas.

Riedel, L. 1951. The refrigerating effect required to freeze fruits and vegetables. *Refrigeration Engineering* 59(7): 670.

Riedel, L. 1956. Calorimetric investigation of the freezing of fish meat. *Kaltetechnik* 8: 374.

Schwartzberg, H. G. 1976. Effective heat capacities for the freezing and thawing of food. *J. Food Sci.* 41(1): 152.

Smith, J. G., Ede, A. J., and Gane, R. 1952. The thermal conductivity of frozen foodstuffs. *Modern Refrigeration* 55: 254.

Staph, H. E., and Woolrich, W. R. 1951. Specific and latent heats of foods in the freezing range. *Refrigeration Engineering* 59(11): 1086.

Thompson, D. R., Hung, Y. C., and Norwig, J. F. 1984. The influence of raw material properties on the freezing of sweet corn. In *Engineering and Food,* Vol.

1, (Ed.) McKenna, B. M. Elsevier Applied Science Publishers, London.

Watt, B. K., and Merrill, A. L. 1975. *Composition of Foods.* Agricultural Handbook No. 8, U.S. Dept. of Agriculture, Washington, D.C.

Wilson, H. E. 1986. Numerical simulation of individual quick freezing of spherical foods. M.S. Thesis. University of California, Davis, CA.

12

Developments in Microwave Food Processing

Richard E. Mudgett

University of Massachusetts
Amherst, Massachusetts

INTRODUCTION

Home microwave oven penetration in the United States is providing an unusually favorable climate in the food industry for research in food formulation and packaging to develop new or improved microwavable products. While the use of large-scale microwave processes is increasing rapidly, recent improvements in the design of high-powered microwave ovens, reduced equipment manufacturing costs, and trends in electrical energy costs offer a significant potential for developing new and improved industrial microwave processes.

Fallout from this consumer-driven research for home products will benefit industrial development, because microwave heating effects at small or large scales are related to electrical and thermal properties that are determined by product composition, structure, temperature, and the processing frequency. The current strategy for consumer product development is based on consumer appeal and convenience in terms of product sales, while the strategy for industrial development is based on product quality and manufacturing costs. The major challenge in either case is to optimize product heating rates and temperature uniformity with respect to the particular processing objective, a goal of no less challenge in conventional food processing.

Home microwave ovens are operated in batch mode at a single frequency, i.c., 2450 MHz, while industrial ovens are operated at 915 or 2450 MHz in batch and continuous modes. Industrial ovens often combine microwave and conventional heat sources to provide surface temperatures needed for browning and crisping, to reduce surface microbial counts, or to accelerate moisture removal. Despite their obvious advantages, however, microwave convection ovens are not commonly used by consumers, due to their higher costs and complexity.

While from the above comments it may seem that consumer product development and industrial process development require completely different approaches, the basic approach in meeting either objective is much the same and requires basic understanding of the electrical and thermal interactions affecting heat and mass transfer between the food product and the oven. These interactions are defined in terms of the electrical and thermal properties of foods and are very much affected by oven design. While many food processors employ empirical methods in product and process development, there is a growing recognition that models based on electrophysical and electrochemical concepts combined with numerical methods for transient analysis of heat and mass transfer can be of significant value in product or process development. In this context, the sections that follow review the fundamental basis of microwave interactions with foods, their electrical and thermal properties, some microwave modeling concepts, the current state of development for both home and industrial microwave ovens, and, finally, possible future developments in microwave food processing.

FOOD PROPERTIES

Microwave Interactions with Foods

Although microwave interactions with foods depend heavily on salt and moisture content, it is generally useful to consider interactions for low moisture foods separately from those of intermediate or high moisture ones, since their mechanisms of interaction may be quite different. At the molecular level, microwaves are seen to interact with chemical constituents of intermediate and high moisture foods by several mechanisms, as seen in the model of Fig. 12.1 (Mudgett, 1974). Proteins, lipids, and regions near colloidal surfaces are relatively inert and do not interact significantly with microwaves. Proteins are partly soluble and partly insoluble, with ionizable surface regions that may tightly bind water or salts to give rise to zeta potentials and double layer effects associated with free surface charge (Mudgett and Westphal, 1989). Lipids are also hydrophobic except

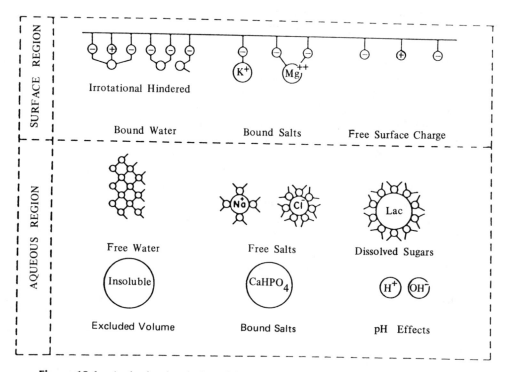

Figure 12.1 A physicochemical model for microwave interactions with foods. (*From Mudgett, 1974.*)

for ionizable carboxyl groups of fatty acid residues and do not interact appreciably with microwaves in the presence of aqueous ions, because these are much more interactive and selectively absorb microwave energy. There are some direct interactions with colloidal food constituents at frequencies below the microwave region, but the bulk aqueous regions of foods are the major sites for microwave interaction at intermediate and high moistures. However, there are probably indirect interactions between hydrocolloids and water also affecting microwave behavior at microwave frequencies in ways that are not yet clear.

The principal mechanisms of interaction in multilayer and capillary regions of moist foods are: (1) rotation of water molecules, i.e., dipoles with centers of positive charge on hydrogen atoms and negative charge on oxygen atoms, and (2) conductive migration of ions, i.e., dissolved salts, in an electrical field that reverses its direction billions of times each second, to dissipate energy. The rotation of tightly bound water dipoles in monolayer regions is sterically hindered in rotation, as seen in Fig. 12.1. Tightly bound water also represents a small fraction of the total water molecules in intermediate and high moisture products based on typical bound water levels in foods ranging from 5–10 grams of bound water/100 grams of dry solids (Karel, 1975). Some of the free water molecules are also bound by the dissolved salts in hydration sheaths. At lower moisture contents, the salts become more concentrated and precipitate as their concentrations exceed saturation levels, thus limiting ionic conductivities.

Microwaves also interact with alcohols and alcoholic moieties of dissolved sugars and polysaccharides by dipole rotation. Alcohols, such as ethanol, are pure polar solvents whose molecules are hydrogen bonded in the liquid state and which are freely miscible with water. Sugars and noncellulosic polysaccharides are more or less soluble in water to an extent that may depend on the number of free aldehydic and hydroxyl groups in solution available for hydrogen bonding. Low alcohol or sugar levels in aqueous mixtures do not modify the interactions of microwaves with water and dissolved ions to any great extent, but at higher concentrations do alter the frequency response of water in such mixtures. For example, frequencies at which microwaves show maximum interaction in ethanol–water mixtures are between those of pure water and ethanol (Buck, 1965). This is attributed to hydroxyl–water interactions that stabilize hydrogen bonding of free water. Similar effects have been shown for high concentrations of sugars and starch in water, but appear to be negligible at low concentrations (Roebuck et al., 1972).

Surprisingly, although microwaves do not appear to interact with lipids and colloidal solids in intermediate or high moisture foods, they interact strongly with these constituents in the absence of moisture as evidenced by calorimetric measurements of olive oil and bone-dry food solids and also by ignition and/or charring of food solids by microwaves following moisture removal. Calorimetric measurements of olive oil and dry food solids show levels of energy absorption that cannot be due to free water and ion activity. While the mechanistic basis for energy absorption in these materials is not clear, microwave interactions are known to result from rotational modes, rather than vibrational, electronic, or nuclear modes of interaction characteristic at lower or higher frequencies, and are related to permanent and induced dipole moments (Pomeranz and Meloan, 1987). Perhaps the major difficulty in accounting for these interactions in a quantitative way is that it is not yet possible to estimate net dipole moments in complex food mixtures.

Electrical Properties of Foods

The electrical properties affecting energy transmission and absorption in foods have been reviewed in detail (Mudgett, 1986a). These properties are tabulated for a wide variety of food measurements in a recent review by Kent (1987). The most fundamental of these properties are the relative dielectric constant and dielectric loss factor, which reflect the ability of an insulating material to store or dissipate energy in an electrical capacitor, respectively. The vector sum of these properties is called the relative permittivity of the material (Von Hippel, 1954a):

$$\kappa^* = \kappa' - j\kappa'' \qquad (1)$$

and their ratio, a measure of the material's ability to generate heat, its loss tangent:

$$\tan \delta = \kappa''/\kappa' \qquad (2)$$

Moist foods are not good insulators, since they contain high concentrations of aqueous ions that selectively absorb microwave energy and dissipate heat due to interactions resulting in "molecular friction," e.g., the release of energy in disrupting hydrogen bonds between water molecules. The dielectric constant and loss factor of most foods depend primarily on their moisture and salt contents and are predicted with reasonable accuracy by electrochemical models that treat the food as a binary mixture of an aqueous ionic solution and food solids with phase properties calculated from the chemical composition and temperature of the product and the processing frequency.

Dielectric properties of the aqueous ionic phase are predicted by the Debye model for pure polar solvents, i.e., water, as modified to reflect depression of the dielectric constant for pure water through ion hydration and elevation of the dielectric loss through ionic conductivity (Collie, et al., 1948; Hasted et al., 1948):

$$\kappa_i' = \frac{(\kappa_s - 2\delta C - \kappa_0)}{1 + (\lambda_s/\lambda)^2} + \kappa_0 \qquad (3)$$

$$\kappa_i' = \frac{(\kappa_s - 2\delta C - \kappa_0)(\lambda_s/\lambda)}{1 + (\lambda_s/\lambda)^2} + \frac{\Lambda C}{1000 \, \omega\varepsilon_0} \qquad (4)$$

Values of the static dielectric constant and critical frequency for water in these equations decrease with temperature, as seen in Table 12.1, and result in decreasing values of both the dielectric constant and loss of water as the temperature is increased. However, equivalent conductivities in Eq. (4) increase with temperature. The dielectric loss of an aqueous ionic solution has a dipole loss term, due to the rotation of water molecules, with negative temperature dependence and a conductivity term, due to the migration of ions, with positive temperature dependence. The net effect of these terms on total dielectric loss of the aqueous ionic phase at any temperature is seen to depend on its dissolved salts concentration. The conductive loss term also increases as the process frequency decreases. The dielectric constant and loss of solid phases in foods are similar to

Table 12.1   Static Dielectric Constant and Critical Wavelengths of Water

Temperature (°C)	Static constant	Critical wavelength
0	88.2	3.34
10	84.2	2.39
20	80.4	1.80
30	76.7	1.39
40	73.1	1.12
50	69.8	0.91
60	66.6	0.76
70	62.1	0.61

Source: Collie et al., 1948.

those of ice, and maybe assumed to be close to $K^* = 3 - j0$ based on literature data (Von Hippel, 1954b) and to be invariant with frequency and temperature.

Predicted dielectric constant and loss values for a distribution of binary phase properties by phase volume fraction and designated as the Distributive model:

$$\kappa_m^* = \kappa_s^* v_s + \kappa_c^* (1 - v_s) \tag{5}$$

are shown in Table 12.2 at frequencies of 915 and 2450 MHz and temperatures of 0–100°C. The moisture contents for each of the products in their native state are similar, but free salt concentrations are known to vary significantly with moisture content. Predicted dielectric constants and losses are shown at lower moisture contents and higher salt concentrations, since measurements of sorbed freeze-dried potatoes show that the salts are concentrated at reduced moistures below saturation levels (Mudgett et al., 1980). Indicated salt concentrations at these moistures are based on the assumption that the salt concentrations are inversely proportional to the product's native moisture content.

Predicted values of the dielectric constant are similar at both frequencies for comparable moistures, with small variations due to water binding by salts. They decrease at higher temperatures and reduced moistures, due to temperature dependence of the static dielectric constant and critical frequency of water and increasing solid contents. Dielectric loss values increase as temperature is increased at 915 MHz and are generally higher at 915 MHz than at 2450 MHz, due to conductive losses. However, loss values at 2450 MHz decrease initially as temperatures are increased due to the negative temperature dependence of the dipole loss and then increase with temperature due to the positive temperature dependence of the conductive loss term in Eq. (4). That is, dipole losses are dominant at low temperatures and conductive losses are dominant at higher temperatures, with temperature crossover points depending on free salt content.

Table 12.2 Distributive Model Values of Dielectric Constant and Loss Factor for a Typical Meat, Fruit, and Vegetable

Product	Moisture (%)	Salts (M)	Temperature (°C)	915 MHz K′	915 MHz K″	2450 MHz K′	2450 MHz K″
Lean beef	75	0.15	0	70.7	19.9	67.6	20.4
			25	63.6	28.8	62.7	17.5
			50	56.3	40.9	56.1	18.7
			75	49.0	49.0	48.9	22.5
			100	41.6	73.0	41.5	28.1
	50	0.23	0	52.3	19.4	50.0	16.7
			25	47.1	29.8	46.4	16.1
			50	41.8	43.2	41.6	18.7
			75	36.3	59.5	36.3	23.5
			100	30.9	78.6	30.9	30.0
	20	0.56	0	24.7	17.4	23.7	10.8
			25	22.3	28.4	22.1	12.9
			50	19.9	41.7	19.8	16.9
			75	17.4	58.8	17.4	22.7
			100	15.0	80.8	15.0	30.6
Potato	80	0.10	0	74.5	16.2	71.2	19.7
			25	67.1	21.5	66.1	15.2
			50	59.4	29.7	59.2	14.7
			75	51.7	39.9	51.6	16.7
			100	43.9	51.9	43.9	20.2
	50	0.16	0	52.7	15.4	50.4	15.3
			25	47.5	22.4	46.9	13.4
			50	42.2	31.9	42.0	14.5
			75	36.8	43.5	36.7	17.6
			100	31.3	57.0	31.3	22.0
	20	0.40	0	25.2	13.7	24.2	9.5
			25	22.8	22.0	22.5	10.6
			50	20.4	32.3	20.3	13.4
			75	17.9	45.0	17.9	17.5
			100	15.4	60.7	15.4	23.1
Apple	85	0.05	0	77.9	11.8	74.5	18.7
			25	70.3	13.1	69.2	12.4
			50	62.3	16.6	62.0	10.0
			75	54.2	21.6	54.1	9.9
			100	46.1	27.6	46.1	11.2
	50	0.09	0	53.1	10.8	50.8	13.7
			25	48.0	14.0	47.3	10.3
			50	42.6	19.0	42.5	9.7
			75	37.2	25.4	37.2	10.8
			100	31.8	32.9	31.8	13.0
	20	0.21	0	25.7	8.7	24.7	7.7
			25	23.4	13.1	23.1	7.3
			50	20.9	18.8	20.9	8.3
			75	18.5	25.7	18.5	10.4
			100	16.0	33.8	16.0	13.4

Table 12.3 Distributive Model Values of Dielectric Constant and Loss
Factor for a Simulated Potato Product with 50% Void Volume

Product	Moisture (%)	Salts (M)	Temperature (°C)	915 MHz K′	915 MHz K″	2450 MHz K′	2450 MHz K″
Potato puff	50	0.16	0	26.8	7.7	25.7	7.6
			25	24.2	11.2	23.9	6.7
			50	21.5	15.9	21.5	7.2
			75	18.8	21.7	18.8	8.8
			100	16.1	28.5	16.1	11.0

The distribution model can also be expanded to predict dielectric constant and loss values for products with significant volume fractions of air by the addition of a term based on the relative permittivity of air ($\kappa^* = 1 - j0$). This is illustrated for a hypothetical potato product of 50% moisture content with an air void volume fraction of 0.5 in Table 12.3. As expected, the predicted values are about one-half those of the potato at 50% moisture content shown in Table 12.2.

Predictions of dielectric constant and loss based on a nonlinear model using a "form factor" to distribute binary phase properties by phase volume fraction obtained from the modified Fricke model:

$$\kappa_m^* = \frac{\kappa_c^*[\kappa_s^*(1 + Xv_s) + \kappa_c^*(1 - v_s)X]}{[\kappa_c^*(X + v_s) + \kappa_s^*(1 - v_s)]} \tag{6}$$

give slightly better argreement with dielectric measurements in the literature for some foods (Mudgett 1986a). For example, a form factor of 1 (spheres) is more typical of fruits and vegetables and a form factor of 2 (cylinders) more typical of meats and fish. However, the equation is not very useful, since the form factor, i.e., the ratio of major to minor axes for particles suspended in a continuous phase, is difficult to define in structural terms for the aqueous and solid phase of most food products.

Predictive models for products with high soluble carbohydrate levels, e.g., sugars and starches, are not presently available, although they would be very useful in formulating and processing microwavable bakery products. However, it is known that the dielectric loss in aqueous mixtures of sugars and starches is synergistic, i.e., the loss in mixture is higher than the loss of either pure component (Roebuck et al., 1972). The synergistic effect in carbohydrate–water mixtures appears to be analogous to the effect seen in ethanol–water mixtures (Buck, 1965), which is attributed to stabilization of hydrogen bonding patterns that results in shifting the critical frequencyof mixtures to an intermediate frequency between the critical frequencies of pure alcohol and water. It seems likely that synergistic loss effects of carbohydrate–water mixtures are similar in nature and result from the stabilization of hydrogen bonding patterns due to interactions between the hydroxyl groups of carbohydrates and water dipoles.

Microwave Energy Transfer

A plane wave in free space striking the surface of a dielectric material at some angle of incidence is partly transmitted and partly reflected back to the energy source, as shown in Fig. 12.2 (Von Hippel, 1954a). The transmitted energy is refrefracted toward the normal plane of incidence, as described by Snell's law:

$$\psi = \sin^{-1}(\eta/\eta_0) \sin \phi \tag{7}$$

and is then absorbed within the material in accordance with Lambert's law:

$$P_z = P_0 \, e^{(-2\alpha z)} \tag{8}$$

For a plane wave in free space, energy transfer is determined by a material's intrinsic impedance:

$$\eta = \eta_0/(\kappa^*)^{1/2} \tag{9}$$

which determines the material's transmission and reflection coefficients, i.e., fraction of incident energy transmitted and reflected at the surface:

$$T = \frac{2\eta \cos \phi}{\eta \cos \phi + \eta_0 \cos \psi} \tag{10}$$

$$R = \frac{\eta \cos \phi + \eta_0 \cos \psi}{\eta \cos \phi + \eta_0 \cos \psi} \tag{11}$$

The reflection of energy at the surface gives rise to standing wave patterns that result from interaction of the incident and reflected waves, with nodes (voltage maxima) and antinodes (voltage minima) whose ratios are defined by the voltage standing wave ratio in a loaded cavity:

$$VSWR = \frac{1 + |R|}{1 - |R|} \tag{12}$$

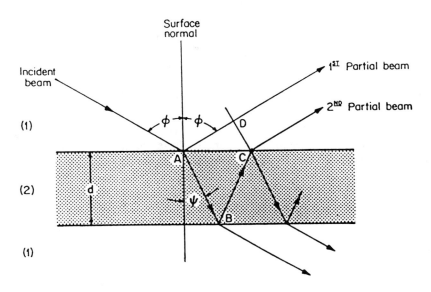

Figure 12.2   Energy transmission, reflection and refraction at food surfaces. (*From Von Hippel, 1954a.*)

Energy transfer from a power source to the food load in a microwave oven is apparently more complicated than the above discussion would suggest, since the amount of energy coupled by the load varies significantly with load volume and dielectric properties, as seen in energy coupling water and olive oil based on temperature rise measurements with time from the equation (Mudgett, 1986b):

$$\frac{dT}{d\theta} = \frac{14.3\ P'}{\rho C_{p?}} \tag{13}$$

Results of these measurements are shown in Fig. 12.3. In the 750 watt home oven used for these measurements, energy transfer efficiencies were calculated from the following equation based on an empirical volumetric coupling coefficient of approximately 0.004 for oil and 0.008 for water:

$$P_0 = P_m(1 - e^{-kv}) \tag{14}$$

Variations in energy transfer efficiencies of water with load volume based on calorimetric measurements for a 625 watt oven show coupling coefficients from

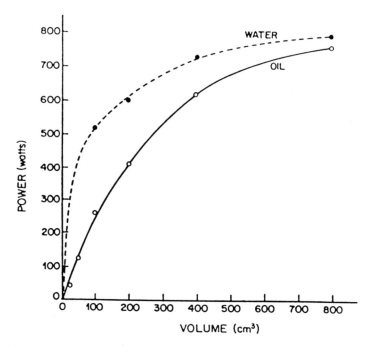

Figure 12.3 Energy transfer efficiencies for oil and water as a function of load volume, (*From Mudgett, 1986a.*)

Eq. (14) of 0.002–0.017 for load values of 2000 and 50 mL, respectively (O'Meara, 1989), suggesting that coupling efficiencies among home ovens from different manufacturers may vary significantly with load volume. This has been a major problem in the development of microwavable products for home oven use because of the differences in heating times required from one manufacturer's oven to another and even among different models by the same manufacturer. It is not a major problem for most industrial ovens, because energy transfer can be maximized by matching the loaded cavity impedance to that of the magnetron based on VSWR measurements (Krieger, 1989).

Microwave Energy Propagation

Microwave energy propagation in foods is determined by the product's complex propagation factor (Von Hippel, 1954a)

$$\gamma = \alpha + j\beta \tag{15}$$

which is also related to its dielectric permittivity:

$$\gamma = j\omega(\varepsilon^*\mu^*)^{1/2} \tag{16}$$

The real part of the propagation factor is the product's attenuation factor, which is obtained from the equation:

$$\alpha = \frac{2\pi}{\lambda_0} \left[\frac{\kappa'[(1 + \tan^2\delta)^{1/2} - 1]}{2} \right]^{1/2} \tag{17}$$

The imaginary part of the propagation factor is the product's phase factor, which is obtained from the equation:

$$\beta = \frac{2\pi}{\lambda_0} \left[\frac{\kappa'[(1 + \tan^2\delta)^{1/2} + 1]}{2} \right]^{1/2} \tag{18}$$

As indicated above, microwave energy transmitted at the surface of a material is refracted in accordance with Snell's law (Eq. 7) and then absorbed in accordance with Lambert's law (Eq. 8) as an exponential function of depth from the surface based on the material's attenuation factor. The depth below the surface at which the field strength in a material, i.e., voltage gradient, is $1/e$ of the field strength at the surface is designated as a material's penetration depth. The penetration depth is the reciprocal of the material's attenuation factor:

$$Z = 1/\alpha \tag{19}$$

The power penetration depth of the material, i.e., the depth below the surface at which the unattenuated power is $1/e$ of the power transmitted at the surface, is:

$$Z_p = 0.5Z \tag{20}$$

It is also useful to define a half-power penetration depth, i.e., the depth from the surface at which half of the transmitted power is not attenuated:

$$Z_L = 0.35Z \tag{21}$$

The fractions of energy absorbed and unabsorbed as a function of reduced sample depth, i.e., depth from the surface of incidence divided by penetration depth, are shown in Fig. 12.4.

Predicted values of reflection coefficients and power penetration depths by the distributive model based on dielectric constants and losses for lean beef, potato, and apple given in Table 12.2 are shown in Table 12.4 for frequencies of 915 and

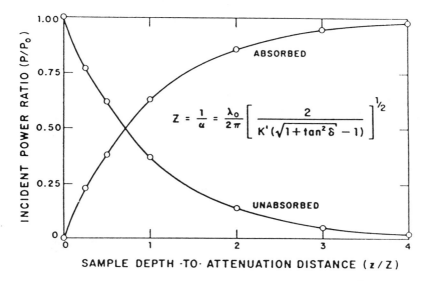

Figure 12.4 Fraction of absorbed and unabsorbed energy as a function of reduced depth from food surfaces. (*From Von Hippel, 1954a.*)

2450 MHz and temperatures of 0–100°C. Reflection coefficients are quite similar over the range of frequencies, moistures, and temperatures, reinforcing the notion that energy transfer efficiencies are primarily based on volumetric coupling coefficients that are nearly constant with frequency and temperature and that are consistent with multiple transmission and reflection of energy at product surfaces, i.e., energy reflection at product surfaces to cavity walls and then back to product surfaces in multiple reflections at decreasing energy levels. Penetration depths are significantly greater at 915 MHz than 2450 MHz for low temperatures, but there is relatively little difference in their values at either frequency for higher temperatures. Parenthetically, the reflection coefficient of dried food solids is about one-half that for products of higher moisture content at each frequency, and their penetration depths at 915 MHz (39 cm) are about twice those at 2450 MHz (15 cm) for all temperatures, based on the intrinsic impedance calculated from the permittivity of food solids.

The imaginary part of the propagation factor, the phase factor, affects the phase velocity (velocity of energy propagation) and space period (wavelength of the propagated energy) within the product, as seen in the equations:

$$v_s = \omega/\beta \qquad (22)$$

$$\lambda = 2\pi/\beta \qquad (23)$$

Table 12.4 Distributive Model Value of Reflection Coefficient and Penetration Depth for a Typical Meat, Fruit, and Vegetable

Product	Moisture (%)	Salts (M)	Temperature (°C)	915 MHz R	915 MHz Z	2450 MHz R	2450 MHz Z
Lean beef	75	0.15	0	0.79	4.5	0.79	1.6
			25	0.79	3.0	0.78	1.8
			50	0.79	2.0	0.77	1.6
			75	0.79	1.5	0.76	1.2
			100	0.80	1.1	0.75	0.9
	50	0.23	0	0.76	3.9	0.76	1.7
			25	0.76	2.5	0.75	1.7
			50	0.77	1.7	0.74	1.4
			75	0.79	1.3	0.74	1.0
			100	0.80	1.0	0.74	0.8
	20	0.56	0	0.69	3.1	0.67	1.8
			25	0.71	2.0	0.67	1.5
			50	0.74	1.4	0.67	1.1
			75	0.77	1.1	0.68	0.8
			100	0.80	0.9	0.71	0.6
Potato	80	0.10	0	0.79	5.6	0.79	1.7
			25	0.79	4.0	0.78	2.1
			50	0.78	2.8	0.77	2.1
			75	0.78	2.0	0.76	1.7
			100	0.78	1.5	0.75	1.3
	50	0.16	0	0.76	5.0	0.76	1.8
			25	0.76	3.3	0.75	2.0
			50	0.76	2.3	0.74	1.8
			75	0.77	1.6	0.73	1.4
			100	0.78	1.3	0.72	1.0
	20	0.40	0	0.69	4.0	0.67	2.1
			25	0.70	2.5	0.67	1.8
			50	0.72	1.7	0.66	1.4
			75	0.75	1.3	0.67	1.0
			100	0.78	1.1	0.68	0.8
Apple	85	0.05	0	0.80	7.8	0.80	1.8
			25	0.79	6.7	0.79	2.6
			50	0.78	5.0	0.78	3.1
			75	0.77	3.6	0.76	2.9
			100	0.76	2.7	0.75	2.4
	50	0.09	0	0.76	7.1	0.76	2.0
			25	0.75	5.2	0.75	2.6
			50	0.74	3.7	0.74	2.6
			75	j0.74	2.6	0.72	2.2
			100	0.74	2.0	0.71	1.7
	20	0.21	0	0.68	6.2	0.67	2.5
			25	0.68	4.0	0.66	2.6
			50	0.68	2.8	0.65	2.2
			75	0.70	2.0	0.64	1.7
			100	0.72	1.6	0.64	1.3

While phase velocities of a traveling wave in foods are from 1/2 to 1/10 those of their velocities in free space, energy propagation at microwave frequencies is instantaneous as compared with conventional heat transfer rates. Propagation wavelengths in foods are also from 1/2 to 1/10 those in free space for high and low moisture products, respectively. This could be of significance in terms of harmonic mode propagation (energy transmission at multiples of the fundamental frequency) because of reduced wavelengths and energy levels at higher harmonic frequencies.

Heat Transfer Considerations

While microwaves generate heat within foods by instantaneous energy transfer from an electromagnetic field, heat is also transferred within the food product to regions of lower temperature by Fourier conduction and between the oven and product surfaces by natural or forced convection. Conventional ovens that use external energy sources may also involve radiative heat transfer. Rates of heat penetration in conventional heating are determined by conditions for internal conduction and surface convection reflected by the thermal diffusivity of the productivity:

$$\alpha_t = k/\rho C_p \tag{24}$$

and the overall heat transfer coefficient at the product's surface, which is primarily determined by air flow conditions, ambient temperature, and relative humidity of the oven cavity. The relative magnitudes of a product's thermal conductivity and overall heat transfer coefficient are reflected in the Biot modulus (Holman, 1976):

$$Bi = hS/k \tag{25}$$

Low values of the modulus indicate that internal heat transfer resistance is low, as compared with surface convection resistance, and vice versa. Another ratio, the Fourier modulus, compares a characteristic product dimension, i.e., radius of a cylinder or sphere and slab half-thickness, with an approximate thermal penetration depth for a given heating time:

$$F_0 = \alpha_t \theta/S^2 \tag{26}$$

These moduli determine the rate of heat penetration by conventional heating and whether heat penetration is conduction- or convection-limited.

Thermal diffusivities of foods are determined primarily by product moisture content, chemical composition, porosity, and temperature and have been predicted with reasonable accuracy for some products by empirical models as a function of moisture content and temperature (Singh, 1986):

$$\alpha_t = 0.088 \times 10^{-6} + (\alpha_{tw} - 0.088 \times 10^{-6})W \tag{27}$$

$$\alpha_t = [0.05376W + 0.00028(T + 273)] \times 10^{-6} \tag{28}$$

Literature values of thermal properties for many food products are given in a recent review (Polley et al., 1980). Semi-solid foods, such as fatty meats and frozen entrees, vary in thermal diffusivity from one region to another, leading to differential heating effects similar to those in microwave heating that are due to variations in dielectric properties between different regions.

One of the recent trends in microwave food processing is an increase in the use of combination ovens with microwave and conventional elements, particularly for industrial applications. Heat transfer modes are particularly important in such applications, but also need to be considered in ovens using only microwave energy sources, since heat generated within the food product by interactions of microwave with food constituents is also transferred within the product and at product surfaces during microwave heating and in thermal equilibration after a microwave oven is turned off. The trend toward combination ovens is mostly due to evaporative heat losses at product surfaces in cool microwave cavities that limit surface temperature close to the boiling point of water. This inhibits nonenzymatic browning and crisping reactions occurring at much higher surface temperatures and decreases surface microbial inactivation rates due to cooling effects of surface moisture evaporation.

MODELING MICROWAVE PROCESSES

Microwave Heating Simulations

Microwave time-temperature profiles have been simulated with 5–10% accuracy in 95% moisture agar slabs and cylinders at salt concentrations of 0.015 and 0.10 M of varying size by finite-difference equations with an internal heat generation term based on distributive model values of dielectric constant and loss in each shell volume as a function of salinity and temperature (Swami, 1982). Similar results are reported in simulations of heating patterns in rice flour and corn starch at moistures of 20–40% (Ofoli and Komolprasert, 1988) and frozen foods (Ohlsson, 1983; Taoukis et al., 1987). Concepts of microwave process modeling by numerical methods are also reviewed in the recent literature (Talmon et al., 1982;

Risman, et al., Ohlsson, 1988; Tong and Lund, 1989). These models generally employ finite-difference or finite-element methods and, to date, are limited to homogenous food products, although the method of Taoukis et al. (1987) is based on mathematical transformations that treat thawing as the propagation of a phase front.

One of the complications in using models for home ovens is that transmitted energy levels are not easy to determine, since transfer efficiencies vary with oven design and cavity loading, i.e., product volume and moisture content. This is not a problem in most industrial ovens, since their cavities can be "tuned" to match the food load by regulating the voltage standing wave ratio (Krieger, 1989). This feature is not available in relatively inexpensive home ovens that need to be accurately calibrated as a function of load volume and composition.

A further limitation in some of the models being used is that the electrical properties of the product are not modeled as a function of local temperature or chemical composition. Several models assume that these properties are constant in each volume element during the heating period. This assumption is not valid since food properties vary significantly with local temperature and with local moisture and free salt concentrations due to surface moisture loss and moisture equilibration, unless the product is packaged in an effective moisture barrier. Despite such limitations, mathematical modeling based on physical and chemical food properties offers an attractive alternative to trial and error methods now used by many food processors for microwave product and process development.

Conceptual Models for Microwave Heating Effects

Generalized conceptual models for microwave heating of homogeneous foods in finite-difference forms are shown in Figs. 12.5, 12.6, and 12.7 (Mudgett, 1985) for slabs, cylinders, and spheres. Electromagnetic energy from the magnetron is delivered to the microwave cavity by a coaxial transmission line or a waveguide. Some of the energy strikes food surfaces directly, while some strikes the cavity walls, where it may be absorbed to some extent, reflected to the product in a series of multiple reflections, or reflected back to the magnetron, depending on cavity loading.

Energy incident at product surfaces is partially transmitted and partially reflected, depending on the product's transmission and reflection coefficients [Eqs. (10) and (11)], which are determined by its intrinsic impedance [Eq. (9)] with subsequent attenuation of the transmitted energy within the food product by Lambert absorption [Eq. (8)]. Energy transmitted to the product is refracted toward the direction normal to the surface of incidence by Snell's law [Eq. (7)], which determines its trajectory within the product. Energy reaching the side of the product opposite to its surface of incidence is then seen to be partly reflected within the

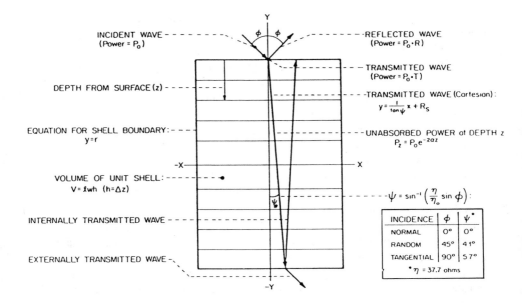

Figure 12.5 Electrophysical model for energy transfer in a slab. (*From Mudgett, 1985.*)

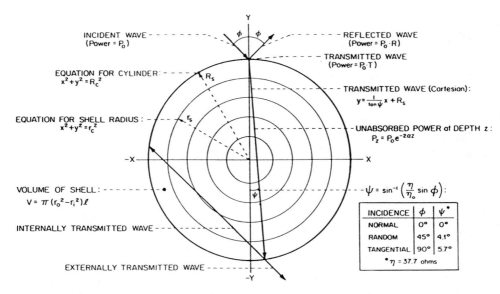

Figure 12.6 Electrophysical model for energy transfer in a cylinder. (*From Mudgett, 1985b.*)

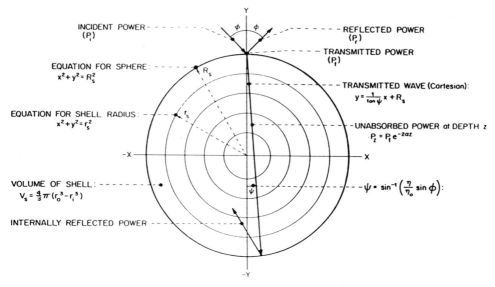

Figure 12.7 Electrophysical model for energy transfer in a sphere. (*From Mudgett, 1985*).

product and partly transmitted back to the cavity, where it is conceived to restrike the product in a multiple series of reflections at successively lower energy levels. Whether or not this is true, there is no way to determine the overall reflection coefficient without measuring the voltage standing wave ratio and empirical coupling coefficients obtained through oven calibration need to be used for modeling, unless the oven is equipped with a VSWR indicator.

Heat Transfer Equations

Heat transfer in microwave processing may be represented by the same equations as those in conventional heating except that the coupling of microwave energy introduces an internal heat generation term similar to that required for the analysis of exothermic chemical reactions. For foods of conventional geometry, one-, two-, and three-dimensional equations may be used for microwave heating of spheres, cylinders, and slabs with the heat generation term modeled in terms of food electrical and thermal properties as functions of chemical composition of the product, temperature, and the processing frequency:

$$\frac{\partial^2 T}{\partial r^2} + \frac{2}{r}\frac{\partial T}{\partial} + \frac{q'}{k} = \frac{1}{\alpha t}\frac{\partial T}{\partial \theta} \tag{29}$$

$$\frac{\partial^2 T}{\partial z^2} + \frac{\partial^2 T}{\partial r^2} + \frac{1}{r}\frac{\partial T}{\partial r} + \frac{q'}{k} = \frac{1}{\alpha_t}\frac{\partial T}{\theta} \tag{30}$$

$$\frac{\partial^2 T}{\partial x^2} + \frac{\partial^2 T}{\partial y^2} + \frac{\partial^2 T}{\partial z^2} + \frac{q'}{k} = \frac{1}{\alpha_t}\frac{\partial T}{\partial \theta} \tag{31}$$

These may be expressed in finite-difference of finite-element forms and solved for transient heating conditions subject to appropriate boundary conditions for heat and mass transfer.

The internal heat generation term may be approximated in finite-difference analysis by the average power absorbed in each differential element of volume within the product as a function of depth from the surface during discrete time intervals expressed in units that are dimensionally consistent with the units employed for the electrophysical properties and temperature of the product for each volume element. Power absorption in each volume shell is then determined by subtracting the power remaining at the inner shell boundary from that at the outer shell boundary by means of Lambert absorption, with an attenuation factor based on shell dielectric properties calculated by the distributive or modified Fricke models. Power per unit volume is then calculated by dividing the power absorbed in each shell by the shell volume. As previously indicated, power at the surface of the product must be determined for home ovens as a function of moisture content.

Heat transfer at product surfaces is seen to depend on oven air temperature, water vapor pressure at the surface, water vapor pressure in air, and air flow regimes in the microwave cavity. A procedure employed to estimate overall heat transfer coefficients at product surfaces involves measuring the voltage across the oven blower under load, disconnecting the magnetron, and placing the product in the cavity. The blower is then operated at the measured voltage and product weight loss determined at fixed time intervals. This procedure provides an air velocity and flow regime identical to that in oven operation without microwave heating effects. Dry and wet bulb temperatures of cavity air are then measured to determine the partial pressure of water vapor in air. The drying rate in the constant rate period is then used to determine the overall mass transfer coefficient by means of the equation:

$$K_y = M'/(p - p_o) \tag{32}$$

The vapor pressure of water at the surface is then estimated from the product's moisture content, based on moisture isotherm data. For high moisture products, water activity (equilibrium relative humidity) is assumed to be close to 100%. The overall heat transfer coefficient is then estimated from the equation:

$$h = K_y \Delta H_v^o (p - p_o)/(T_d - T_w) \tag{33}$$

and convective losses in heat transfer from the equation:

$$q_c = hA(T_s - T_a) \tag{34}$$

Microwave Heating Patterns

Some unusual heating patterns are seen in microwave food processing that may lead to nonuniform temperature distributions, several of which are also seen in conventional heating processes. A unique pattern in microwave heating results from cavity impedance mismatch effects that cause interactions between incident and reflected waves within the cavity that may result in voltage standing wave ratios that affect energy distribution at product surfaces and lead to "hot and cold spots" within the product due to variations in surface power density per unit area. Differences between voltage maxima and minima are much greater for larger impedance differences at high moisture, as shown in Fig. 12.8, than for smaller impedance differences at low moistures.

Such effects are minimized in batch ovens by "mode-stirrers" that distribute energy at the surface more uniformly or by rotating the product on a turntable in the cavity. In continuous industrial processes, the effects are minimized by movement of the product through the field. While standing wave patterns may be generated in some products from interactions between internally transmitted and reflected energy, this seems unlikely in intermediate or high moisture products of reasonable thickness, since most of the energy is attenuated before reaching the surface opposite that of energy incidence. Internal standing wave patterns are more likely to be generated in low moisture products with low attenuation factors of small thickness. However, wavelengths in low moisture products are less than one-half those in free space and temperature differences in microwave heating are reduced to some extent by thermal equilibration.

Nonuniform heating effects also result from energy transmission at more than one surface of incidence, i.e., slab and cylinder edges receive energy from two and slab corners from three directions. These effects are avoided by shielding the product with a reflecting material such as aluminum foil to restrict energy transmission to a single direction.

Although surface heating patterns similar to those of conventional heating (i.e.,

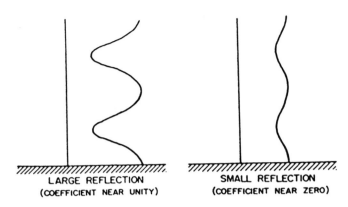

LARGE REFLECTION
(COEFFICIENT NEAR UNITY)

SMALL REFLECTION
(COEFFICIENT NEAR ZERO)

Figure 12.8   Effect of intrinsic impedance on standing wave magnitudes. (*From Mudgett, 1986b.*)

temperatures near the surface are higher than center temperatures), center heating patterns are also seen in which center temperatures are higher than those nearer the surface. These patterns have been shown to depend on product geometry, thickness and free salt content, and the processing frequency. Effects of salt content and diameter are demonstrated by temperature measurements in agar cylinders at 2450 MHz, as seen in Figs. 12.9 and 12.10 (Mudgett, 1986b). Surface heating patterns were seen in 7.2-cm diameter cylinders with 0.1 and 0.3 M salt concentrations, while center heating patterns were seen for a water cylinder of the same diameter, as shown in Fig. 12.9. Center heating was also found for a 4.6-cm cylinder at 0.1 M salt content, but not for a cylinder of 0.1 M salt content with a diameter of 10.2 cm, as seen in Fig. 12.10. Center heating was probably first seen in nutrient agar cylinders of low salt content by Copson and Decareau (Copson, 1962) at 915 MHz, while surface heating of the cylinders was seen at 2450 MHz. In terms of electrical properties, center heating effects are likely due to low attenuation factors resulting in higher penetration depths, which lead to power densities nearer the center higher than those closer to the surface, due to less rapidly decreasing power densities than those volume elements nearer the center for low attenuation factors. While neither pattern is ideal in food processing, both center and surface heating effects might be combined by exciting the microwave cavity at both frequencies to obtain more uniform temperature profiles.

One of the major problems in microwave heating occurs in heterogeneous food products with substantially different phase properties. For example, microwave tempering or thawing of frozen foods leads to "runaway heating" effects, which result from selective heating of unfrozen brine pockets (Mudgett et al., 1979),

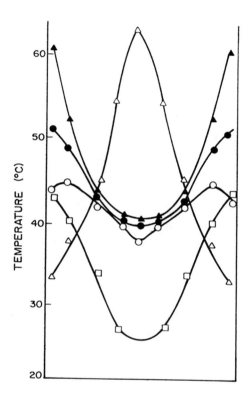

EXPERIMENTAL CONDITIONS

SYMBOL	TIME(min)	T_0(°C)	POWER LEVEL	SALTS(M)	AGAR%	GELATIN%	BEAKER	END SHIELD	CYLINDER RADIUS(cm)
○	1.5	24	9	0.1	5	–	–	+	3.6
●	1.5	24	9	0.1	5	–	+	+	3.6
△	1.5	24	9	–	5	–	–	+	3.6
▲	1.5	24	9	–	5	15	+	+	3.6
□	1.5	24	9	0.3	5	–	–	+	3.6

Figure 12.9 Effect of conductivity on center heating in agar cylinders. (*From Mudgett, 1986b.*)

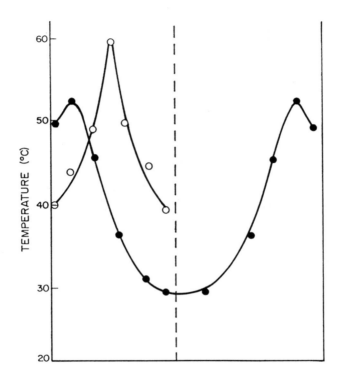

EXPERIMENTAL CONDITIONS

SYMBOL	TIME(min)	T_o(°C)	POWER LEVEL	SALTS (M)	AGAR%	END SHIELD	CYLINDER RADIUS (cm)
●	3	24	9	0.1	5	+	5.1
○	1	24	9	0.1	5	+	2.3

Figure 12.10   Effect of diameter on center heating in agar cylinders. (*From Mudgett, 1986b.*)

unless the oven is operated intermittently to provide thermal equilibration by internal heat conduction. Home ovens provide duty cycles with a time base that varies from one manufacturer's oven to another. O'Meara (1989) points out that 50% power levels could be obtained with a 1-minute time base in a 30-sec on and 30 sec off duty cycle or a 1-sec time base with a 1/2 sec on and 1/2 sec off duty cycle and that neither of these duty cycles is expected to have the same heating effect as full-time operation at 50% power.

Differential heating effects are also seen for a number of semi-solid foods, such as pizzas and jelly doughnuts, and in liquids with large solid inclusions such as meat and vegetable soups. For soups, the conductivity of the liquid phase is much higher than the suspended solids phase. In jelly doughnuts, the liquid jelly phase has a high sugar content that absorbs microwave energy more efficiently than the low moisture dough phase. These effects are explained by differences in attenuation factors. That is, energy transmitted to the phase with a higher attenuation factor is highly absorbed before it reaches the phase with a lower attenuation factor. However, energy transmitted to the phase with the lower attenuation factor is lightly absorbed and a significant fraction of the energy reaching the phase boundary is transmitted through the boundary, as demonstrated by differential heating effects seen in composite water and salts agar cylinders (Mudgett and Nash, 1980).

Thermal Inactivation Effects

Finite-difference models for microwave heating have also been developed with numerical integration of microbial lethalities for semisolid (Keenan, 1983) and liquid (Foley, 1985) foods based on models for time-temperature profiles. The models were based on the generalized nth-order kinetic model for thermal inactivation given by Labuza (1980):

$$\frac{dC}{d\theta} = -kC^n \tag{35}$$

which, in conjunction with the Arrhenius equation for temperature dependence of specific rate constants, can be used to represent the inactivation of microbial pathogens or spoilage organisms and nutrient or quality, e.g., flavor, texture and color, retention based on measured thermal resistances:

$$k = k_o \exp\left[\frac{E_a(T - T_o)}{RTT_o}\right] \tag{36}$$

Thermal resistance values for microbial contaminants and nutrients of foods are given by Harris and Karmas (1975), Jay (1986), Lund (1977), Stumbo (1973), and Thompson (1982) for a large number of model systems and food products.

The use of such models for predicting thermal inactivation effects needs to be based on numerical integration of incremental time-temperature profiles at product locations with minimum lethalities (temperatures) to ensure microbial safety in pasteurization and sterilization processes, since temperatures are generally not uniform and dwell times at each temperature are much shorter than those in conventional heating. While other numerical integration methods may be more suitable (Pachner, 1984), Simpson's method was chosen by Keenan (1983) as a preliminary basis for modeling microbial lethalities in agar cylinders of varying diameter and salinity seeded by a strain of *Escherichia coli* with measured thermal resistances over a wide range of pasteurization temperatures. Comparison of predictions with experimental lethality measurements at various depths from the surface showed reasonable agreement with first-order kinetics at 20% power levels, suggesting thermal equilibration during the off periods in the oven's duty cycle, and with second-order kinetics at full power, suggesting selective energy absorption by the organism due to salt concentration from the suspending medium. Surface lethalities were lower than predicted, presumably due to evaporative cooling effects. Center heating was also seen at low salt concentrations in a cylinder of small diameter. For the range of experimental conditions examined, the models predicted the locations of minimum temperature and lethality.

Continuous-flow microwave processes are of special interest for liquid food products because of their potential for rapid and uniform heating and reduced heat-exchanger surfaces fouling, as compared with conventional heating methods. Reduced fouling by fluid products can be obtained in microwave processing by using chemically resistant polymers that are also "transparent" to microwaves. Microbial lethalities were simulated in apple juice seeded with *Es. coli* in a continuous-flow pasteurization process by Foley (1985) by means of Romberg integration using the IMSL program DCADRE (Rice, 1983). Resisdence-time distribution effects were also considered on the axial dispersion model of Levenspiel (1972). Predicted and measured lethalities were in reasonable agreement at low and intermediate flow rates, with or without correction for residence-time distribution effects. The further refinement of these models would be useful in developing pasteurization and sterilization processes based on computer-aided design. Additional aspects of microwave pasteurization and sterilization processes have been reviewed by Mudgett and Schwartzberg (1980).

HOME MICROWAVE OVENS

Due to the increasing number of women working outside the home, there has been an increasing consumer demand for foods that offer greater convenience and time savings in preparation, which has led to a current home oven market penetration level of more than 80%. This demand is providing a strong marketing incentive to the food manufacturing and foodservice industries to consider innovative approaches in food formulation and packaging to develop new and improved microwavable food products for home oven use. Many of the factors being considered in microwavable product development for home ovens have been reviewed recently by IFT's Expert Panel on Food Safety and Nutrition (Mudgett, 1989). However, there are some areas of concern in developing products for home ovens that may be useful to discuss in more detail.

Oven Calibration

Because of the large variations in oven design among home oven manufacturers, both the food manufacturer and the user are faced with a staggering array of maximum power levels, energy transfer efficiencies, and reduced power time bases. During the past five years, there has been a tremendous increase in the number of home ovens imported. At present, 79% of home ovens sold in the United States are manufactured by foreign corporations: of these, 34% are manufactured in the United States, and 66% are imported primarily from Japan, Korea, and Singapore (Buffler, 1988). There is not yet a standardized technique in the United States for oven power measurements in ovens, which range from 175 to 700 watts in maximum power level. Since there are now at least 58 million ovens in the United States with unlabeled power levels, it would be useful to develop a technique which would allow consumers to calibrate their own ovens. Buffler has suggested a reasonable method to measure maximum power levels, based on the time to boil a cup of water, that would permit users to correct recipes written for a different oven power level.

The lack of a standardized calibration procedure also places a heavy burden on food manufacturers in product development and requires product performance evaluation in test kitchens with ovens that are proposed to be standardized by one or two liters of water by temperature rise measurements (Schiffman, 1988). However, energy transfer efficiencies vary extensively with load volume and moisture content (Mudgett, 1986b). Oven calibration curves based on temperature rise measurements of oil and water as a function of volume and provided by oven manufacturers for each home oven model would greatly simplify microwavable product development by food manufacturers in standardizing recipe development for the many models currently available to home users. Such a procedure would involve measuring energy coupling coefficients of water and oil in each oven and determining an overall oven coupling coefficient by distributing the coefficients of oil and water by phase volume fraction. Overall coefficients so obtained would then be used in Eq. (14) to determine fractions of maximum power coupled by the product as a function of load volume and moisture content. This procedure has been used to calibrate several home ovens and is useful in estimating time-temperature profiles for a wide range of product volumes and moisture contents by microwave heating models currently available. Coupling coefficients for a 715 watt oven calibrated by this method were found to vary from 0.004 at low moistures to 0.008 at high moistures, suggesting that 25% of maximum oven power is coupled for a 50-mL load volume and that a minimum load volume of about 250 mL would be required to couple 75% of maximum oven power. A minimum load volume of 500 mL would be required to couple 95% of the maximum power for this oven.

It would be clearly useful for the oven manufacturer to calibrate each model's performance prior to sale than to require each food manufacturer and user to do so for the many different models currently available. Calibration data should be supplied with the oven in an operating manual that could be used by the food manufacturer in product development and by the user to obtain more satisfactory heating performance. A more accurate oven calibration procedure would also reduce the current level of dissatisfaction with home oven products by many users and would lead to further expansion of the profitable home oven market for both electronic and food manufacturers (Cronk, 1988).

An additional area for consideration by oven manufacturers is optimization of the time base used to reduce power levels in home ovens, because time bases used for power reduction vary from one manufacturer's oven to another (O'Meara, 1989). This is seen to depend on the effects of thermal equilibration during the off-period of the oven's duty cycle and could be investigated by currently available finite-difference models for continuous microwave heating modified to include various time bases, i.e., duty cycle time lengths, and on-off periods.

Food Formulation

There is an increasing emphasis in microwavable product development on the use of starch, cellulose, alginates, alcohols, oils, and emulsifiers to obtain more uniform heating and use of browning and crisping agents to overcome effects of low surface temperatures. For example, microcrystalline cellulose is used as a non-nutritive bulking agent in the diet or as a functional ingredient (Carroll, 1989). Functional properties of cellulose gels of interest in product development include emulsion stabilization, foam stabilization, ice crystal control, syneresis control, and moisture migration control at high processing temperatures. Carroll points out that crystalline cellulose is used in sandwich dressings and cake sauces for improving thermal stability, emulsion stability, flow control, and texture of microwavable foods, in addition to other advantages. She also reports the use of cellulose gels to enhance temperature and emulsion stability or to control ice crystal growth in microwavable frozen food products such as chocolate cake sauce, sandwich spread, omelets, pancakes, cheese and white sauces, brownies, creamed chicken and cream soup base, noting that imbibition of water by cellulose gels inhibits the formation of large ice crystals.

Emulsifying agents are also used in microwave food formulation to stabilize moisture retention by interaction with fats and oils, proteins, and starches to prevent moisture migration and improve product texture (Connerton, 1989). The selection of an emulsifying agent is seen to depend on its functionality, its thermal behavior, and structural integrity during microwave heating. There is, however, a question as to whether fats and oils absorb much microwave energy in products with significant moistures, since their dielectric loss is much lower than that of water and aqueous ions. Concepts of relative intrinsic impedance suggest that the aqueous phase of foods selectively absorbs most of the energy and that fats and oils are nearly "transparent" to the electrical field in such products. This is open to question, though, since pure fats and oils do couple microwave energy and are known to heat quite rapidly, due to relative densities and heat capacities, to higher temperatures than water in microwave heating due to their higher boiling points. It does seem likely that emulsifiers that link hydrophobic and hydrophilic regions of food constituents alter the dielectric properties of food constituents through interactions which increase the number of dipoles interacting with the electrical field. This is shown to depend on the hydrophilic/lipophilic balance (HLB) of the emulsifier used based on data presented by Connerton showing that the time to heat partially hydrogenated soybean oil loaded with various emulsifiers to 88°C in a microwave oven decreased with increasing HLB values.

The use of hydrocolloids such as alginate, starch, xanthan gum, carrageenan, guar gum, and carboxymethylcellulose in formulating microwavable foods such as desserts, salad dressings, beverage mixes, ice cream, cheese spreads, toppings, sauces, and gravies is well established (Shukla, 1989). But as Shukla observes, effects of their interactions with microwave energy are not well understood. He does raise some interesting questions that seem well worth investigating at the molecular level in terms of water binding, polymer–polymer and solute–polymer interactions, and, perhaps most importantly, the effects of viscosity on dipole rotation and ion migration, the major mechanisms of microwave interaction with foods. It seems likely that viscosity would increase water relaxation and ion migration times. Yet, there is little information in the literature on viscosity effects in microwavable foods or on the effects of hydrocolloid interactions in foods at the molecular level. Shukla does show some data on binding of water by starch, casein, pectins, and alginate indicating maximum bound water levels from 0.26 to 0.55 g/g of hydrocolloid and relatively imobilized water levels from 0.20 to 0.58 g/g, which are significant in terms of effects on dielectric behavior of hydrocolloids, as seen by the synergistic loss effect observed in pregelatinized starch (Roebuck et al., 1972) due to stabilization of hydrogen bonding. Suffice it to say, considerably more investigation is needed on the effects of hydrocolloid and emulsifier interactions in foods and their impacts on dielectric behavior.

Packaging Considerations

There is increasing concern about the use of susceptors to provide localized heating in products such as microwave popcorn and pizzas, because of the high temperatures generated by these highly interactive materials and their effect on the possible migration of plasticizers and antioxidants from polymers used in food packaging to the food product (IFT, 1988). Susceptors in current use include polyester film bags, paperboard trays, and plastic materials laminated with ferrite or aluminum particle films that can result in temperatures up to 450–500°F and that have been found to clog pores in packaging materials, cause off-flavors in foods, and migrate to the product. One study by a packaging company found temperatures higher that 675°F for aluminum films despite FDA approval of polyester film temperatures to temperatures of 275°F and browning and crisping temperatures from 300–400°F (Anonymous, 1988). FDA regulations on packaging of foods and beverages treat chemical substances added directly to foods the same as those added indirectly by migration from food packages. This has stimulated recent research to develop improved polymers with plasticizers that are not as easily leached, plasticizer reformulation with silicone oil, methyl siloxane, epoxidized soybean oil, polyesters and other chemical compounds; and surface treatment of packaging films to induce cross-linking and decrease plasticizer migration at film surfaces. More positively, the use of susceptors has given higher popcorn yields, browning and crisping of surfaces in wrapped products, and crispier microwave pizzas by providing additional heat at locations that are not easily heated by microwaves. It also seems possible that susceptors could be designed at reduced temperatures by controlling concentrations of the active material, suspending the active material in a different medium, or the use of novel susceptor materials. The development of more effective barriers to polymer migration is also seen as an alternative in limiting migration.

Materials such as aluminum foil are also used in food packages to reflect microwave energy in portions of composite products. A frozen dessert package with layers of ice cream, a sauce, and a brownie uses a metalized lid to shield the top layer of ice cream and a metalized electrically conductive material on the sides of the container to shield the ice cream and underlying sauce layer, while the bottom layer of brownie is unshielded (O'Meara, 1988). Shielding is also used to eliminate edge and corner heating effects in slabs and cylinders and to prevent microwave energy absorption in compartmented trays with one or more food components that are not to be heated. Despite arcing problems seen for metals with sharp or rough edges due to electrical charge dissipation, the use of properly designed aluminum foil and metalized shields is well accepted by the industry, although consumers are having some difficulty in overcoming earlier warnings against the use of metals in a microwave oven.

Safety Considerations

It is generally believed that the effects of microwave interactions with foods are solely thermal in origin, although there are still some reports suggesting the possibility of athermal effects (Mudgett, 1989). However, there are some potential hazards against which precautions should be taken in home ovens, as summarized by Schiffman (1989), related to undisclosed hot spots, generation of steam, boilover, volcanic eruption from superheating, frying in oil, and fires. According to Schiffman, the most common safety hazard is in excessive heating that is hidden from and may result in injury to the consumer. Aside from the possible deleterious effects on product quality, he notes that temperatures at the surface may be significantly lower than the product's interior, because of evaporative cooling losses, and that there may be localized "hot spots" within the product due to high field intensities resulting from standing waves or to center heating effects. An example of center heating is given as baby foods heated in small cylindrical jars, where the glass jar remains cool and center temperatures may be high enough to scald an infant's mouth. Steam buildup in a sealed popcorn bag which did not vent properly was also cited in scalding a child's eyelids when the bag was opened. Similar steam hazards are expected from unusually tight containers, lids, or plastic wrapping seals, unless covers are first loosened and plastic wraps are pierced prior to microwaving.

Boilover problems are seen for liquids such as water and coffee. In one scenario, coffee that is quiescent after heating erupts when it is removed from the oven. In a second scenario, water that is quiescent after heating erupts when a teaspoon of instant coffee is added. Schiffman attributes this to bubble nucleation and coalescence within the liquid due to dissolved gases that result from localized superheating of the liquid, i.e., temperatures above the boiling point of water. "Volcanic" eruption from superheating effects have been experimentally verified by Buffler and Lindstrom (1988). Schiffman also notes that viscous products such as hot cereals, stews, and chili permit larger bubbles that can cause volcanic effects and boilover of the product.

He also notes that pure oils absorb significant amounts of microwave energy and can be heated to temperatures near the flash point, presenting a hazard in microwave frying in terms of spattering and boilover caused by wet foods and handling the hot oil after frying, and suggests that microwave frying is too hazardous for home oven use. He also cites numerous reports of fires within home ovens. These include microwave popcorn, which he attributes to faulty packaging materials, drying of herbs such as organo and thyme, and the use of metallic twist ties. While the mechanism for microwave energy absorption by oils an dry food solids is not clear it is known that these materials are heated by microwaves to high temperatures. It therefore seems reasonable to believe that food solids continue to absorb energy following moisture removal and can, when heated too long, reach spontaneous ignition temperature, as seen by charring and ignition of some dried food solids.

INDUSTRIAL MICROWAVE OVENS

Processes in Commercial Use

Although high-powered microwave ovens have been slowly accepted by the food industry for large-scale unit operations, an increasing number of processes have become commercially successful. Decareau and Peterson (1986) report some 250 microwave installations in operation at that time, about 200 for tempering of frozen meats and fish, 16 for precooking bacon, poultry and meat patties, 30 for drying pasta, onions, and snack foods, 5 for vacuum concentration of fruit juices, and 3 for pasteurizing bread and yoghurt. Decareau (1985) estimates annual growth of these processes at 1.0–1.5 MW/year with annual growth of 2.5 MW/year for microwave food processing on a worldwide basis. Microwaves are often combined with conventional heating sources, with microwaves for internal heating, and conventional sources to obtain surface browning and crisping, lower surface microbial counts, or accelerate drying.

Tempering Processes

Typical commercial processes (Table 12.5) are described in a benchmark summary of microwave applications in the food and beverage industry (Sanio and Michelussi, 1988). The most successful of these is in tempering frozen foods, a process in which the frozen product is heated to a temperature just below the freezing point and then allowed to fully thaw at low temperature. Tempering is commonly done in batch and continuous processes at 915 MHz and at power levels from 25 to 120 kW without conventional heating. Continuous tempering systems also use 2450 MHz power sources, with circulation of refrigerated air around the product to limit edge and corner heating at shallower penetration depths. Some advantages cited for these processes include: (1) easier slicing, (2) lower building and maintenance costs of refrigeration, (3) reduced spoilage and drip losses, (4) more flexibility in scheduling inventory control, and (5) simple cleaning operations and improved sanitation, in addition to space reductions. Tempering times of 5

Table 12.5 Typical Commercial Applications

Application	Frequency (MHz)	Power (kW)	Tube size (kW)	Conventional heat
Tempering				
Batch	915	30	30	None
Continuous	915	80	40	None
Pasta drying	915	30–50	30–50	Hot air
Precooking				
Bacon	915	50–300	50	Hot air
Poultry	2450	50–80	2.5	Steam
Meat patties	2450	30	2.5	None
Vacuum drying	2450	40	—	Infrared

Source: Adapted from Sanio and Michelussi (1988).

minutes without unpacking the product and with little or no drip loss are reported, as compared with processing times from 2–5 days and drip losses up to 5% of product weight by conventional thawing methods. There are about 200 tempering installations around the world which are cost-effective based on reduced spoilage and drip losses alone. Tempering is done at 915 MHz by intermittent operation permitting thermal equilibration to avoid runaway heating effects obtained in continuous microwave heating. Throughputs of 6000 pounds per hour of frozen hamburger are reported for a 90 kW unit at 915 MHz, with throughputs for a 120 kW unit up to 10,000 pounds per hour.

Atmospheric Drying Processes

Atmospheric during operations are reportedly in use at 26 installations for products such as pasta, tomato paste, onions, rice, snack foods, and bacon bits. These systems are operated at 915 MHz and are combined with hot-air blowers to increase convective moisture transfer at product surfaces. Microwave drying is cost-effective at moistures below 20%, above which it is more economical to use conventional drying methods, where primary energy costs are approximately one-third less than conventional dryers. For pasta drying, process times are reduced from 8.0 to 1.5 hours, fuel consumption is reduced 20%, and operating costs by 20–25%, with improvements in product color and texture, reduced bacterial loads, and less case-hardening. A 30 kW unit with hot air at 82–93°C and 15–20% relative humidity operated at about 100 feet per minute is reported to dry one ton of pasta per hour. Other advantages cited are similar to those of tempering.

Precooking Processes

Precooking of bacon, meat patties, poultry, battered fish products, bacon bits, and crumbs combine microwaves with steam and/or hot air. In a two-stage bacon process at 915 MHz, microwaves are used to remove water and fat, after which hot air at 71°C is passed over the bacon to remove surface moisture, with a throughput of about 500 pounds per hour and a 40% yield, while a three-stage process with steam and hot air gives a throughput of more than 1000 pounds per hour. Other advantages cited in microwave bacon processing are increased yield (about 25–40%), an increase in the number of slices per pound, and reduction in preparation time by institutional users. In one poultry precooking process, microwaves are used with saturated steam to reduce *Salmonella* contamination, with processing of breasts and thighs in an 80 kW unit and legs and wings in a 50 kW unit with 2.5 kW power modules. The heavier breasts and thighs need more energy and longer processing times and are generally processed at 915 MHz for greater penetration depths, while wings and legs and may be processed at 2450 MHz. Throughputs of more than 1 ton per hour are cited with processing times ranging from 12 to 16 minutes. Advantages cited include reduced moisture loss and higher yields, less product shrinkage, and savings in labor costs.

One continuous microwave unit precooks meat, poultry, and fish portions, in addition to bacon, meat patties, sausage links, pizza toppings, and meatballs. Precooking of meat patties by the Swedish "Inpro" process in a 40 kW unit a 2450 MHz yields a ton of meat patties per hour and browns the patties between temperature-controlled Teflon-coated aluminum platens prior to passage through a microwave tunnel (Decareau, 1985). The process is reported to be widely used in both Sweden and Japan (Sanio and Michelussi, 1988).

Other Commercial Processes

Commercial processes successful on a more limited scale include the Swedish "Scanpro" blanching process, which dips peeled whole potatos in ascorbic acid to reduce enzymatic browning, followed by vacuum packaging and microwaving of the vacuum-packed potatoes in a 30 kW, 2450 MHz tunnel with a throughput of some 1300 pounds per hour (Decareau, 1985). Products blanched by microwaves include mushrooms, fruit, corn, and other vegetables (Sanio and Michelussi, 1988).

Microwave bakery processes have been successful in applications combined with conventional heating in simultaneous or sequential operations (Schiffman et al., 1981). Donuts in proofing and frying operations that require special flours and dough conditioners to control dough rheology, structural integrity, and fat retention are proofed in 4 minutes versus 40–60 minutes in conventional donut proofing and frying times reduced by 20% (Schiffman et al., 1971) with some 24 industrial users.

Microwave and radio-frequency drying processes are commonly used for finishing of cookies and biscuits, with process time reductions from 50% (Schiffman et al., 1981) to 67% (Pei, 1982). While microwave processes are not now used in bread baking, a bread pasteurization process for 2450 MHz, 48 kW unit is used to inhibit mold growth, and a similar process is used to make bread crumbs at several locations (Decareau and Peterson, 1986).

Microwave pasteurization processes are reported for breads, potatoes, fish, fruits, vegetables, and salads in at least 50 locations, primarily in Sweden, Denmark, Germany, and Switzerland. Bread is processed in 2450 MHz, 48 kW units at throughputs up to more than a ton per hour, potatoes in 2450 MHz, 30 kW units at throughputs up to 1500 pounds per hour with 6-minute process times, and other products at 2450 MHz with throughputs of about 500 pounds per hour. According to Schiffman (1989a), microwave pasteurization processes are expanding rapidly due to expansion of refrigerated food products in plastic trays for both reheating home products and microwave pasteurization in the plant. He also observes that at least 10 installations for vacuum concentration of fruit juices and drying of herb and plant extracts, yeast and vegetable powders are reported in Europe by 2450 MHz, 48 kW units coupled with a heat pump vacuum concentrator that uses 2800 watt-hours to produce about 2 pounds of flakes from a feed of 10% solids (Sanio and Michelussi, 1988).

Processes Under Development

Microwave unit operations currently in various stages of development include bread baking, freeze-drying, setting meat emulsions, meat coagulation, puffing, rendering, and roasting (Sanio and Michelussi, 1988). Some of these are now in limited commercial operation but represent little installed power (Decareau and Peterson, 1986). Microwave tunnels of 80 kW at 2450 MHz with jet impingement and infrared sensing for process control are reported in use at the American Institute of Baking in Kansas. Microwave freeze-drying has been studied at the United States Army Natick Research, Development and Engineering Center and at North Carolina State University, with two pilot-scale units in current operation for meats, fruits, and vegetables and favorable cost projections for commercial processing. The Natick pilot units are 3–12 kW at 2450 MHz. A commercial coffee process may already be in operation in Switzerland. A semicommercial process to set sausage emulsions is in operation in one plant in Switzerland, at 2450 MHz for small diameters and 915 MHz for large diameters in a continuous tunnel. Many commercial installations for meat coagulations are reported in Japan, Norway, and Sweden, with the throughput of a 40 kW installation of more than one ton per hour.

Oyster shucking in a 1.5 kW installation at 2450 MHz is reported to open the shells at the rate of 21 oysters per minute. Puffing of marine snack foods at 8 kW and 2450 MHz with a throughput of some 100 pounds per hour and vegetables and fruit products at 25 kW and 915 MHz are also cited (Sanio and Michelussi, 1988). One installation for fat rendering (Bird, 1979) using 115 kW at 2450 MHz is now in operation, and continuous units at 915 and 2450 MHz have been obtained from an American manufacturer by the USDA. At least one installation of 40 kW at 2450 MHz is used in Japan to roast fish, with a throughput of some 4000 slices per hour, and cocoa beans roasted in Switzerland have throughputs from about 150 to 220 pounds per hour. The Swiss process is also used to roast coffee beans. The beans are roasted in 5–10 minutes with increased yield and reduced smoke pollution. Finally, liquid and semisolid food products foods in plastic pouches have been sterilized in pilot-scale units at the United States Army Natick Research and Development Center (Ayoub et al., 1974), the Swedish Food Institute, and the Alfa-Laval Company in Sweden (Stenstrom, 1974a, b, c) and is expected by Decareau and Peterson (1986) to be commercial within the next few years. Microwave sterilization equipment is currently available at 915 and 2450 MHz at power levels from 5–25 kW (Sanio and Michelussi, 1988).

Economic Considerations

The relatively slow acceptance of industrial microwave food processes has been, historically, related to unfavorable equipment and energy costs, as compared with conventional heating processes, and to a lack of a theoretical basis for electrical interactions of microwave energy with foods and their relationship to microwave heating performance. Recent research on the dielectric behavior of foods and models for microwave heating is now available and is providing an information base for improved microwave process design. Similarly, trends in microwave equipment and energy costs have become much more favorable for the development of new and improved microwave food processes. As costs of gas and oil rise due to their depletion and alternative energy sources such as coal and nuclear energy are used to generate electrical power, energy cost differentials are less significant and other economic considerations become more critical. A projection of United States energy requirements by the year 2000 (Anonymous, 1981) shows a significant increase in energy needs, which will be largely be met by increased use of coal and nuclear fuels, as shown in Table 12.6. In addition, capital costs of industrial ovens have also been reduced by modular designs using magnetrons of high reliability manufactured in bulk quantities at low cost. Current costs of installed microwave equipment in the 100–200 kW range are about $2000/kW. On this basis, the annual costs of a 100 kW unit of modular design at 2450 MHz in New England would be $42,469 (Table 12.7) based on very

Table 12.6   Meeting United States Energy Needs

Source	Energy (quads)	
	1979	2000
Domestic oil	21	18
Domestic gas	19	18
Domestic coal	15	40
Nuclear energy	3	12
Renewables	5	10
Imports	17	10
Total	90	108

Source: Anonymous, 1981.

conservative assumptions of energy costs, installed capital costs and maintenance costs and capital recovery, but not including labor costs and neglecting cost reductions in space savings, product yields, etc. Based on power levels and throughputs for processes reported by Sanio and Michelussi at 2450 MHz and modular annual costs shown in Table 12.7, cost per pound of product processed at 5000 hour plant operation was calculated, as shown in Table 12.8. Relative costs are generally less than one cent per pound, except for vacuum, drying, puffing and cocoa roasting, which may be scaled up or down by a load factor for annual operating hours. As indicated, the assumed modular costs are very conservative based on values assumed for capital recovery and energy costs and could be more or less depending on the microwave unit processor. It is again emphasized that capital costs of installed microwave equipment in the 100–200 kW range are about $2,000 per kW. More sophisticated methods of cost evaluation are described by Edgar (1986) based on differential cash flow and investment analysis by simple payback, net present value, and internal rate of return methods, which can be more accurately applied in microwave process evaluation.

CONCLUSION

Microwave food processing in the last decade, made significant advances, which are now being fueled by the dramatic increase in home oven use based on consumer

Table 12.7   Approximate Costs/Pound of Product at 2450 MHz Based on 25 kW Modular Microwave Unit Costs

Process	kW	Throughout (lb/hr)	Annual costs ($)	Cost/lb ($)
Meat tempering	40	1,600	67,950	0.0084
Onion drying	25	10,000	42,469	0.0008
Potato blanching	30	1,320	50,963	0.0077
Meat coagulation	40	2,200	67,950	0.0062
Inpro process	40	2,200	67,950	0.0062
Bread pasteurizing	48	2,200	81,540	0.0074
Vacuum dry juice	48	107	81,540	0.1524
Puffed products	8	88	13,590	0.0309
Roasting cocoa	5	66	8,154	0.0247
Fat rendering	115	5,000	195,397	0.0078

Table 12.8   Basis for 25 kW Modular Microwave Unit Costs

Cost factor	Basis
Industrial power rate	$0.15/Kwh
Equipment conversion efficiency	70%
Installed equipment costs	$100,000
Magnetron replacement costs	$1,000
Equipment life	10 yr
Annual operation	5,000 hr
Annual Cost Estimates:	
crf-10 yr-20%	$23,850
Magnetron replacement	1,000
Energy cost	17,619
Total annual cost	$42,469

demand for new and improved convenience products. This demand is creating research opportunities in food formulation and packaging that seem likely to stimulate both home and industrial oven development resulting in new and improved microwave food products and processes. Although there are still many unresolved problems, further research and development efforts by food equipment manufacturers, processors, and academicians is certain to result in a new wave of prosperity for food and electronics manufacturers in conjunction with the expanded database on microwaves generated during the past decade.

One of the problems that still persist in home ovens is the need for better methods for measuring oven power. This should be addressed by the equipment manufacturing and food processing communities. An important goal for academic research is to gain a more basic understanding of the subtle interactions among hydrocolloids, emulsifiers, and water, primarily a task for food scientists, and to translate this information into models for microwave processing, primarily a task for food engineers. Microwave research is increasingly being addressed by the food industry on an in-house basis, with little support from government-sponsored microwave research, although it would appear to be in the national interest to develop energy conservative methods to ensure a safe and nutritious food supply.

As to the future, expansion of microwave food processes seem likely to be stimulated by fundamental and applied research during the past decade, which has led to a new understanding of microwaves and foods. Some recent innovations in equipment design, food formulation, and packaging have also made a big impact on the microwave market. These include the use of jet impingement of hot air to increase heat transfer, infrared and humidity sensors to test product doneness, the use of combined microwave and conventional heat sources, the use of food ingredients that improve thermal stability and extend product shelf life, the development of multistage processes, the application of susceptors to provide localized heating, the use of shielding to obtain more uniform heating, and the development of modeling concepts for microwave heating based on food properties and combined heat and mass transfer. Other innovations are sure to follow and may include feedback or adaptive control systems to maximize energy coupling efficiency, computerized "cookbooks" for programming meal preparation, the use of multiport feed systems to increase heating uniformity, the development of dedicated processing systems with high energy efficiencies and the development of new techniques in food formulation and packaging that will eliminate some of the current problems in product and process design. In the more remote future, solid-state microwave ovens of high reliability, light weight, and high energy efficiency may become a reality, with frequency-agile sources that provide more uniform heating. This is now being considered by the United States Army Natick Laboratories Advanced Systems Concepts Directorate for future space travel (Mudgett, 1988) and could lead to a new generation of home and industrial microwave ovens.

SYMBOLS

A	Heat transfer area (cm^2)
Bi	Biot number (N.D.)
C	Free salts concentration (equiv $liter^{-1}$)
C	Concentration (variable)
C_p	Specific heat at a constant pressure (cal $g–1$ C^{-1})
d	Thickness of dielectric (cm)
e	Naperian constant: 2.71828 (N.D.)
E_a	Energy of activation (cal mol^{-1} κ^{-1})
F_o	Fourier number (N.D.)
h	Heat transfer coefficient (cal sec^{-1} cm^{-2} C^{-1})
h	Height of slab (cm)
ΔH_v^o	Latent heat of vaporization (cal mol^{-1})
k	Volumetric coupling coefficient (N.D.)
k	Thermal conductivity (cal sec^{-1} cm^{-1} C^{-1})
k	Thermal denaturation constant at temperature T (min^{-1})
k_o	Thermal denaturation constant at temperature T_o (min^{-1})
k_y	Mass transfer coefficient (cm^2 sec^{-1})
l	Length of slab (cm)
M'	Mass transfer rate (g sec^{-1})
n	Reaction order (N.D.)
p	Water vapor pressure at surface (kPa)
p_o	Vapor pressure of pure water (kPa)
P_i	Power incident at dielectric surface (W)
P_m	Power coupled by matched load (W)
P_o	Power coupled by unmatched load (W)
P_r	Power reflected at dielectric surface (W)
P_t	Power transmitted at dielectric surface (W)
P_z	Power not attenuated at depth z from surface (W)
P'	Power per unit volume of dielectric (W cm^{-3})
q'	Rate of internal heat generation (cal sec^{-1} cm^{-3})
r	Radial coordinate (cm)
r_o	Shell radius of cylinder (cm)
r_s	Shell radius of sphere (cm)
R	Reflection coefficient (N.D.)
R	Universal gas constant: 1.987 (cal mol^{-1} κ^{-1})
R_c	Radius of cylinder (cm)
R_s	Radius of sphere (cm)
T	Temperature (variable)
T	Transmission coefficient (N.D.)

T_a	Ambient temperature (°C)
T_d	Dry bulb temperature (°C)
T_o	Reference temperature (K)
T_s	Surface temperature (°C)
T_w	Wet bulb temperature (°C)
v_s	Volume fraction of suspended phase in mixture (N.D.)
v_s	Phase velocity in dielectric (cm sec^{-1})
V	Product volume (cm^3)
VSWR	Voltage standing wave ratio (N.D.)
w	Width of slab (cm)
W	Water content (wt%)
x	Rectangular coordinate (cm)
X	Form factor: ratio of ellipsoid major/minor axes (N.D.)
y	Rectangular coordinate (cm)
z	Depth from surface of dielectric (cm)
z	Rectangular coordinate (cm)
Z	Voltage attenuation depth (cm)
Z_H	Half-power attenuation depth (cm)
Z_p	Power attenuation depth (cm)

Greek Letters

α	Attenuation factor (cm^{-1})
α_t	Thermal diffusivity of product (cm^2 sec^{-1})
β	Phase factor (cm^{-1})
γ	Propagation factor (cm^{-1})
δ	Loss angle (rad)
δ'	Average hydration number (N.D.)
∂	Partial derivative operator (N.D.)
ε^*	Absolute dielectric constant (farad cm^{-1})
ε_o	Dielectric constant of free space: 8.854×10^{-14} (farad cm^{-1})
η	Intrinsic impedance of dielectric (ohm)
η_o	Intrinsic impedance of free space: 377 (ohm)
θ	Processing time (sec)
κ_o	Optical dielectric constant of polar solvent (N.D.)
κ_s	Static dielectric constant of polar solvent (N.D.)
κ'	Relative dielectric constant (N.D.)
κ_i'	Relative dielectric constant of ionic solution (N.D.)
κ_m'	Relative dielectric constant of mixture (N.D.)
κ''	Relative dielectric loss (N.D.)

κ_i' Relative dielectric loss of ionic solution (N.D.)
κ_m'' Relative dielectric loss of mixture (N.D.)
κ^* Relative permittivity (N.D.)
κ_c^* Relative permittivity of continuous phase (N.D.)
κ_m^* Relative permittivity of mixture
κ_s^* Relative permittivity of suspended phase (N.D.)
λ Wavelength of measurement (cm)
λ Wavelength in dielectric (cm)
λ_o Wavelength in free space (cm)
λ_s Critical wavelength of polar solvent (cm)
Λ Equivalent conductivity of ionic solution (mho cm^2 equiv^{-1})
μ^* Absolute permeability (henry cm^{-1})
ρ Product density (g cm^{-3})
ϕ Angle of energy incidence at surface (rad)
ψ Angle of energy refraction in dielectric (rad)
ω Angular frequency (rad sec^{-1})

REFERENCES

Anonymous 1981. Report of U.S. Energy Research Advisory Board, Research and Development Panel, Washington, DC.

Anonymous 1988. Food and Drug Administration looking into susceptor packaging, *Microwave World* 9: 11.

Ayoub, J. A., Berkowitz, D., Kenyon, E. M., and Wadsworth, C. K. 1974. Continuous sterilization of meat in flexible pouches. *J. Food Sci.* 39: 309–313.

Bird, L. L. 1979. Rendering of material such as meat. U.S. Patent 4,168,418.

Buck, D. E. 1965. The dielectric spectra of ethanol-water mixtures in the microwave region. Ph.D. thesis, Massachusetts Institute of Technology, Cambridge, MA.

Buffler, C. 1988. A simple home test to determine microwave oven power output. *Microwave World* 9: 5–8.

Buffler, C., and Lindstrom, T. 1989. Experimental evidence of water erruption caused by super-heating. *Microwave World* 9: 10–11.

Carroll, L. e. 1989. Hydrocolloid functions to improve stability of microwavable foods. *Food Technol.* 43: 96–100.

Collie, C. H., Hasted, J. B., and Ritson, D. M. 1948. The dielectric properties of water and heavy water. *Proc. Royal Soc. of London* 60: 145– 160.

Connerton, S. D. L., and Shuleva, J. J. 1989. How fats, oils, and emulsifiers function in microwave food formulations. Paper presented at MW Foods '89, 2nd Intl. Conf. on Formulating Food for the Microwave Oven, March 14–15.

The Packaging Group Inc., Milltown, NJ.

Copson, D. A. 1962. *Microwave Heating.* AVI Publishing Co., Westport, CT.

Cronk, C. 1988. Microwave oven standardization and the consumer. *Microwave World* 9: 4.

Decareau, R. V., and Peterson, R. A. 1986. *Microwave Process Engineering.* Ellis Horwood Ltd., Chichester, England.

Edgar, R. 1986. The economics of microwave processing in the food industry. *Food Technol.* 40: 106–112.

Foley, A. M. 1985. Modeling a continuous microwave pasteurization process. M.S. thesis, Univ. of Massachusetts, Amerhurst, MA.

Harris, R. S., and Karmas, e. 1975. *Nutritional Evaluation of Food Processing.* AVI Publishing Co., Westport, CT.

Hasted, J. B., Ritson, D. M., and Collie, C. H. 1848. Dielectric properties of ionic solutions, Parts 1 and 2. *J. Chem. Phys.* 16: 1–21.

Holman, J.R. 1976. *Heat Transfer* 4th ed. McGraw-Hill Co., New York.

IFT. 1988. Migration of toxicants, flavors and odor-active substances from flexible packaging materials to foods. A Scientific Status Summary by the IFT Panel on Food Safety and Nutrition. *Food Technol.* 40: 109.

Jay, J. J. 1986. *Modern Food Microbiology,* 3rd ed. Van Nostrand Reinhold Co., New York.

Karel, M. 1975. Physiocochemical modification of the state of water in foods. In *Water Relations in Foods,* (Ed.) R.B. Duckworth, Academic Press, New York.

Keenan, M. 1983. Prediction of thermal inactivation effects in microwave heating. M.S. thesis, University of Massachusetts, Amherst, MA.

Kent, M. 1987. *Electric and Dielectric Properties of Food Materials.* Science and Technology Publishers, London.

Krieger, B. 1989. Personal communication.

Labuza, T. P. 1980. The effect of water activity on reaction kinetics of food deterioration. *Food Technol.* 34: 36–41,59.

Levenspiel, O. 1972. *Chemical Reaction Engineering.* John Wiley and Sons, New York.

Lund, D. B. 1977. Design of thermal processes for maximum nutrient retention. *Food Technol.* 31: 71–78.

Mudgett, R. E. 1974. A physical chemical basis for prediction of dielectric properties in foods at ultra-high and microwave frequencies. Ph.D. thesis, Massachusetts Institute of Technology, Cambridge, MA.

Mudgett, R. E., 1985. Modeling microwave heating characteristics. In *Microwaves in the Food Processing Industry,* (Ed.) Decareau, R. V., Academic Press, New York.

Mudgett, R. E. 1986a. Electrical properties of foods. In *Engineering Properties of Foods.* (Ed.) Rao, M. A. and Rizvi, S. S. H. Marcel Dekker, New York.

Mudgett, R. E. 1986b. Microwave properties and heating characteristics of foods.

Food Technol. 40: 84–93.

Mudgett, R. E. 1988. A systems analysis of solid-state microwave oven feasibility for military space feeding applications. Final report to the U.S. Army Natick Research, Development and Engineering Center's Advanced Systems Concepts Directorate, December.

Mudgett, R. E. 1989. Microwave food processing, A Scientific Status Summary by the IFT Panel on Food Safety and Nutrition. *Food Technol.* 43: 117–126.

Mudgett, R. E., Goldblith, S. A., Wang, D. I. C., and Westphal, W. B. 1980. Dielectric behavior of a semi-solid food at low, intermediate and high moisture contents. *J. Micr. Pwr.* 15: 27–36.

Mudgett, R. E., Mudgett, D. R., Goldblith, S. A., Wang, D. I. C., and Westphal, W. B. 1979. Dielectric properties of frozen meats. *J. Micr. Pwr.* 14: 209–216.

Mudgett, R. E. and Nash, J. 1980. Phase impedance mismatch effects in dielectric heating of composite food systems. *Proc. Int. Micr. Pwr. Inst.* 15: 126–129.

Mudgett, R. E., and Schwartzberg, H. G. 1982. Microwave food processing: pasteurization and sterilization—a review. *AIChE Symp. Ser.* 218, 78: 1–11.

Mudgett, R. E., and Westphal, W. B. 1989. Dielectric behavior of an aqueous cation exchanger. *J. Micr. Pwr. E.^{2}* 24: 33–37.

O'Meara, J. P. 1988. Using science for consumer products: an ISMI-CAS connection. *Microwave World.* 9: 6–11.

O'Meara, J. P. 1989. Variable microwave power: a dilemma for the microwave oven user. *Microwave World* 10: 12–15.

Ofoli, R. Y., and Komolprasert, V. 1988. On the thermal modeling of foods in electromagnetic fields. *J. Food Proc. and Preserv.* 12: 219–241.

Ohlsson, T. 1983. Fundamentals of microwave cooking. *Microwave World* 4: 4–9.

Ohllson, T. 1988. Dielectric properties and microwave processing. Paper presented at the NATO Conference *Food Properties and Computer-Aided Engineering of Food Processing Systems.* Porto 17–21, October.

Pachner, J. 1984. *Handbook of Numerical Analysis Applications.* McGraw-Hill, New York.

Pei, D. C. T. 1982. Microwave baking—new developments. *Bakers Dio.* 56: 8,10,32.

Polley, S. L., Snyder, O. P., and Kotnour, P. 1980. A compilation of thermal properties of foods. *Food Techol.* 34: 76–90.

Pomeranz, Y., and Meloan, C. E. 1987. *Food Analysis: Theory and Practice,* 2nd ed. Van Reinhold Nostrand Co., New York.

Rice, J. R. 1983. *Numerical Methods, Software and Analysis: IMSL Reference Edition.* McGraw-Hall Co., New York.

Risman, P. O., Ohlsson, T., and Wass, B. 1987. Principles and models of power density distribution in microwave oven loads. *J. Micr. Pwr. E^2* 22: 193–198.

Roebuck, B. D., Goldblith, S. A., and Westphal, W. B. 1972. *J. Food Sci.* 37: 199–204.

Sanio, M. R., and Michelussi, I. 1988. *Microwave Aplications in the Food and Beverage Industry.* Ontario Hydro, Toronto.

Schiffman, R. F. 1988. A standardized microwave power test for food companies. *Microwave World.* 9: 3.

Schiffman, R. F. 1989a. Food product safety problems due to microwave heating. Paper presented at MW Foods '89, 2nd Int. Conf. on Formulating Food for the Microwave Oven, March 14–15. The Packaging Group, Inc., Milltown, NJ.

Schiffman, R. F. 1989b. New trends in consumer microwavable food products. *Waveguide* 1: 3.

Schiffman, R. F., Mirman, A. H., and Grillo, R. J. 1981. Microwave proofing and baking bread utilizing metal pans. U.S. Patent 4,271,203.

Schiffman, R. F., Roth, H., Stein, E. W., Kaufman, H. B., Hochhauser, A., and Clark, F. 1971. Application of microwave energy to donut production. *Food Technol.* 25: 718–722.

Shukla, T. P. 1989. Hydrocolloids for microwavable food formulations. Paper presented at MW Foods '89, 2nd Int. Conf. on Formulating Foods for the Microwave Oven, March 14–15. The Packaging Group, Inc., Milltown, NJ.

Singh, R. P. 1982. Thermal diffusivity in food processing. *Food Technol.* 36: 87–91.

Stenstrom, L. A. 1974a. Method and apparatus for treating heat-sensitive products. U.S. Patent 3,809,844.

Stenstrom, L. A. 1974b. Heating of products in electromagnetic fields. U.S. Patent 3,809,845.

Stenstrom, L. A. 1974c. Heat treatment of heat-sensitive products. U.S. Patent 3,814,889.

Stumbo, C. R. 1973. *Thermobacteriology in Food Processing,* 2nd ed. Academic Press, New York.

Swami, S. 1982. Microwave heating characteristics of simulated high moisture foods. M.S. thesis, University of Massachusetts, Amherst, MA.

Talmon, Y., Davis, E. A., Gordon, J. and Davis, H. T. 1982. Temperature profiles in heated frozen foods. In *Food Process Engineering,* AIChE Symp. Ser. 218, 78: 76–82.

Taoukis, P., Davis E. A., Davis H. T., Gordon, J., and Talmon, Y. 1987. Mathematical modeling of microwave thawing by the modified isotherm migration method. *J. Food Sci.* 52: 445–463.

Thompson, D. R. 1982. The challenge in predicting nutrient changes during food processing. *Food Technol.* 36: 97–108.

Tong, C. H., and Lund, D. B. 1989. Modeling thermal moisture profiles in foods during microwave heating. Paper presented at MW Foods '89, 2nd Int. Conf. on Formulating Foods for the Microwave Oven, March 14–15. The packaging Group Inc., Milltown, NJ.

Von Hippel, A. R. 1954a. *Dielectrics and Waves.* MIT Press, Cambridge, MA.

Von Hippel, A. R., Ed. 1954b. *Dielectric Materials and Applications.* MIT Press, Cambridge, MA.

13

Robotics in Food Processing

J. Peter Clark

Epstein Process Engineering, Inc.
Chicago, Illinois

INTRODUCTION

Robotics is the art and science of using programmable machinery to replace human labor. While robots have become relatively common in many areas of manufacturing, they have been relatively uncommon in the food industry so far (Dingle, 1985). When a computerized literature search was performed in 1980, there were no publications found that combined the words food and robotics in the title. When a similar search was performed in 1988, there were only three relevant references identified, and two of these could not be found. There have been more frequent discussions in the trade press, but the overall conclusion is that the food industry has not yet applied robotics to any great extent.

This paper focuses, therefore, on some of the basic principles of robotics and related technologies, describes some of the few applications to date in the food industry, and identifies some of the future opportunities.

The distinction must be made between hard automation and robotics. The food industry has been highly automated in the sense that specialized machinery has been developed to replace human labor in such areas as filling and sealing packages, cartoning and palletizing, and large-scale movement of materials. In contrast, robots are more versatile than the typical specialized automatic machine in that they can be reprogrammed relatively easily to do a variety of tasks. On the other hand, an industrial robot is relatively slow in its action. This is the price paid for the versatility because each motion of the robot must be controlled by a computer. Commercially available industrial robots are relatively expensive and are not typically designed for operation in a food plant environment.

There are several types of devices that are considered industrial robots. The most common is an arm (Fig. 13.1). As shown in Fig. 13.1, the robot really involves a system in addition to the arm. The system includes a controller, which is a substantial industrial computer, a teach-pendant with which the arm is instructed, and a video display terminal and disc drive on which performance data may be stored. Figure 13.2 is a close-up of the arm showing the degrees of freedom at the six joints which permit movement of the arm in three-dimensional space. Robot arms differ in the number of joints or degrees of freedom and in the power mechanism. Both electrical and hydraulic robot arms are commercially available. A variety of tools can be placed on the end of the arm. The most common in industrial manufacturing are welding and spray paintheads. However, it is easy to imagine cutting and gripping tools also being attached.

An alternate approach to the arm is the XYZ pick and place device, which uses linear paths in three dimensions to move a tool or objects.

A large version of the pick and place device is the high rise stacker crane found in automatic storage and retrieval system (AS/RS) warehouses. These cranes are restricted in their movement but are computer controlled to store and retrieve pallets and other objects making very efficient use of storage volume. (Pleban et al., 1985).

A different type of robot that has found fairly wide application in the food industry is the automatic guided vehicle (AGV), which can replace a person-driven forklift. Typically, an AGV follows a path determined by a wire in the floor of a warehouse or factory and can pick up and deliver materials from storage and work stations. In addition to saving labor, AGV often is more gentle in handling fragile materials, such as bright stacked cans (Ellis, 1985).

Finally, certain vision systems might be considered as a type of robot which has found application in the food industry (Somers et al., 1987; Jeffcoat et al., 1987). Vision systems are discussed in greater detail in Chapter 14.

APPLICATION OPPORTUNITIES FOR ROBOTS

In general, the conditions that make application of a robot attractive include jobs or environments that are dangerous for humans, jobs that are boring and repetitious,

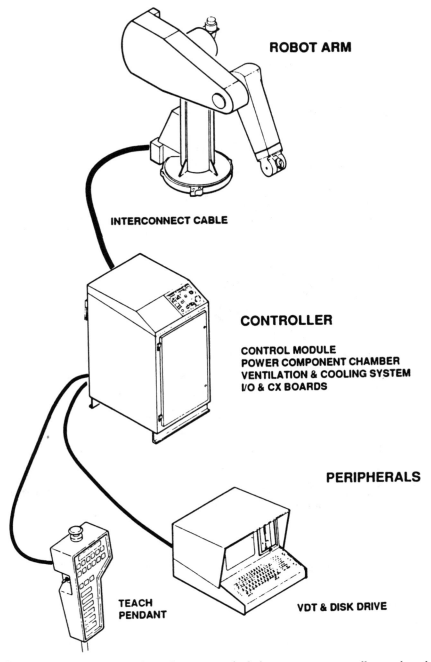

ROBOT ARM

INTERCONNECT CABLE

CONTROLLER

**CONTROL MODULE
POWER COMPONENT CHAMBER
VENTILATION & COOLING SYSTEM
I/O & CX BOARDS**

PERIPHERALS

**TEACH
PENDANT**

VDT & DISK DRIVE

Figure 13.1 Components of a robot system include an arm, a controller, and such peripherals as a teach pendant and display, all connected by appropriate cables. *(Courtesy of Westinghouse Electric Corporation.)*

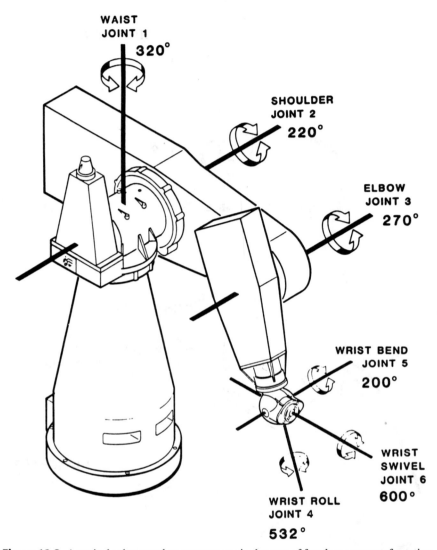

**WAIST
JOINT 1
320°**

**SHOULDER
JOINT 2
220°**

**ELBOW
JOINT 3
270°**

**WRIST BEND
JOINT 5
200°**

**WRIST
SWIVEL
JOINT 6
600°**

**WRIST ROLL
JOINT 4
532°**

Figure 13.2 A typical robot arm has as many as six degrees of freedom or axes of rotation with varying degrees. *(Courtesy of Westinghouse Electric Corporation.)*

jobs where relatively little judgment is required, jobs that require relatively little skill, and, finally, jobs where people may be a source of contamination. In other words, robots are most likely to replace people performing tasks that are relatively unpleasant and may be difficult positions to fill.

Some specific areas of the food industry that meet the criteria for attractive applications of robots include meat packing, cookie and candy assortments, storage and retrieval of pallets of products, transportation of packaging material and finished goods, sorting of ingredients and finished goods, and quality assurance laboratories.

Meat packing is one of the most dangerous industries in the United States. Labor turnover is high, and the tasks of disassembly have been "engineered" so that workers have relatively few, but repetitive tasks to perform. This in turn has led to the repetitive motion injury syndrome and some large fines to companies where workers have suffered. In principle, robots should find some application and there have been some published expressions of interest (Courter, 1985; Charlier, 1988). Following is the abstract of Courter's publication:

> A robot which breaks up pork carcasses is being developed in Australia. A commercial version is expected in six months. The dexterity is already available. The sensors to separate meat from fat are still lacking. Two approaches are being considered. A robot can measure a few basic parameters, and rely on the similarity between pigs for correct action, or all the parameters can be measured. The robot must be fast to be economical. The Australian system uses a remote pointer which feeds information into a micro processor, which in turn directs a knife. 250 carcases can be cut an hour. An alternate technology might use image digitizing.

This application sounds very interesting, but no further information has been found.

In spite of the seemingly obvious opportunity and the published expression of interest and intent, it appears that the meat industry has in fact not progressed very far in the application of robots. Some obstacles include the relatively low margins of the meat packing business, the relatively low research and developments budgets and staffs of the industry, and the lack of commercially available equipment designed for the hostile environment of a meat plant. For sanitation reasons, most meat processing facilities are cooled to about 50°, which leads to nearly 100% relative humidity. In addition, meat processing equipment is routinely washed down at least once a day with hot water. These conditions are very demanding for mechanical equipment, which must resist the resulting corrosion and challenge to electrical systems.

In spite of the obstacles and relative lack of progress so far, meat packing remains one of the great opportunities for robots in the food industry.

Relatively simple pick and place devices have been used for special packing applications such as cookie and candy assortments. While such packing operations are far less dangerous than meat packing, there have been instances of worker injury due to repetitive motions in cookie bakeries. In addition, humans are a source of contamination, which would be eliminated if a machine were used instead. One obstacle so far is the relatively slow speed of most robotic machinery as compared with humans. On the other hand, robots do not tire, take coffee breaks, and one machine can work around the clock if necessary. (Cronshaw 1983).

Automatic storage and retrieval systems (AS/RS) are relatively common in the food industry, especially in snack foods, ice cream, and frozen foods. An AS/RS typically has a higher capital investment cost than a conventional warehouse, but it can have a smaller footprint because it can be much taller than a conventional warehouse. A frozen storage AS/RS eliminates the need to have people working in the discomfort of a low temperature environment. The computer-controlled management of storage space and the highly specialized cranes and rack systems permit very high utilization of the storage volume in such a warehouse. In particular, first in, first out inventory management can be achieved even with very large numbers of stock keeping units (SKU). The design and construction of an AS/RS is very demanding especially if the environment is low temperature. The floor must be very flat and the racks must be precisely installed so that the crane can accurately insert the loads. The floor must be protected against frost heaving as this will cause misalignment of the racks and crane. Lubricants for motors and conveyors must be properly chosen for the environments. When these issues are properly addressed, automatic storage and retrieval systems perform well and justify their considerable cost.

Automatic guided vehicles (AGVs) are also fairly common in the food industry and are best used whenever transportation paths are long and can be relatively well defined. Movement of pallets of finished goods from palletizers to storage and of packaging material from storage to use points are the most common applications. In addition to wheeled vehicles like driverless forklift trucks, there are other automated transportation systems. A common one that has been applied in food plants is a "smart" monorail on which self-propelled vehicles move, picking up and dropping off containers as instructed by a central computer. Switches and sidings are laid out to permit such a system to service multiple stations with a relatively simple circuit.

Palletizing is a special type of robotic application which is quite common in the food industry. Typically, cases are accumulated on a feed conveyor and deposited in a predesignated pattern to make a tier and eventually a full pallet load. It is possible, but not common, to combine a palletizer and a unitizer such as a stretch wrapper. Usually these two operations are separate. More sophisticated palletizers can accumulate cases of different products and build several pallets at the same time by shuttling the incomplete pallets back and forth. One application is known of an industrial robot arm being used as a palletizer in a plant where cases of product differ widely in size and shape but are relatively light in weight. (Most industrial robot arms have a relatively small lifting capacity.)

Vision systems, which would share with robots the need for high powered computing, are increasingly applied in the food industry. For example, ingredients such as carrots or potatoes can be pumped in water through sections of clear pipe where they are brightly illuminated and off-spec pieces are detected optically. A diversion valve following the inspection eliminates the off-color material. Such a system handles large quantities of ingredients at a substantial savings in human labor. Image analysis systems in which a video image is digitized and processed by a computer are applied in several food plants and will have increasing applications in the future. For example, such a system can inspect metal cans for roundness, pizzas for accurate placement of condiments, candy bars for correct shape and frozen meals or entrees for presence of all components. A vision system alone may just collect data but also can be connected to a mechanical device to eliminate or divert out-of-specification materials. Because of the sophistication of the computations and the versatility of such a system, it deserves inclusion as a form of robotics.

One final application of robotics in food processing is the use of small arms and pick and place devices in the laboratory (Swientek, 1986).

Small robot arms can be programmed to prepare samples and execute multistep analytical procedures such as are routinely performed in food plant laboratories. This is repetitive and boring work for humans, which can lead to mistakes in many cases. The robot will perform the same task reliably and continuously around the clock. Since modern analytical instruments almost always have electronic controls and even self-contained computers, linking the instruments and the manipulator is relatively straightforward.

ISSUES AND OBSTACLES

The food industry is a relatively high speed and high volume industry compared with other types of manufacturing. In contrast, typical "intelligent" automation devices have relatively slow speeds such as 15 cycles per minute. When the products of food processing are relatively concentrated (in cases or pallets), such a speed may be adequate; earlier in the process it is not. Thus, the major applications of robots in food processing have come in finished goods material handling or packaging materials handling.

The general category of tooling, that is appropriate devices at the end of a robot arm, has not been developed very far in the food industry. Such things as saws, knives, and gripping devices must be developed that are reliable, clean, sanitary, and precise.

Some of the most creative applications of robotics will come with the integration of sensing devices such as vision and image analysis with robotic manipulators. This would permit the removal of meat from bones, the separation of fat from lean meat, the placement of irregular objects in trays, and the adaptation to variable conditions. Both the sensing devices and the algorithms need to be developed for food application.

Increased computing power for robotic applications is highly likely. The cost of computers constantly declines, and the ability to apply the power is becoming more widespread.

Finally, as previously mentioned, there is a need to adapt commercially available robotic devices to the food plant environment so as to maintain sanitation and survivability of the machinery. Robots have been built and used in clean environments, but not in the high humidity wet environment typical of a food plant.

In summary, robotics has not been widely applied in the food industry. There are few references outside the trade press and these appear to be mostly speculative. Direct contacts with suppliers and research centers elicited little interest in research on robotics for the food industry. The industry itself is probably going to have to perform such research.

Application of robots in the food industry remains a great opportunity for creative engineering.

ACKNOWLEDGMENTS

Mary T. Murphy of Andersen Consulting Company arranged for the literature search. Figures were provided by Westinghouse Electric Corporation.

REFERENCES

Charlier, M. 1988. Robots to be used in slaughterhouses. *Wall Street Journal,* Dec. 27: 82.

Courter, E. 1985. Meat processing robots. *Robot/X News* 4(4): 8.

Cronshaw, T. 1983. Putting on the style. *Industrial Robot* 69.

Ellis, R. E. 1985. Automated guided vehicles move packaged milk. *Food Processing,* April: 84–85.

Jeffcoat, R. L., Captain, K. M., and Hudson, D. M. 1983. Machine vision in industry. *Particulate and Microbial Control,* Jan/Feb.

Lingle, R. 1985. Robotics: thought for food, *Prepared Foods* 154(12): 61–64.

Pleban, J. P. and Ellis, R. F. 1985. Automatic storage retrieval systems. *Food Processing,* Jan: 78–79.

Somers, J. D., Frank, E., and Swientek, R. J. 1987. Optical sorting. *Food Processing,* Jan: 78–79.

Swientek, R. J. 1986. Emerging analytical technologies in the food R&D laboratory. *Food Processing,* Dec: 128–130.

14

Integration of Computers in Food Processing

Israel Saguy

The Pillsbury Company
Minneapolis, Minnesota

Leon Levine

Leon Levine & Associates, Inc.
Plymouth, Minnesota

Stephen Symes and Enrique Rotstein

The Pillsbury Company
Minneapolis, Minnesota

INTRODUCTION

In recent years, the staggering double-digit growth of computer systems established a \$188 billion world market (Anonymous, 1989a). The sale of personal computers reached \$28.5 billion in the United States alone (Anonymous, 1989b) and are still growing at a rate of 12–17% a year. The astonishing reduction of cost and, simultaneously, the increased computer power measured by MIPS (million instructions per second) is mind boggling. For instance, by 1993 the cost of hardware to process one MIPS is predicted to reach \$140, \$6,000, and \$54,000 for personal computers, minicomputers, and mainframe, respectively. These numbers represent an amazing improvement of performance/cost compared to 1981 figures. Also, a \$5,000 desktop with 20 MIPS may soon outperform a mainframe which not too long ago cost \$2.9 million (Anonymous, 1989a).

Computer sales are slowing down in the United States but growing in Europe and Japan. Overall sales grew 10% in 1988, 43% faster in Europe than in the United

States. Based on these figures, European sales could surpass America's by the mid-1990s (Anonymous, 1989a). One of the most signficant changes in recent years is network communications. An increasing number of business customers and corporate users are turning networks into the fastest growing segment of the computer industry. Worldwide sales of network hardware and software jumped 85% in 1988 to $4.8 billion. Furthermore, 15% of the 40.1 million PCs in U.S. businesses are linked together. By 1992, 47% of the 60.1 million desktop computers in use will be networked. Coca-Cola Foods is one example where nearly 1000 PCs are already networked (Anonymous, 1989c).

The food industry and its counterparts face three major challenges: the capability to ensure products of consistently high quality to satisfy consumer needs and demands, the ability to produce cost-effective products, and the flexibility to respond rapidly and economically to the changes occurring in the marketplace. Traditional technologies utilized in the food industry are insufficient to meet the challenges and needs of the future. Perhaps only computer technology can simultaneously meet these challenges and provide the flexibility and adaptability dictated by market forces.

Assessing the overall expenditure by the food industry on computers is very difficult if not impossible. Our computerized literature search resulted in very scarce information. Only one report shows that in 1988 the food and beverage industry have increased their expenditures by 15.7% compared to 1987. This rate is one of the lowest in the U.S. (Anonymous, 1988a).

New developments in computer and communications technology are leading to the integration of virtually all manufacturing functions. Artificial intelligence (AI), expert systems, computer-aided design (CAD), computer-aided manufacturing (CAM), and computer-integrated manufacturing (CIM) are only a few examples of this evolution.

Although the penetration of computer technology into the food industry is lagging, its impact nevertheless is already significant. Unlike other American businesses, only a few food industries suffer from lack of technology. Companies are flooded with an astonishing, even bewildering, array of computers, telecommunications gear, and "factories of the future." Yet the ability of the food industry to utilize and integrate this technology in daily operations is more important, and may be considered its Achilles' heel.

The best use of computer technology results not from the unbridled acquisition of leading-edge hardware, but from a solid understanding of technological processes, and from their innovative and disciplined use. This basic and scientific understanding is paramount for computer integration. Furthermore, it is critical to look beyond short-range cost/benefit analysis only, since companies that let simple financial arguments dictate how they reach their strategic decisions fail to act on the right questions and ultimately choke on the dust of more farsighted competitors (Brody, 1988).

This chapter highlights only a few topics of computer integration in food processing. They are chosen to explain and illustrate some of the basic principles of modeling, simulation, optimization, sensing, and process control.

MODELING, SIMULATION, AND OPTIMIZATION OF INDUSTRIAL PROCESSES

Modeling of a process is the application of engineering sciences and knowledge to describe mathematically a process and its performance. Simulation is the use of this model to assess different scenarios. This is done to gain understanding, to evaluate alternatives, and to answer specific questions. When the simulation is performed with the aim to maximize (or minimize) a certain process objective, we describe the task as optimization. This is usually done using some type of mathematical algorithm to guide the search towards the desired optimum.

Modeling, simulation, and optimization can take place before the actual system is built or to improve an existing system. In general, the more complex the system and the greater the number of interactions between its constituents, the more valuable it is to use these techniques. The reasons for adopting these techniques are multiple and include: improving quality, improving yields, predicting operability, developing control strategies, cutting costs, evaluating alternatives, changing operational modes, analyzing risk, debottlenecking, formulating start-up, shut-down, and change-over policies, and doing parametric sensitivity studies.

The model can be static, focused on steady state operation, or dynamic, focused on transient state. It can be fully deterministic, not allowing for variation, or probabilistic with time-varying parameters (stochastic).

Modeling and simulation are no newcomers to the engineering profession. What makes these techniques particularly valuable nowadays is the widespread use of computers. They bring the power to handle large quantities of information and solve sets of hundreds and thousands of equations in a very short time. Therefore, once the models are set and validated, we can "test drive" all kinds of hypotheses and play "what if" scenarios in the time it takes to introduce the fractional data changes required.

Principles of Modeling

Modeling is an important approach to a number of problems in the food industry. The modeling procedure draws upon techniques developed utilizing multidisciplinary inputs such as engineering, mathematics, statistics, numerical analysis, and computer software expertise. Nevertheless, its successful implementation and application to the real problems occurring in food science and technology require, first and foremost, intimate familiarity with the system and the problem being analyzed. Hence, it is important to the food practitioner to be well informed about the various tools available, such as mathematical techniques, statistical tools, etc. These techniques are widely documented in various textbooks (e.g., Saguy, 1983) and articles, and will not be discussed in further detail here.

This short introduction highlights some of the general principles of modeling and their application to food processing.

The mathematical formal way of describing any process is given by one or more equations expressed by the following functional relationship (using a vector notation):

$$y = f(x, P, e) \tag{1}$$

where y is the dependent variable vector; x is the independent variable vector; P is the vector parameter, and e represents the error vector, which includes both the inherent and measured error.

The model may be a linear and/or nonlinear set of equations either algebraic, differential, or partial differential that describes dynamic behavioral aspects of the system being analyzed. This set is referred to as the mathematical model, and is generally solved by applying a numerical scheme (algorithm) in a computer program. These different model classes, required to describe complex systems such as food processing, are quite complicated in many cases.

As the process becomes more complex and requires increased sophistication and automation, there is a need for an analytical approach to problem solving. Typically, there are several models for the same system, each suitable for solving a particular problem. In both of the two broad classifications—steady state and dynamic models—the degree of detail required depends on the problem to be solved as well as on the nature of the data available.

To date, no general formulation is available to produce the "proper" model, nevertheless, we utilize the procedure outlined in Fig. 14.1 to demonstrate these steps (Saguy and Karel, 1987):

Planning.   The first and perhaps most crucial step in model building is that of problem definition. Technical problems normally occurring in the food industry are very diverse and extremely complex, thus, it is up to the food practitioner to establish and define exactly the problem and its objective. Adequate problem definition and pertinent assumptions are essential and invaluable in outlining a path from the problem to the solution in any application. Adequate problem definition is even more important for real complex applications such as simulation and optimization of food processes which follow different control and production policies. As there are no general rules that can be applied for choosing the "right" goals and adequate budget, management must address and resolve these issues.

Development.   The second step of modeling involves writing the appropriate

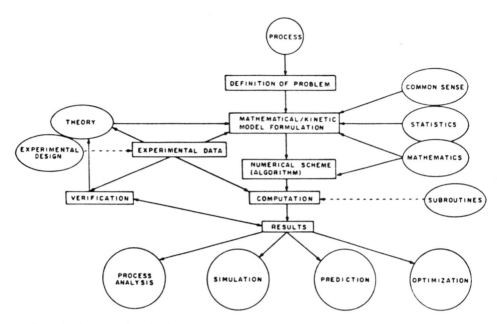

Figure 14.1   Procedure for formulation of kinetic/mathematical models. (*Reprinted from Saguy and Karel, 1980, with permission.*)

equations describing the process and the theory that governs the phenomena. If such a theory is not known, one postulates an empirical model and tests its validity during the verification step. After a proper model is chosen, data are gathered and the parameters of the model as well as its adequacy are estimated, utilizing computer-aided statistical and mathematical tools. Special attention is needed to find the specific model most consistent with the data, to estimate how well the model fits the data, and to see if it actually is consistent with the data representing the phenomena being analyzed. It is worth noting that the model may be as "good" as the assumptions made and the data gathered and should be used in most cases within the range of actual data collected; extrapolation beyond the range of the data used to establish the model is not recommended. Experimental design and sequential analysis are fundamental requirements for obtaining "good" data in the most economical way.

Verification and maintenance.   The validity of the theory and/or the experimental model must always be verified by comparing the results to a second set of data, which is not a part of the model building and is used only for verification needs. It is worth emphasizing that no model should be utilized before adequate and thorough verification is successfully carried out. Issues such as stability of the derived parameters, lack of fit, and applicability of the model should be resolved.

Whenever theoretically driven models are unavailable, several empirical models are thoroughly tested using robust statistical tests specially designed for this purpose.

The three-step model-building procedure outlined below includes three major blocks, namely, planning, development, verification and maintenance, and is graphically summarized in Fig. 14.2.

The derived model is a set of mathematical equations describing the data gathered. This always raises the questions of what data to collect and how many data points are needed. The data collected facilitate these requirements (Rand,

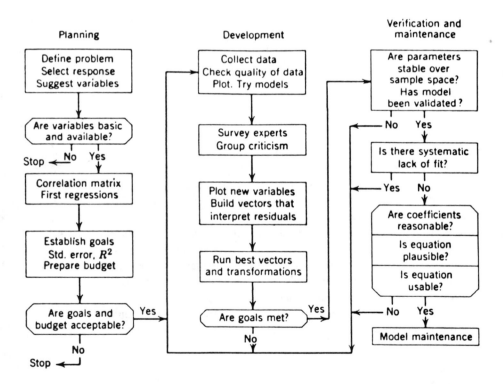

Figure 14.2 Summary of the model-building procedure. (*From Draper and Smith, 1981, with permission.*)

1983): data should cover evenly the region of interest; enough distinct data points are needed so that the model of the system is overdetermined; and there should be replicates to provide an estimate of error.

Clearly, modeling takes significant time and effort and is not an easily acquired skill. Yet, once established, the model is extremely useful for simulation and optimization. To fully benefit from the investment in developing a model, simulation is exploited to address a spectrum of problems and issues such as different manufacturing scenarios, optimization, quality improvements, control, etc.

Simulation is normally based on this sequence:

1. **Define**—Goals and objectives; assumptions; model; and data gathering.
2. **Mathematical modeling**—Algorithm; verification; validation; and documentation.
3. **Implementation**—Analysis of data output, scenarios and manufacturing conditions; documentation of results; basic understanding; and user training.

The above steps indicate that payback is possible only after a significant up-front investment is made in basic understanding of the process, data gathering, and analysis.

On the surface, simulation is considered as simply the computerized creation of an analog of a real manufacturing process. Yet, as pointed out previously, the complexity of food processes mandates that simulation be a highly selective process, fully structured in its methodology and driven by pertinent data. This means that simulation has to be managed very carefully. Naturally, simulation does not stand apart from other technologies and cannot be managed in isolation. In many cases it is connected to computer-aided design (CAD), manufacturing resource planning (MRP II), or just in time (JIT) (Malone, 1988).

Simulation methodologies have not changed dramatically in recent years. Nevertheless, the specially designed simulation software computer languages have changed a great deal. For instance, the GPSS/H is a descendant of GPSS, the 25-year-old general-purpose simulation language. Other software includes SI-MAN (flow chart interface), TESS, SLAM, and MAP/1. The new software is simpler, more modular, and easier to maintain. Also, the software has become more natural (less math and code), more menu-driven, and more communicative with higher animation and graphics. The proliferation of personal computers has significantly affected this field of simulation software, creating an even more user-friendly environment.

The Macroscopic and the Microscopic Approach

In the simulation process, there is a wide spectrum of alternatives, from the plant regarded as a single unit to zooming in on the detailed modeling of a particular piece of equipment. Therefore, it is important to decide early just what the scope of the simulation subject will be. Consider, for instance, the apple processing plant illustrated in Fig. 14.3. We call the macroscopic approach the practice of taking all processes in the figure as "black boxes," for which we only have input-output information. Typically, this includes flowcharts in and out, energy consumption, or production labor. This is straightforward information, which results from material and energy balances. When complemented with cost data, it provides an overall picture of the plant operation, thus allowing us to play "what-if" scenarios or to apply optimization algorithms.

The advantage of the macroscopic approach is that one can have a model running in a period of a few weeks. All information is in itself groundwork for further modeling, and the conceptualization of the plant as a whole always helps to develop understanding. The information obtained is most useful especially when a large number of interactions within the process exist and the process nature is more complex. A typical result of a macroscopic simulation is shown in Fig. 14.3, for which we need to determine the optimal product mix resulting in maximum profit for the plant. Similarly, one obtains the optimum operation pattern for minimum energy consumption, minimum waste, optimal raw material variety mix, etc.

There is a limitation to the macroscopic approach. Because of the black box nature of the unit operations and processes, one cannot depart too much from those operating conditions for which the data are generated. Consider as an example the issue of waste in a cheese depositor, which is part of a pizza manufacturing plant (Fig. 14.4). Point A corresponds to the current operating production rate for which we obtain the data. We do not have a model for the depositor; it is a black box. Thus, we assume in running our macro simulation that the depositor follows a linear relationship expressed by line 1. The true operating line is line 2, and therefore our model is more unrealistic the more we depart from point A.

The microscopic approach is to model each one of the individual pieces of equipment, so as to simulate its performance under changing conditions. This is a task akin to process equipment design, with exclusion of detailed mechanical and electrical design. What we need are equations describing equipment behavior in terms of relevant variables. In the case depicted in Fig. 14.4, it is an empirical equation describing line 2. The aroma recovery unit in Fig. 14.3 requires a model based on the theory of vacuum distillation. The model of one unit can be the sole objective of the simulation effort, because we need more understanding, or we contemplate working under drastically different operating conditions. On the other hand, when all units of a production plant are modeled, we move into a higher hierarchy of simulation. Here the macro- and microscopic approaches come together, and we have the ability to realistically simulate all kinds of scenarios with a high degres of confidence in the validity of our results.

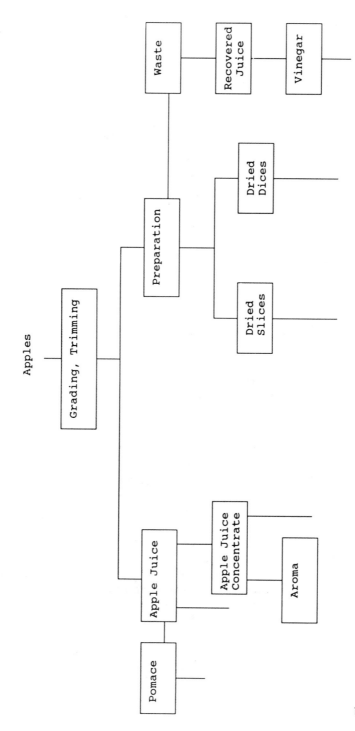

Figure 14.3 Flowchart of an apple-processing plant.

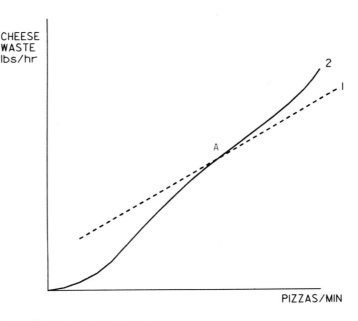

Figure 14.4 Waste in a pizza depositor.

Linear Programming as a Macroscopic Tool

We illustrate the principles involved in linear programming (LP) with an example. Consider that the quality (q) of a product has been proven to be a function of the concentration of sugar (x) and acid (y). The empirical function is:

$$q = \frac{2}{3}x + y \qquad (2)$$

We wish to make decisions related to quality. For that reason, Eq. (2) is called "objective function." It is subject to the following constraints:

$$x \geq 0; \qquad y \geq 0; \qquad y \leq x + 0.2; \qquad x \leq 0.8 - y \qquad (3)$$

This means that we cannot have a negative sugar or acid content, that the acid content must be lower than the sugar content by at least 0.2, and the sugar content cannot be greater than 0.8 − y. Figure 14.5 illustrates the constraints, defining a region (shaded in the figure) made up of all feasible points, i.e., all points which fall within the four constraint conditions given by Eq. (3).

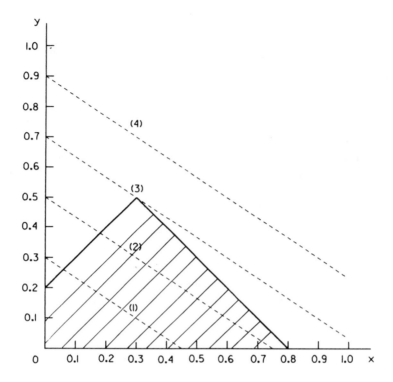

Figure 14.5   Optimal LP resolution of a quality problem.

We can rearrange Eq. (2):

$$y = q - \frac{2}{3}x \tag{4}$$

to show that the quality objective of q is the intercept of the linear function $y = f(x)$. In Fig. 14.5 lines (1), (2), (3) and (4) correspond to intercepts, i.e., quality indices, 0.3, 0.5, 0.7, and 0.91, respectively. It is obvious from Fig. 14.5 that the maximum value of g compatible with the constraints is $q = 0.7$, corresponding to $x = 0.3$ and $y = 0.5$. This illustrates the main feature of LP: the maximum of the objective function is a feasible point located at one of the vertices of the polygon constructed by drawing the problem contraints.

In practice we have much more than two variables and we need to use matrix algebra to solve the problem, but the essential characteristics remain the same.

Linear programming problems are normally solved using the simplex technique. This is a two-phase procedure. Phase 1 finds an initial basic feasible solution if one exists. Phase 2 uses these values as the starting point and either finds the solution or reveals the information that the minimum is unbounded. There are a number of computer packages which solve LP problems, such as LINDO (Schrage, 1981).

Example: Apple Juice Concentrate Plant. Figure 14.6 shows an apple juice concentrate plant. The problem is studied by Bandoni et al. (1988). The plant crushes apples, which are fed to five parallel continuous belt presses. The resulting pomace is rewetted for 30 minutes at 60°C and further pressed in a Bucher-Guyer discontinuous press. Both juice streams are fed to the aroma stripper, the head product going to the aroma recovery column and the tail product to clarification. The clarified juice is concentrated to 72° Brix in a triple-effect evaporator. We will explore the use of linear programming for optimizing this fixed process arrangement. The objective function is the profit, z, as given by:

$$z = \sum_{j=1}^{N} c_j \, x_j \tag{5}$$

where x_j are the independent variables—x_1, fruit input, x_2, pomace not subject to rehydration, and x_h, rehydration water—and c_j are coefficients, which include cost and sales price of different items. The plant sells two products: aroma and juice concentrate. The constraints are of the type:

$$x_j \geq 0 \tag{6}$$

and

$$\sum_j A_{ij} \, x_i \leq b_i \tag{7}$$

These are capacity constraints; A_{ij} are the coefficients that, when multiplied by x_j, result in the capacity requirement for equipment i. The model is essentially macroscopic and constructed on the basis of a material balance. The only detailed modeling corresponds to the rehydration and processing and obtains an empirical equation relating soluble solids recovery with the rehydration percentage. Table 14.1 shows the result of running the LP model. It provides data on the value of all the flow rates as depicted in Fig. 14.6. Once these data are obtained and compared with the corresponding equipment capacities, one can do a bottleneck analysis. Operational capacity requirements are compared with actual equipment capacities. The result is shown in Table 14.2, as a percentage of the latter. We see that the clarification section is the current bottleneck of the plant. If this section were expanded, the next bottleneck would be the discontinuous press, then the boiler, and so on. The influence of changes in costs, prices, and capacities is studied by a modified LP technique, as indicated by Bandoni et al. (1988).

To analyze the influence of apple variety on plant performance, the LP program

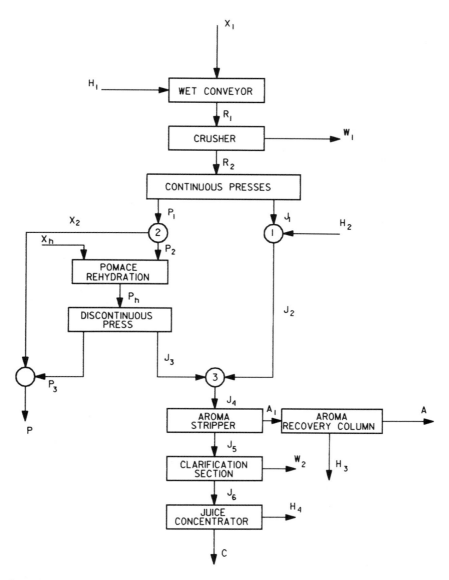

Figure 14.6 Flow sheet for apple juice concentrate plant. (*From Bandoni et al., 1988, with permission.*)

Table 14.1 Optimum Net Profit and Operating Flow Rates

Net profit	$z = 131,300$ \$ month^{-1}
Fresh apple input	$x_1 = 4912.9$ tons month^{-1}
Unextracted pomace	$x_2 = 0.0$ tons month^{-1}
Rehydration water	$x_h = 785.9$ tons month^{-1}
Pulp	$R_2 = 4911.8$ tons month^{-1}
Continuous press juice	$J_1 = 2976.5$ tons month^{-1}
Pomace from continous press	$P_1 = 1964.7$ tons month^{-1}
Rehydrated pomace	$P_h = 2750.6$ tons month^{-1}
Discontinuous press juice	$J_3 = 1111.4$ tons month^{-1}
Total pomace	$P = 1636.2$ tons month^{-1}
Juice feed to stripper	$J_4 = 4090.2$ tons month^{-1}
Stripped juice	$J_5 = 3600.0$ tons month^{-1}
Clarified juice	$J_6 = 3420.0$ tons month^{-1}
Juice concentrate	$C = 619.0$ tons month^{-1}
Aroma concentrate	$A = 34.4$ tons month^{-1}
Steam consumption	1967.9 tons month^{-1}
kg fresh apple/kg concentrate	7.94

Table 14.2 Bottleneck Analysis

Equipment	Percentage of full capacity
Mill	56.8
Continuous presses	58.5
Discontinuous press	95.5
Aroma stripper	65.5
Boiler	82.0
Clarification section	100.0
Multiple effect evaporator	54.8

can be run at different levels of soluble solids, representing the different varieties available. Figure 14.7 shows the results, indicating that the higher the soluble solids content, the greater the yield and profit, whereas more raw material is required for other products (C is tons of concentrate per month). Yet another approach is to consider that apples may be scarce or in surplus from one season to another. To consider the implications, the LP model is solved with preimposed x_i values, below and above the optimum. Figure 14.8 shows the resulting optima. The net profit z line shows, as it should, decreases to the left and right of the previously found optimum $x_i = 4,912.9$ tons/month. Beyond $x_1 = 6,174$ tons/month it is necessary to discard some of the pomace without further treatment. As for the percentage rehydrated, below $x_1 = 4,458$ tons/month, its optimal value is 60%, becoming progressively less as the tonnage of apples increases above this value. It becomes zero at the level at which untreated pomace must be disposed.

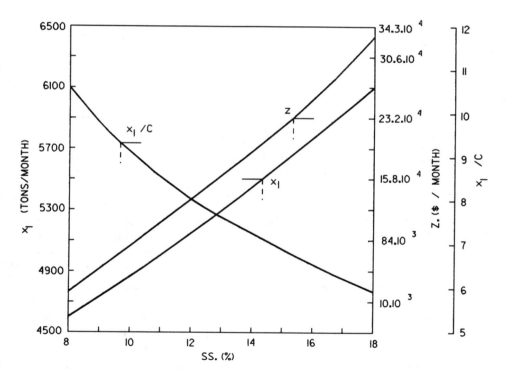

Figure 14.7 Influence of soluble (SS) on optimal performance. (*From Bandoni et al., 1988, with permission.*)

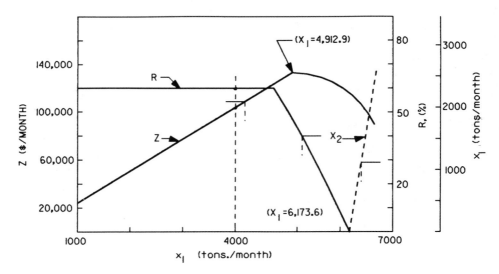

Figure 14.8 Optimum values of rehydration percentage (R) and of pomace disposed of for different raw material input rates, (*From Bandoni et al., 1988, with permission.*)

Much more insight can still be gained with respect to optimal plant operation, but the above is enough to illustrate the power of the method.

Example: Quality optimization in aseptic processing.   We further demonstrate the utilization of linear programming for quality optimization in aseptic processing. This example is based on the paper by Saguy (1988).

Aseptic processing and packaging refers to a technique whereby food is commercially sterilized outside of the package, aseptically filled, and sealed in a previously sterilized package. The process utilizes high-temperature–short-time (HTST) or ultrahigh-temperature (UHT) processing methods. Aseptic processing offers an avenue to optimize overall food quality. Due to public safety considerations, the sterilization effect on microbial inactivation of low-acid foods usually is evaluated only during the holding process.

To derive the optimal process, microbial and enzyme inactivation, cooking quality, and retention of heat-labile constituents are considered. The attributes chosen in this case are two different quality constituents which express two main modes of quality loss during processing. Chlorophyll and thiamin retention are such attributes.

The first step is the formulation of the objective function and constraints in a linear form. Hence, the equations expressing microbial inactivation (F), cook-value (C), enzyme inactivation (E), and quality (Q) are linearized by taking their logarithm. The linear formulation follows:

Choose the holding time, x_1, and the temperature, x_2, to maximize the objective function, Z, defined as the maximum retention of both thiamin and chlorophyll:

$$Z = x_3 + x_4 \tag{8}$$

where x_3 and x_4 are defined as in Eq. (12). Subject to these constraints:

Microbial inactivation

$$x_1 + 1/z_F \cdot x_2 \geq b_1 \tag{9}$$

where $x_1 = \log (t)$ (time = time in sec); x_2 = temperature, °F; z_F = the number of degrees required for the thermal inactivation curve to traverse one log cycle and taken as 18°F.

Cook-value

$$x_1 + 1/z_C \cdot x_2 \geq b_2' \text{ minimum}$$

and

$$x_1 + 1/z_C \cdot x_2 \leq b_2' \text{ maximum} \tag{10}$$

where $z_C = 59.4$°F.

Enzyme inactivation

$$x_1 + 1/z_E \cdot x_2 \geq b_3 \tag{11}$$

where $z_E = 50$°F.

Quality (thiamin and chlorophyll)

$$x_1 + 1/z_Q \cdot x_2 + x_3 = b_4$$

$$x_1 + 1/z_Q \cdot x_2 + x_4 = b_5 \tag{12}$$

where $z_Q = 39.6$ and 69°F for thiamin and chlorophyll, respectively; x_3, x_4 = $-\log[\ln(Q/Q_o)]$; and Q, Q_o are the initial and final concentrations of thiamin and chlorophyll, respectively.

Using the linear notation derived above, the tableau representing the linear programming problem and the values of the b parameters are listed in Table 14.3. These data derive from the following constraints:

1. A minimum of Fo = 5 min is needed for sufficient spore inactivation.
2. A cook-value greater than 1 min and lower than 2 min is desired. These values are within the range recommended for vegetables.
3. Peroxidase is chosen as the constraint for enzyme inactivation. A $D_{250} = 1$ min is used, representing the resistance of a typical enzyme.
4. For practical purposes, the holding time is within 20 and 120 sec.
5. The holding temperature is 250–300°F.

The right-hand-side values expressing the above constraints after the logarithmic transformation are also included in Table 14.3.

The optimal solution derived by linear programming is summarized in Table 14.4 (the effect of 1·D and up to 3·D values are tested) and in a simplified two-dimensional chart in Fig. 14.9. The latter indicates that for 1·D enzyme

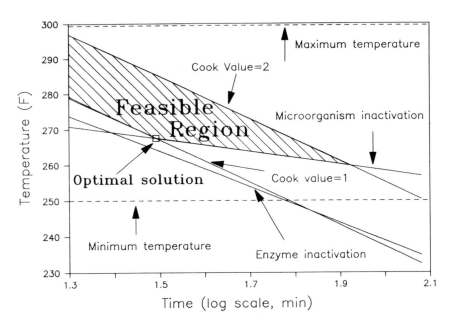

Figure 14.9 Optimization of an aseptic process.

Table 14.3 Initial Simplex Tableaux Containing the Linear Programming Coefficients for the Optimization of HTST Processing of Peas

Eq. no.	Task	Z	x_1	x_2	x_3	x_4	Constraint	Right-hand side
1	Objective	1	0	0	1	1	—	
	Microorganism inactivation (12·D)	0	1	0.0556	0	0	$\geq$	16.366
2	C-value($\geq$1)	0	1	0.0168	0	0	$\geq$	5.987
3	C-value($\leq$2)	0	1	0.0168	0	0	$\leq$	6.288
4	Enzyme inactivation (3·D)	0	1	0.0200	0	0	$\geq$	7.255
5	Thiamin	0	1	0.0253	1	0	=	9.928
6	Chlorophyll	0	1	0.0145	0	1	=	6.160
7	Time ($\geq$20 sec)	0	1	0	0	0	$\leq$	1.301
8	Time ($\leq$2 min)	0	1	0	0	0	$\leq$	2.778
9	Temperature (max.)	0	0	1	0	0	$\leq$	300
10	Temperature (min.)	0	0	1	0	0	$\geq$	250

Table 14.4 Optimal Solution Derived by Linear Programming

| Enzyme inactivation (min) | Linear programming results | | | | Optimal derived values | | | |
	x_1	x_2	x_3	x_4	Time (sec)	Temp. (°F)	Thiamin (%)	Chlorophyll (%)
1.0·D	1.49	267.5	1.67	0.79	31	267.5	97.9	85.0
1.5·D	1.30	282.7	1.48	0.76	20	282.7	96.7	84.0
2.0·D	1.30	288.9	1.31	0.67	20	288.9	95.3	80.7
2.5·D	1.30	293.8	1.19	0.60	20	293.9	93.8	77.8
3.0·D	No feasible solution				—	—	—	—

inactivation, the optimal solution is defined by the binding constraints of microorganism inactivation and cook-value. The optimal solution at this point is 31 seconds and 267.5°F. Under these conditions the thiamin and chlorophyll retention are 97.9 and 85.0%, respectively (Table 14.4).

Additional results indicate that for the system considered, from 1·D up to 2.5·D of enzyme inactivation is possible. Retention of thiamin and chlorophyll is 97.9–93.8% and 85.0–77.8%, respectively. These values are very high and indicate that indeed quality attributes may be optimized.

The constraint imposed on the minimal time is active for values higher than 1.5·D enzyme inactivation. This constraint is of practical importance, and its actual value should be evaluated for each system. Data presented in Table 14.4 also indicate that 3·D inactivation of peroxide in the holding tube is not possible. This finding manifests that for vegetable products the risk of peroxidase regeneration at high-temperature sterilization is a serious consideration and possible alternatives should be evaluated.

The optimal solution also indicates that peroxidase inactivation is the active constraint at 1.5–2.5·D. Hence, microorganism resistance may not be the active constraint for HTST processes. These data are in agreement with enzyme inactivation studies where some enzymes are shown to be extremely resistant and thus may represent a severe problem.

Detailed Process Simulation and Optimization Processing

To further demonstrate some of the principles of process simulation and optimization that should start with modeling as discussed above, it is helpful to give a few specific examples rather than to attempt to build a theory of modeling methodology.

Example: Potato blanching. Modeling of potato blanching is studied by Kozempel and co-workers (Kozempel et al., 1988) in the context of their effort to develop a potato flake process simulator. To model the blancher we need a material and energy balance and enough data to decide what kind of mixing behavior occurs in the equipment: back-mix, plug, etc. Consider a blancher such as the one shown in Fig. 14.10. For full back-mixing of fluid:

$$C_w = \frac{w\,M\,C_{p1}\,(1 - \alpha)}{w\,M\,(1 - \alpha) + W_{of}} \tag{13}$$

$$\alpha = 8/(\pi^2)\,exp(-\pi^2 D t/L^2) \tag{14}$$

where D is the effective diffusivity of soluble solids; t, blanching time; L, slice thickness; all other terms are defined in Figure 14.10.

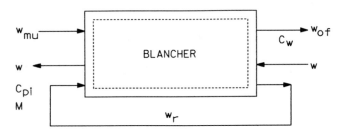

Figure 14.10 Schematic presentation of a potato blancher (w_{mu} = lb/hr make up water; w_{of} = lb/hr overflow; w = lb/hr potatoes; w_r = lb/hr recirculating water; M = lb/water/lb potato; C_{pl} = lb solute concentration/lb of potato, C_w = lb soluble solids/lb water).

Equation (13) provides a functional relationship between the operating variables. We use it to simulate how the blancher functions. Consider for instance that our concern is to minimize the biological load in the overflow water. Figure 14.11 shows a simulation centered around current operating conditions (C_w = 0.019, x = 1.00). Lines w, t, L, and w_{of} show how C_w changes when the current value of the corresponding variable is multiplied by a coefficient factor, x.

This simplified example shows an aspect of simulation: the search for understanding. It illustrates that we need good hard data of inputs and outputs and a knowledge of the equipment behavior. Given this, we can map the way the equipment will react to a change in variables and compare it with our objectives and constraints.

Example: Combination of microscopic modeling and optimization—optimal drying paths.   An excellent application of microscopic modeling in combination with optimization is the search for optimal drying profiles. The goal of this effort is to maximimze the rate of drying while minimizing or controlling undesirable side reaction(s) which may have a negative effect on finished product quality.

Several articles on this problem have been published (Mishkin et al., 1982, 1983, 1984). These consider the effects of air drying of potato slices on browning and ascorbic acid oxidation.

The problem is quite complex. The rate of drying of the material is defined by Fick's law, which is stated as the following differential equation:

$$\frac{\delta M}{\delta t} = \frac{\delta}{\delta x} \left(D \frac{\delta M}{\delta x} \right) \tag{15}$$

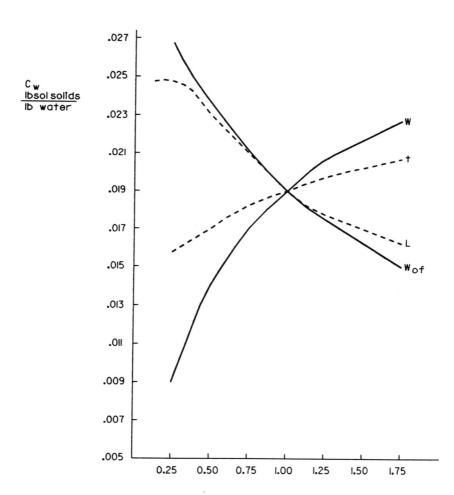

Figure 14.11   Sensitivity analysis.

where M is the product moisture (dry basis); D is the diffusivity; t is time; and x is position.

The diffusivity is a function of temperature and moisture content, so this is a nonlinear, partial differential equation which can only be solved by numerical integration.

Simultaneously with the drying of the potato tissue, browning and ascorbic acid oxidation occur. The rates of thse reactions are defined by: for browning,

$$\frac{dB}{dt} = k_b \qquad (16)$$

for ascorbic acid oxidation,

$$\frac{dC}{dt} = -k_c \cdot C \qquad (17)$$

where B is the browning as measured colorimetrically; C is the ascorbic acid concentration; and k_b, k_c are the appropriate rate constants.

Like diffusivity, the rate constants are a function of both moisture and temperature. To determine the ultimate browning or ascorbic acid retention of the potatoes, Eqs. (16) and (17) must be solved simultaneously with Fick's law (Eq. 15).

Because of the complicated interaction between the rate of drying and the rate of undesirable side reactions, the possibility exists for an "optimal" drying profile, defined as the dryer temperature profile as a function of time. That is one which maximizes the rate of dehydration while subjecting the undesirable side reactions to some constraint and the potato quality to some maximum constraint.

The experimental determination of the "optimal" drying profile is virtually impossible, requiring prohibitively expensive experimentation. As a consequence, this is an ideal application of the use of computer modeling.

Figure 14.12 illustrates the drying profile required to minimize browning. Figure 14.13 illustrates the drying profile required to maximize ascorbic acid retention. The browning study illustrates that it is desirable to have the minimum drying temperature somewhere in the middle of the drying cycle. This is quite surprising as it is not the normal industrial practice.

Example: The use of modeling to assess parameter sensitivity and response of extruders to external influences. Extrusion is a very important food industry unit operation. The characteristics of extruders are very complicated and poorly understood. As a consequence, the extrusion process design often proceeds without adequate understanding of certain important design parameters. One aspect of extruder performance that causes considerable problems for industrial users is

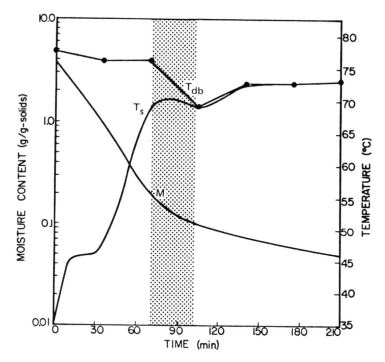

Figure 14.12 Optimal dryer-temperature (T_{db}) control profile for minimizing browning, with resultant potato disk temperature (T_s), moisture (M), path for a 210-min process. Shaded area represents a moisture-sensitive region. (*Reprinted from Mishkin et al., 1983.*)

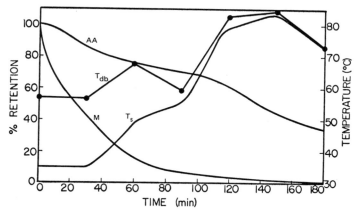

Figure 14.13 Optimal dryer temperature (T_{db}) control profile, and the resultant sample temperature (T_s), moisture content (M), and ascorbic acid concentration (AA) (180-min process). (*Reprinted from Mishkin et al., 1984.*)

extrudate output and performance "surges" without readily apparent causes. If these causes are identified, then the process design can include items to help minimize the problem.

Several recent papers (Roberts and Guy, 1986; Levine et al. 1987a,b) explore the sources of observed instabilities of extruders through the use of mathematical modeling.

The models solve the differential equations which qualitatively describe the fluid mechanics of flow within the extruder as a function of time. The "driving forces" considered are perturbations in the extruder feed rate and extrudate viscosity. Several conclusions are drawn from the outputs of these models about the correct design characteristics of extruders and the accompanying feed and temperature control systems. Since the behavior of the extruders is so complex, without this modeling, rational design choices are not readily made.

The response of an isothermal, single-screw, front-fed extruder to perturbation in the moisture content of the feed material has been considered (Levine et al., 1987a). A typical result is presented in Fig. 14.14. The figure illustrates the response of the output of the extruder to a step change in the ±3% moisture content of the extrudate, for a material having a power law flow index of 0.5, which is a typical value. The parameter, N, represents the steady state output of the extruder as a fraction of maximum extruder output (free discharge). The consequences of the moisture change are quite dramatic and perhaps surprising. For a decrease in moisture content an instantaneous output approximately 60% above the steady state values occurs. A 3% increase in feed moisture results in an instantaneous output which is approximately 40% below the steady state. It is not difficult to see that with regular variation in feed moisture, enormous variations in extruder output are continuously observed.

A similar analysis (Levine et al., 1987b) is done for starve-fed extrusion screws, the normal operating mode of twin screw extruders. The output responds to changes in feed moisture in much the same way as described above for the single-screw extruder. In addition, the change in feed moisture affects the filled length of the starved screw, and the filled length directly impacts the power consumption of the extruder. Figure 14.15 illustrates that changes in moisture can have dramatic transient effects of filled length.

Figure 14.16 (Levine et al., 1987b) shows the response of the extruder to step changes in feed rate to the starve-fed extruder. The figure illustrates that changes in the feed rate transmit to the extruder output very rapidly, indicating that not much dampening of this upset occurs.

Figures 14.14 through 14.16 demonstrate to the system designer that extra expenditures to improve the performance of the extruder feeders is money well spent.

These examples consider an isothermal extruder. When cooking extrusion is considered, the problem is far more complex, because of the changes in extrudate

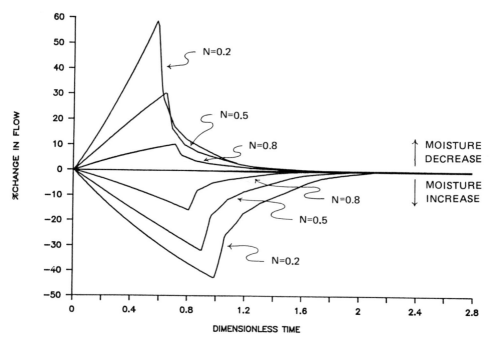

Figure 14.14   Effect of a step change in moisture content (±3%) on extruder output (flow index power law = 0.5). (*From Levine et al., 1987a.*)

viscosity associated with the cooking operation. A recent article (Roberts and Guy, 1986) uses modeling to consider this problem. This study concludes that an improperly designed extruder may be inherently unstable. Hence, a small perturbation in feed rate or viscosity may result in a sustained oscillation in extruder output, torque (power consumption), and die pressure, as illustrated in Fig. 14.17. These authors use the model to explore design parameters which help stabilize extruder operation.

These simulations are quite useful to the extrusion system designer. Experimental determination of the sources and importance of perturbations to extruder behavior is very difficult to obtain. This work, although only partially quantitative, provides clear-cut design directions.

SENSORS, BIOSENSORS, AND MACHINE VISION

A recent report predicts that U.S. end-user demand for smart sensors will grow over the 1988–1993 period at real annual rates of from 9 to 70%, and the total will

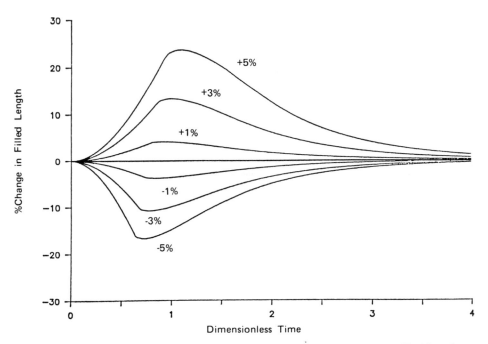

Figure 14.15   Effect of a step change in water content (±5%) on extruder filled length (index power law = 0.5; delivery = 0.4; resistance = 0.25). (*From Levine et al., 1987a.*)

exceed \$2 billion by 1993 (Berler, 1988). A smart sensor is defined as a device consisting of a transducer, a degree of electronic intelligence, and a transmitting or communication means. It includes four main categories: solid state (smart) sensors, biosensors, fiber optics, and vision sensors.

The industries associated with the four product categories are dissimilar. The solid state transducer industry is structured on the conventional transducer manufacturer and is dominated by major corporations at the present time. The semiconductor industry also plays a significant part. The biosensor industry is minuscule at the present time, commercially speaking. Yet, it is expected to grow from \$14.4 million in 1986 to \$365 million in 1991 (Anonymous, 1988b). The biosensor industry consists of small high-tech entrepreneurial research and development firms. The fiber optic sensor manufacturers are mainly electrooptic and fiber optic companies. An entirely different group forms the vision industry, which is one of the most advanced sectors of the sensor categories.

To comply with Deming's quest (Deming, 1982) and to strive to total quality

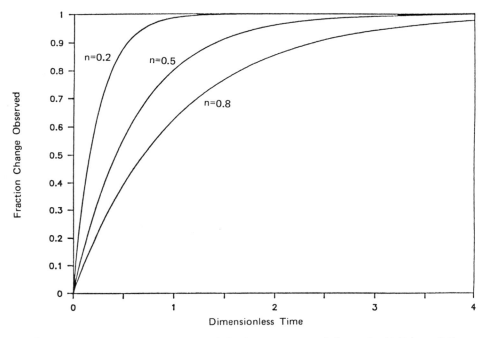

Figure 14.16 Effect of a step change in feed rate on a starve-fed extruder (delivery; - 0.2). (*From Levine et al., 1987b.*)

control (TQC) and statistical process control (SPC), there is an ever-growing need for methods suitable to assess and quantify changes occurring during manufacturing. This ability has tremendous benefits both tangible (e.g., control, efficiency, quality) and intangible (e.g., reduced consumer complaints). Sensing devices are paramount to provide inputs for the various distributed control systems, programmable logic controllers, microcomputers, and other spectra of devices employed to keep the process in balance. While computer technology has leaped to the next generation, sensor technology is lagging behind. Yet, several new technologies are penetrating the food industry. Near infrared (NIR), nuclear magnetic resonance (NMR), optical turbidimeter, and others are already used to provide on-line analysis data. Coupled with artificial intelligence and computer technology, these sensing devices will play a major role in furnishing real-time information for computer-integrated manufacturing (CIM).

The next section is focused on only two main topics: biosensors and machine vision.

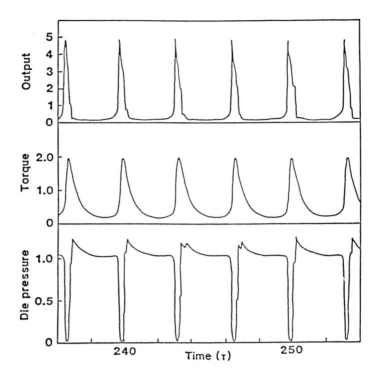

Figure 14.17 Instabilities in an extrusion cooker: a simple model. (*From Roberts and Guy, 1986.*)

Principles of Biosensors and Their Application

Advanced control systems in the food industry today are limited by the availability of on-line measurements. A recent workshop sponsored by the Institute of Food Technologists identified seven most important research needs. The first need is: "Development of innovative technologies for improving food processing systems by various means, but especially to adopt promising biotechnologies, available sensors and robotic controls" (Liska and Marion, 1985). Furthermore, the subcommittee of food process and engineering/automation identified the need to: "Develop sensors and monitoring systems for process control" (Clark et al., 1985). This lack of appropriate sensors and control systems for the food processors is considered one of the major impediments to the development of fully automated food processing and packaging systems. Special emphasis focused on-line/real-time sensing of unique food composition and quality attributes such as flavor, color, and textural characteristics, as well as changes in raw materials.

The comprehensive description of available biosensors, their function and utilization is an immense task (Rechnitz, 1987, 1988; Turner et al., 1987; Janata and Bezegh, 1988) and is beyond our scope. Instead, we briefly review underlying principles of several biosensors and highlight some recent developments in this area.

Types of Biosensors

The use of biological and organic materials as sensors holds promise of revolutionizing many aspects of food technology and may satisfy some of the most needed measurements of on-line food quality attributes.

Biosensors are at the leading edge of sensor research. Worldwide interest in this field is flourishing, and publications on fundamental and applied aspects of biosensors are appearing at a rate approaching 400 papers per year (Rechnitz, 1988). Furthermore, another review dedicated to chemical sensors lists 420 references and reports an astonishing ratio of one review per 3.4 original papers (Janata and Bezegh, 1988). This explosion of information truly reflects the vast interest in this topic and its potential for sensing applications.

A biosensor is a device that responds to the presence of a specific analyte by producing a signal proportional to the concentration of the analyte. A schematic representation of a biosensor is depicted in Fig. 14.18. The biosensor consists of two main parts: receptor and transducer. The receptor is the biological recognition system composed of an immobilized biomaterial which may include one of the following: enzymes, protein, antibodies, DNA, organelles, microbial cell, or whole sections of mammalian or plant tissues. The specificity of the receptor for the analyte may be visualized as a "lock and key" combination. The second part of a biosensor is the transducer, which is an electrical or physical device that converts the effects of the biorecognition reaction into a form which can be electronically amplified. Typical transducers include: potentiometric (voltage), amperometric (current), thermistor (temperature), piezoelectric (force), optoelectronic, field effect transistor (FET), ion selecting field effect transistor (ISFET), enzyme field effect transistor (ENFET), and others. The signal is conditioned and amplified before it is processed by a microprocessor or other output device.

The receptor is normally protected by a membrane that is permeable only to certain molecules in the medium tested. Recent configurations allow the elimination of the membrane by direct adsorption or covalent fixation of the receptor at the transducer surface or by immobilization of the receptor directly onto the transducer. Table 14.5 lists a few typical examples of possible common modes of measurement systems and transducers.

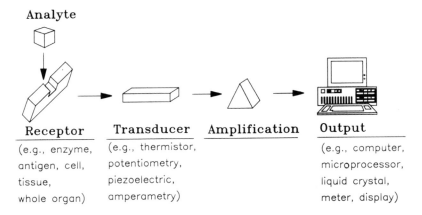

Figure 14.18   A schematic representation of a biosensor.

Table 14.5   Typical Transducers Utilized in Biosensors

Mode of measurement	Transducer
Potentiometry	Ion-selective electrode (ISE), gas selective electrode, Ion-selecting field effect transistor (ISFET), enzyme field effect transistor (ENFET)
Amperometric	Oxygen electrode
Conductimetric	Conductimeter
Optical	Eliptometer, sensitized, semiconductor, fiber optic, photodiode, absorption, fluorescence, reflection
Mass change	Piezoelectric crystal resonance
Calorimetry	Thermistors

The ideal biosensor must be robust, reliable, easy to use, allowing for a single-step preparation and determination. The main advantages of biosensors include high selectivity and specificity, high sensitivity, possibility to detect inactivated enzymes, fast response time, simple to use, comparatively low operation costs, miniaturization, parallel determination of more than one component, and, in some cases, disposability.

Major disadvantages of biosensors are limited availability, instability (physical and chemical environmental properties have marked effect on activity), limited shelf-life, cost, sensitivity to interferences and contamination.

Biosensors are classified either by their transducing systems or by principles of biorecognition. The latter is used in our short description of some primary groups of biosensors.

Chemoreceptors. A chemoreceptor is one involved in the chemical sense of a biomolecule stimulant, such as taste. Obviously, in order to be considered a biosensor the chemoreceptor is coupled with a signal generating device. This coupling is extremely difficult to execute and is far from being resolved.

A typical example of this biosensor is an animal-based system which is capable of responding to levels of 10^{-2} to 10^{-9} M amino acids, while the counterpart living organism is several orders of magnitude more sensitive (i.e., 10^{-15} M). This fast response time and the short period normally needed to reestablish equilibrium (approximately 2–3 milliseconds) is mainly due to two factors: little energy is required to convert the bounded stimulant to an easily detectable electrical signal in the nerve, and the receptor is held far from equilibrium, thus avoiding saturation (Rechnitz, 1988).

Immunological biosensors. Immunosensors promise a particularly wide area of applications. Monoclonal antibodies provide high selectivity and uniformity in binding to numerous biomolecules. Recent advances in this domain provide the first tool for tailor-made sensors for the food industry. A major area of current development is the adaptation of ELISA (Enzyme-Linked ImmunoSorbent Assay) methods (Schmidt, 1988). Normally, ELISA provides a highly sensitive and specific means for measuring a variety of compounds, provided an antibody to the substance of interest is available. ELISA is mainly used within the field of brewing to quantify enzymes added during processing of beer. These include proteases (e.g., papain, ficin, bromelin), starch-degrading enzymes (e.g., α-amylase, amyloglucosidase), and others.

In this application of immunological biosensors, the typical ELISA spectrophotometric measurement of the enzyme and of the indicator is replaced by an electrochemical determination. Electrochemical luminescence is one example of this approach. In practice, however, because of the high molecular weight of the antibodies, it is difficult to couple their reaction with the corresponding antigens to a transducer in such a manner that the resultant signal reflects the antibody–

antigen interaction quantitatively. Perhaps this explains why immunosensors are not widely available.

A recent example combines both immunological technology and a piezoelectric biosensor. The reusable biosensor utilizes an antibody immobilized on a piezocrystal surface. This configuration provides one of the lowest detection limits of any analytical device—about 10^{-12} g (Guilbault and Luong, 1989).

Biocatalytic biosensors.   Biocatalytic biosensors comprise the largest and most established group, including immobilized enzymes, bacterial cells, and plant or animal tissues that may function as biocatalysts. In various biomolecular recognition or conversion reactions, enzymes are needed to convert the desired substance to product/output that can be easily detected. This output is monitored via a spectrum of transducers, which may include electrochemical, fiber optics, thermistors, or other means as outlined in Table 14.5.

Enzyme electrodes exhibit a linear response between analyte concentration of 10^{-1}–10^{-6} M with response times varying between 10 seconds and 30 minutes (Patel and Scott, 1988).

Utilization of living cells such as prokaryotes (bacteria, cyanobacteria) or eukaryotes (algae, fungi, and, more recently, higher animal and plant tissues) offers two real benefits: increased stability and the capability to detect more than one enzyme. Yet, several problems linger. These hurdles include long response and recovery time, tedious and expensive fabrication, short shelf life, and the need to sterilize the biosensor.

An interesting example which combines biosensors, mathematical model building, simulation, and computers is worth noting (Li et al., 1988). In this work biosensors with animal and microbial cells are immobilized close to the tip of a membrane electrode and used for chemical and drug testing. A computer-aided analysis software package is developed to screen biosensor dynamic response. Computer simulation and parameter estimation techniques are utilized to select the "best" model and to describe the biochemical and pharmacological effects of various chemicals and drugs on different cells. The results show that biosensors can be successfully used for screening and to provide useful information about various cell–chemical interactions.

Biosensors for the Food Industry

Biosensor technology offers the food industry a new, rapid type of monitoring and measuring device whose speed, sensitivity, and specificity are crucial for feedback control in modern computer-aided manufacturing and automation. This leads ultimately to an increase in productivity, improved product quality, safety, and cost-effective processes.

The importance of measuring changes occurring in the process is so great that much effort has been invested in the development of on-line nondestructive sensors. However, development of appropriate sensors and biosensors is still a real challenge facing the food industry.

The desirable characteristics of biosensors for the food industry, are as follows (Wren, 1986; Schaertel and Firstenberg-Eden, 1988): relevance of output signal to food quality; accuracy and repeatability; sensitivity and resolution; appropriate dynamic range; speed of response; insensitivity to the environment; amenability to testing and calibration; and low costs.

Theoretically it is possible to design a biosensor for any food application if a proper receptor is available. The most common and useful biosensors are those systems based on biocatalysts. Schaertel and Firstenberg-Eden (1988) listed 90 substances: carbohydrates (e.g., glucose, fructose, lactose, sucrose, starch); amino acids (e.g., L-arginine, D-alanine, L-tryptophan, serine); alcohols (e.g., ethanol); gases (e.g., H_2, CH_4, NH_3, NO, SO_2); enzymes and cofactor (e.g., NAD(P)H, phospholipase A and C), amines (e.g., adenosine, ATP); carboxylic acids (acetic acid, lactic acid, pyruvic acid, fumaric acid); antibiotics (e.g., penicillin); and others. Some of the commercial enzyme biosensors available are listed in Table 14.6.

Two recent projects carried out in the United Kingdom are worth noting. These projects were launched to enable food companies to participate in development and application of sensors for the food industry (Downer, 1988; Patel and Scott, 1988). Of particular interest is a flow-through immunoreactor column coupled to a colorimetric detection device. Antibodies to *Clostridium perfringens* enterotoxin are successfully coupled to preactivated agarose. It is then used to determine the presence of the specific antigen and detected using a second, specific antibody conjugated with horseradish peroxidase.

It should be noted, however, that in spite of the above on-line monitoring capability demonstrated in the lab and in some cases in the plant, the biosensor emergence from research to the marketplace is very slow. Yet, we are convinced that the immense market opportunity is so significant that major efforts will be devoted to the development and evolution of this technology. This will be possible only when a multidisciplinary collaboration between the food industry and other fields of expertise is established. Fundamental knowledge of food chemistry coupled with electrochemical, electromagnetic, optical, and material science is required. This collaboration may provide the only avenue for commercializing sophisticated biosensors.

It is worth noting that the Center for Advanced Food Technology (CAFT), Rutgers University, is in the planning stages for a research project on on-line sensors for food processing systems. The project seeks to conceive new sensor technology for the on-line measurement of moisture and rheological properties in food processing. The objectives focus on three areas of research: optical sensing

Table 14.6 Examples of Commercial Enzyme-Electrode Sensors

Parameter	Company	Linear range (mmol/L)	Sample /hr	Precision (%)	Stability
Glucose	Yellow Springs	1–45	40	<2	300s[a]
Glucose	ZWG	0.5–50	60–90	1.5	>1000s
Glucose	Fuji Electric	0–27	80–90	1.7	>500s
Glucose	Auto Stat	1–40	35	1	—
Glucose	Aplama:G-04	2.5–3.0	20	<3	—
Glucose	Enzalyst	0.5–3.0	—	5	—
Glucose	Seres:Enzymal	0.3–22	60	—	—
Glucose	Provesta	0–10	15	< 5	30D[b]
Lactate	La Roche	0.5–5	20–30	1.5	40D
Lactate	Amron Toyobo	0–8.3	—	<5	>10D
Alcohol	Yellow Spring	0–60	20	2	7D
Alcohol	Provesta	0–10	15	5	30D

[a] = seconds.
[b] = days.
Stability is time or number of samples that can be used before reconditioning the selective electrodes.
Source: From Schaertel and Firstenberg-Eden (1988), with permission.

for determining moisture; new materials synthesis for creating new materials; and rheological sensing systems.

Example: Fish freshness biosensors. To demonstrate the applicability of biosensors in monitoring and measuring the quality of foods, we focus on fish freshness. This unique application is of major importance to the food industry by offering a quick and simple tool to assess quality. Furthermore, it addresses an ever-growing need to provide the consumer the best product possible.

It is a well-established fact that after the death of a living species such as fish, ATP contained in its muscles is degraded sequentially:

$$ATP \rightarrow ADP \rightarrow AMP \rightarrow IMP \rightarrow HxR \rightarrow Hx \rightarrow X \rightarrow U \qquad (18)$$

where ATP = adenosine-5'-triphosphate; ADP = adenosine-5'-diphosphate; AMP = adenosine-5'-monophosphate; IMP = inosine-5'-monophosphate; HxR = inosine; Hx = hypoxanthine; X = xanthine; and U = uric acid.

Changes in hypoxanthine and uric acid could be related to the degree of freshness.

Determination of fish freshness with an enzyme sensor system is described by Karube et al. (1984). The method is based on a new indicator used to assess fish freshness and defined as:

$$K = [(HxR) + (Hx)]/[(IMP) + (HxR) + (Hx)]*100 \qquad (19)$$

In order to determine these compounds,an enzyme sensor system is developed by combining a double membrane consisting of a 5'-nucleotidase membrane and a nucleosidephosphorylase-xanthine oxidase membrane with an oxygen electrode. A small anion-exchange resin column is also connected with the enzyme sensor for separation of nucleotides. Each nucleotide concentration is determined by the current decrease following the addition of certain chemicals. One assay is completed within 20 minutes, and no appreciable decrease of current output occurs during 30 assays. When a sample solution containing 10 mM of IMP is applied, the response is reproducible within 8% of the relative error. Good comparative results were observed between the K values determined by the proposed sensor and by conventional methods. This method is further improved by Watanabe et al., (1986) who uses a multielectrode enzyme sensor system to measure the freshness of fish and to display the freshness pattern by a computer. The apparatus consists of a multielectrode enzyme sensor, relay controller, A/D converter, and micro-computer. The output current of the sensor is read by the computer and converted to digital signals every five seconds. Amounts of hypoxanthine (Hx), inosine (HxR), inosine monophosphate (IMP), and K value are calculated by the computer from this data. Using this technique, freshness of fish is defined as very fresh (K ≤ 10), fresh (K ≤ 40), or not fresh (K ≥ 40). The assay, including sample preparation, takes about 15 minutes.

Watanabe et al. (1987) further improved the method by replacing the immobilized enzyme with a microbial sensor system consisting of immobilized living spoilage bacteria (*Alteromonas putrefaciens*) and an oxygen electrode. The biosensor is applied to the continuous determination of fish freshness. Fish freshness is expressed as the ratio A/B, where A is the current decrease of the microbial sensor obtained by the application of 10–50 mg/L fish extract and B is the current decrease by the equivalent volume of culture medium. One assay can be completed within 13 minutes.

Luong et al. (1988) utilized a Clark-type polarographic probe that is highly specific for detecting hydrogen peroxide and uric acid to measure metabolites as an index of freshness. Three enzymes—xanthine oxidase, nucleoside phos-phorylase, and nucleotidase—are used to convert the nucleotides. The error is reported as 6%. The unit yields a linear relationship between the probe response and hypoxanthine concentration in the range of 0–5 μM. This method has been commercialized by Pegasus Biotechnology (Ontario, Canada), and the unit is called "Microfresh." A comparison of the K values derived by this unit and an

HPLC method reveals high correlation coefficients (e.g., 0.99, 0.94, and 0.98 for cod fillets, whole salmon, and herring, respectively (Fraser, 1989). Furthermore, the unit is utilized to determine the relationship between the K value and the mean sensory scores for salmon and herring. The data indicate a high correlation between these two methods (Fraser, 1989).

A different approach to detect fish freshness by monitoring TMA (trimethylamine) and DMA (diethylamine) concentration with a semiconductive 0.5% (weight) Ru/TiO_2 sensor has been reported (Shimizu et al., 1988). The resistance changes of the sensor are proportional to DMA and TMA concentration in the gaseous headspace. This approach, however, requires further testing.

Even though not following the general description of a biosensor, it is interesting to note a different approach which utilizes a diagnostic test strip assay for determining trimethylamine (TMA) in seafood. The color produced at room temperature is measured visually against a set of six TMA standards. This method compares successfully with sensory, microbial, and chemical assessments of fish freshness (Wong et al., 1988).

Finally, it is quite clear that in spite of these major efforts, the biosensors developed are not fully utilized for on-line measurements. Nevertheless, this research activity is in the right direction and provides the basic knowledge for further developments.

Principles of Machine Vision and Its Application to Inspection and Process Control

Principles. Visual inspection of food is a frequent requirement of process and quality control. The inspection may be qualitative, looking for defects in shape or surface appearance, or quantitative, requiring actual measurement of the food product. Examples of surface feature appearances that may be of interest are such things as the detection of burned spots on baked products, the detection of incomplete coatings, such as in candy bars, the number and count of pepperoni slices on a pizza, the finding of bruises on fresh fruit and vegetables. In addition, visual inspection is an essential part of many food operations, the most obvious example being the trimming of meat products.

These inspection/sorting tasks are quite repetitive, mundane,and boring to a human inspector. As a consequence, the grading accuracy of even well-trained inspectors tends to deteriorate rapidly. If one wishes to operate a plant with "zero defects" or to automate control of the facility, the inability of human inspectors to visually inspect items with consistent accuracy may make these admirable goals unattainable. In addition, the inspector often uses subjective determinations of "acceptable" and "unacceptable." As a result, there are significant sorting differences between inspectors. The application of machine vision (computer-aided inspection) may provide us with the solution to this problem. Ideally, the machine vision system replaces the tiring, inaccurate inspector with an inspector who never tires and whose grading is never affected by subjective judgment. Unfortunately, this goal is more simply stated than attained, but the recent expansion in microcomputer capabilities per unit cost has moved this goal closer to reality.

Figure 14.19 illustrates a machine vision system. The system consists of a presentation system (a conveyor), an image acquisition system (usually a solid state television camera, lights, and optics), an image processing system/controller (computer), and a decision implementer (usually a mechanical actuator). If this were a classical control problem, such as control of the browning of a baked product rather than a straightforward inspection, the computer would interact with a classical control loop driving an oven temperature controller, instead of a mechanical actuator.

While the presentation system often appears to be a simple conveyer, in reality the position of the item of interest and the amount of background noise (stray lighting, conveyor color, extraneous materials, etc.) are important considerations for this aspect of system design.

The image acquisition device is generally a solid state black and white television camera. These cameras contain an array (usually two-dimensional) of charge-coupled devices which have about the same degree of resolution as a television monitor. Color systems are possible, but are just becoming commercially available. The processing of color images poses considerably more computational complexities than one faces with black, grey, and white images. In place of the

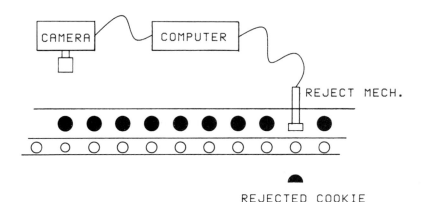

Figure 14.19 Schematic presentation of the major components of a computer vision system.

PROCESSOR/CONTROLLER

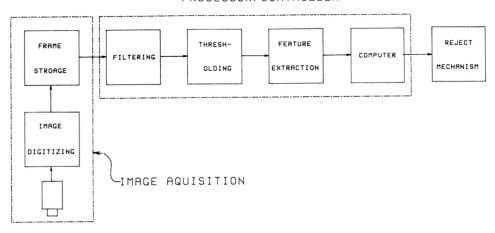

Figure 14.20   Schematic presentation of a computer vision system.

conventional camera, some systems use a linear array camera, which acquires a single line of the image at a time. However, we concentrate this discussion on the conventional black and white television camera.

Figure 14.20 illustrates the machine vision system in more detail than was illustrated in Fig. 14.19. In addition to the camera, the image acquisition system consists of an image digitizer and frame (image) storage. Image digitization is necessary because this is the mode of computer processing and the camera outputs an analog signal. Image digitization converts the camera's analog signal into a "frame" of individual elements (pixels) within the computer, as illustrated in Fig. 14.21. Each pixel contains a grey level value. The number of grey levels of resolution is defined by the number of bits of information stored in each pixel, as shown in Table 14.7.

Note that the number of grey levels chosen to describe the image and the number of pixels selected to break the image define much of the machine vision problem.

To further highlight these facts, let us consider the data in Table 14.8. The data illustrates one of the key problems associated with machine vision and image analysis: it consumes a prodigious quantity of memory and computational capacity. For example, a 256 × 256 pixel image with 256 grey levels requires 64K of memory. This is about the resolution of a normal television monitor. Add to this that the simple act of changing each white pixel to black and black to white requires *at least* 64K calculations. It is not difficult to see that even a moderate amount of

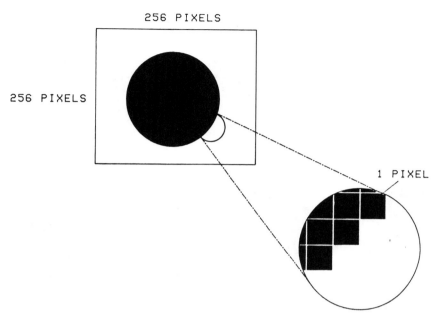

Figure 14.21 Image digitization into pixels.

Table 14.7 Available Grey Level Information

Bits/Pixel	No. of grey levels
1	1 (black or white)
4	16
8	256

Table 14.8 Information Required to Store an Image Containing Different Pixels and Grey Levels

Image size (pixels)	Required bytes of information for grey levels			
	16	64	128	256
64 × 64	512	1k	2k	4k
128 × 128	2k	4k	8k	16k
256 × 256	8k	16k	32k	64k
512 × 512	32k	64k	128k	512k
1024 × 1024	128k	512k	1M	2M

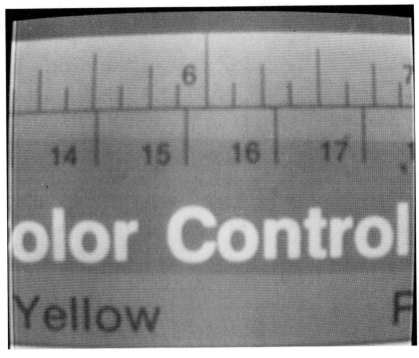

a

Figure 14.22 Effect of various degrees of image digitization on resolution.
(a) 256 Grey levels, 256 x 256.

image manipulation requires extensive calculations. To perform these tasks in real time requires specially designed hardware.

The choice of picture resolution, that is, the number of grey levels and pixels, becomes a compromise. The higher the level of resolution chosen, the finer the detail that can be extracted from the image. However, as Table 14.8 illustrates, this increased resolution is paid for in exponentially increasing memory and computational requirements. One pays for this in both capital cost and speed capability (images per second) of the machine vision system. Most problems probably do not require greater than 256 × 256 resolution. Figures 14.22a–k illustrate the effect of various degrees of image digitization.

In reality, the image digitizing system and frame storage reside in the same "box" as the image processor and controller. It is the image processor/controller that is

b

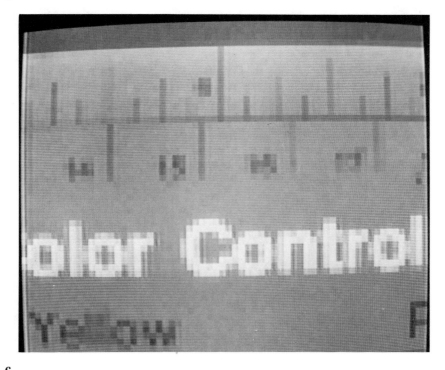

c

(b) 256 Grey levels 128 × 128 pixel; (c) 256 Grey, 64 × 64 pixel;

d

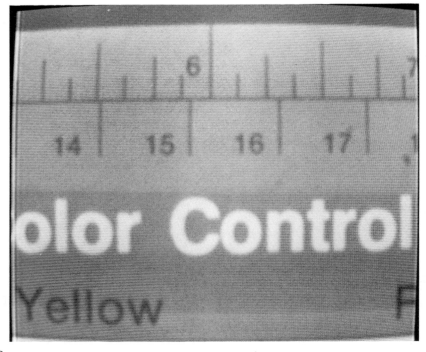

e

Figure 14.22 (continued) (d) 256 Grey, 32 × 32 pixel; (e) 256 Grey, 256 × 256 pixel;

f

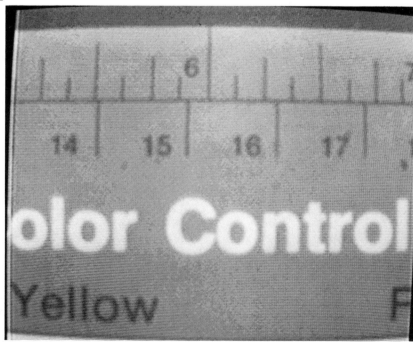

g

(f) 128 Grey, 256 × 256 pixel; (g) 64 Grey, 256 × 256 pixel;

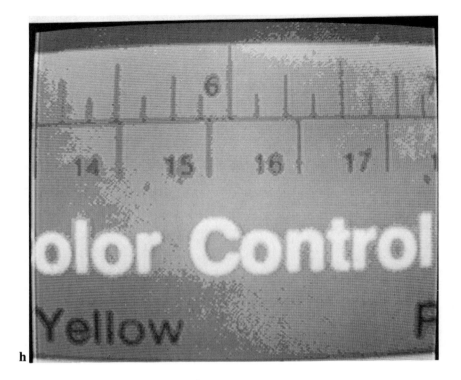

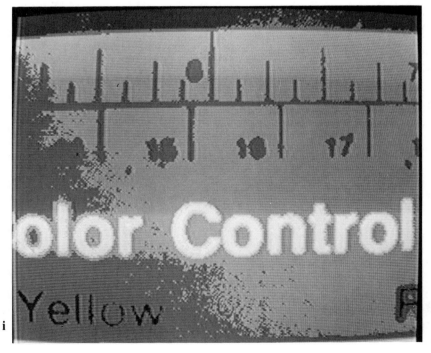

Figure 14.22 (continued) (h) 32 Grey, 256 × 256 pixel; (i) 16 Grey level, 256 × 256 pixel;

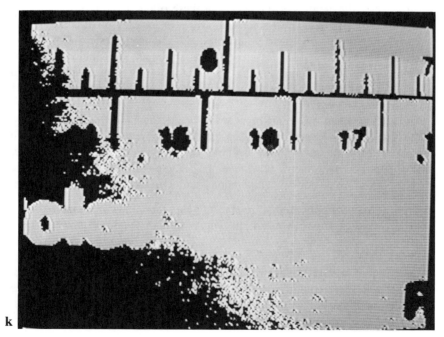

(j) 8 Grey level, 256 × 256 pixel; (k) 4 Grey level, 256 × 256 pixel.

the heart of the vision system. As illustrated in Fig. 14.20, the image processor/controller performs many steps. We discuss each in turn.

Filtering is the act of removing unwanted detail or enhancing desired detail in the image. The literature (Ballard and Brown, 1982; Diamond and Ganapathy, 1982; Baxes, 1984; Suetens and Oosterlinck, 1985; Casansent, 1987) describes many computational algorithms available to perform these tasks. We illustrate the simplest algorithms, all of which are possible to execute at very high speeds.

The most common and simplest filters are simple point transformations, called maps. Maps look at every pixel in an image and transform it based on a predefined rule. Some examples of this might include negation of the image, and removal of specific grey levels within the image. Figures 14.23a–d illustrate the effect of these maps.

The next level of image filters consist of convolutions. These mathematical techniques for filtering the data are much like the techniques used to filter noisy data entering a controller or the separation of a radio signal from its carrier. Refer to the literature for many excellent discussions of the concept of convolution, in general (e.g., Lathi, 1974), and image convolution, in particular (e.g., Ballard and Brown, 1982; Baxes, 1984). As with maps, convolution filters are used to bring out wanted detail or to fade out unwanted detail. Some examples are high and low pass filters and edge detectors (Laplacian operators, etc.). Figures 14.24a–c illustrate several examples of the effects of these filters on an image.

The simple measurement of area, size, shape, or location often is the primary task of feature extraction. Smart programming of the image system allows one to search for almost any type of feature in the binary image. This includes counting the number of objects, identifying the type and location of a defect, etc. From this information, and preset rules, a reject/accept decision is made for the object(s) in the image.

Note that the quantity of data being processed as the image moves from image acquisition through feature extraction is continuously reduced (Fig. 14.25). This successive reduction is absolutely essential when aplying machine vision to a real time (high speed) problem, otherwise the imaging system quickly becomes "bogged down" in numerical computations, and a practical speed is not attained.

Computer vision applications.   In recent years there have been many discussions of practical applications of machine vision to the food industry. Some examples of these are detection of burned cookies/crackers (Gagliardi and Sullivan, 1983), inspection of pouch seal integrity (Gagliardi et al., 1984), measure of thickness of sheeted doughs (McCormick, 1988), detection of stones in nuts (Hudson, 1984), detection of defects in candy bar coatings (Hudson, 1984), pizza crust size and shape (Hudson, 1984), detection of missing components in a multicomponent package (Hudson, 1984), identification of fish species (Anonymous, 1986), detection of bone in meat and poultry (Gerber et al., 1985), identification of poultry parts (Babbit, 1988), detection of bruises on fresh fruit and vegetables (Rehkugler and Throop, 1985; Sarker and Wolfe, 1985; Kanegif, 1987), detection of icing placement on cookies (Gagliardi, 1983), detection of physical damage on corn and soybeans (Gunasekaran et al., 1988), control of bacon slicing (Gagliardi et al., 1984), detection of holes in pizza crusts (Hudson, 1984), and detection of damage on peeled potatoes (Anonymous, 1985).

Most of these applications are relatively straightforward, using the simplest of vision algorithms available to the system programmer. The use of simple algorithms is often necessary in the food industry because of the extremely high inspection rates required. We now illustrate selected applications.

Example: Detection of broken or missing cracker of sandwich cookies.   This example is a very simple application of machine vision. Figure 14.26a illustrates four cookies in different conditions. Note the high degree of contrast between the cookies, the background, and creme filling. As a result, virtually no image enhancement is required. The algorithm to inspect the cookies is simply convert the image to binary black on white (Fig. 14.26b), determine the area of the image, which is simply the summation of the number of white pixels, and compare the area of any cookie to the area of an undamaged cookie (Fig. 14.26c). This vision algorithm is basic to any vision system.

Example: Detection of bruises on fruit.   Like the damaged cookie sample, the detection of bruises on fruit is very straightforward. The computer algorithm is virtually identical. Figures 14.27a–f shows a series of images of bruised bananas and their binary form. It is apparent that the system has no difficulties distinguishing between bruised and unbruised fruit.

Example: Detection of the physical dimensions of a food product.   Clearly, the simple algorithm described above for cookie inspection is applicable to the gross estimation of the size of a food product. This is accomplished by ensuring that the total food product is significantly lighter or darker than the background (conveyer). However, this type of size estimation is often not sufficient. The gross size of an object tells us something about its mass; it tells us nothing about the shape of the object. Knowledge of a food product shape is clearly important. For example, one might wish to know if a cut dough product, such as a tortilla, is truly round, or one may wish to sort/grade an irregularly shaped object by length or width. Clearly, two objects can have the same projected area but very different dimensions. In addition to counting the number of white pixels (area), many systems will also, at the hardware level, perform moment calculations upon the white screen areas. This information determines area, as well as major and minor axis lengths, orientation, and position. These data are used not only for size grading but also to locate a food product for further processing operations (for example, robotic manipulation).

Figure 14.28a displays an image of four carrots. In Fig. 14.28b this image is

Figure 14.23 Effect of maps algorithms on resolution: (a) Original image; (b) Invert image;

c

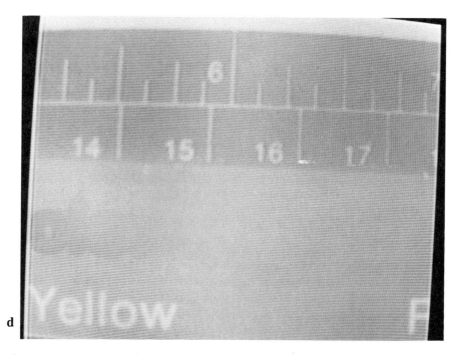

d

Figure 14.23 (continued) (c) Retain only white pixels; (d) White pixels removed.

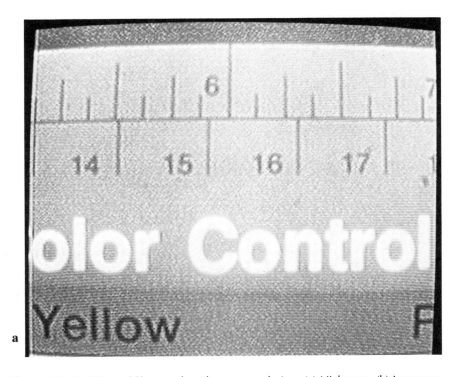

Figure 14.24 Effect of filtering algorithms on resolution: (a) High pass; (b) Low pass; (c) Edge detection.

converted to binary form, while Fig. 14.28c shows the result of a subsequent call to the built-in area/moment calculations. Note that the top two carrots and the bottom two carrots have similar areas, but significantly different lengths and diameters. This information is transmitted to a mechanical system to accomplish size grading.

Example: Determination of fat content in bacon slices and control of package weight. We now illustrate an application that has been suggested to determine fat content of a bacon slice (Gagliardi, 1983). In the processing and packaging of bacon, a net weight target is set. Unfortunately, the fat-to-lean ratio, as is obvious to anyone who has ever purchased this product, varies considerably. Since the density of the fat and lean portions is significantly different, placing a fixed number of slices of uniform thickness into a package does not give "ideal" net weight

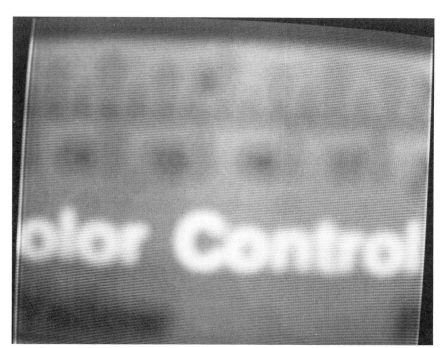

b

c

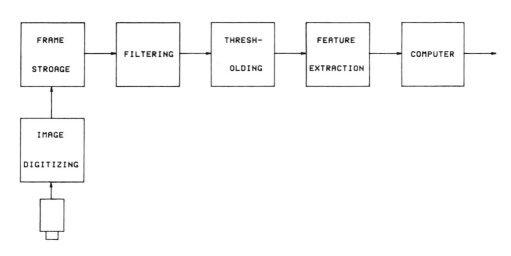

64 kBYTES ⟶ 8 kBYTES ⟶ 256 BYTES ⟶ 1 BIT

Figure 14.25 Quantity of data processed in a typical image acquisition.

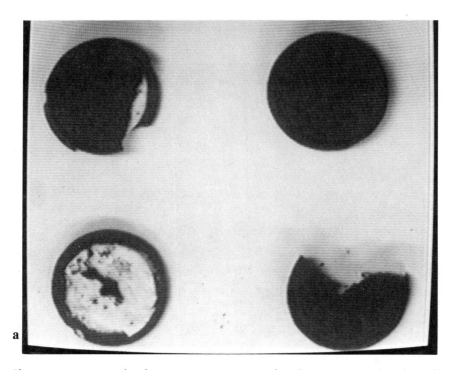

a

Figure 14.26 Example of a computer inspection of cookies: (a) Original cookies; (b) Cookie image converted into binary image of black and white; (c) Comparison of damaged and undamaged cookies area.

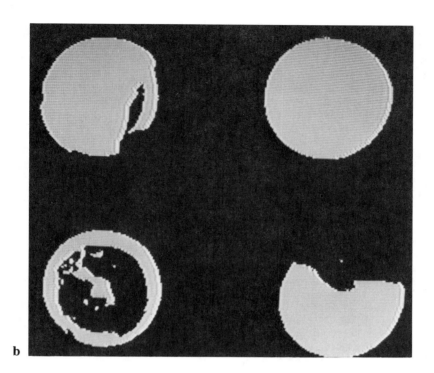

b

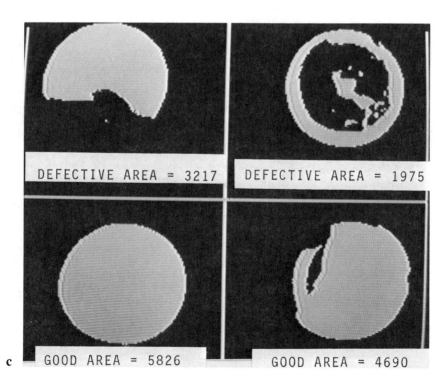

DEFECTIVE AREA = 3217   DEFECTIVE AREA = 1975

GOOD AREA = 5826   GOOD AREA = 4690

c

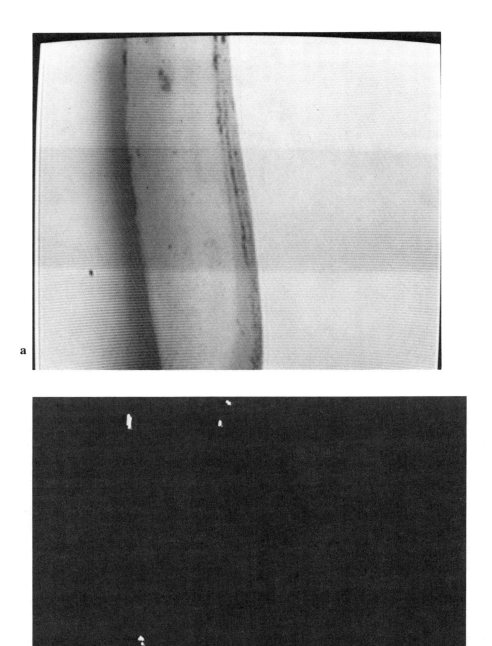

GOOD AREA = 30

Figure 14.27 Example of a computer inspection of a banana: (a) Original image of a good quality banana; (b) Unbruised area;

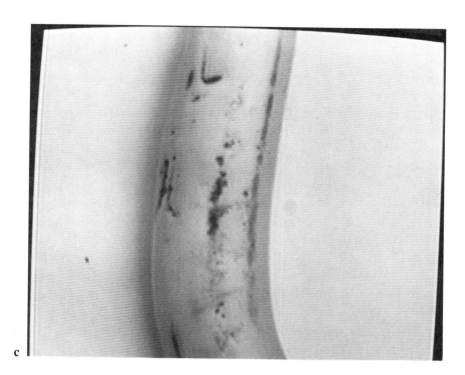

c

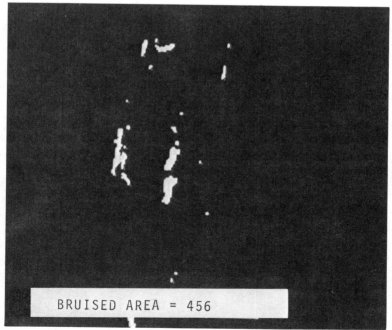

BRUISED AREA = 456

d

(c) Original image of a bruised banana; (d) Bruised area;

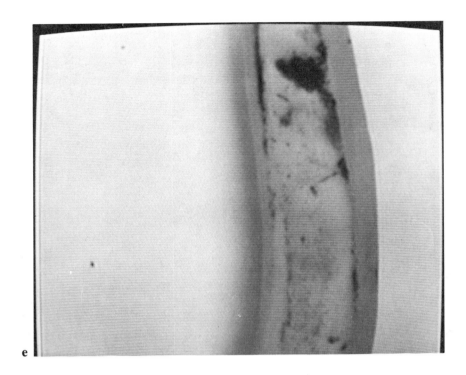

e

BRUISED AREA = 1203

f

Figure 14.27 (continued) (e) Original image of a bruised banana; (f) Bruised area.

control. If one knew the fat-to-lean ratio, and by inference the density of the slice, one would have data with which to automatically modify the thickness of the slices as their density varies. The block diagram of this concept is illustrated in Fig. 14.29.

Machine vision makes the automatic estimation of the fat-to-lean ratio very straightforward. The technique uses the same area measurement algorithms as discussed above, with a slight "twist."

Figure 14.30a shows an image of a slice of bacon on a dark-colored conveyor. Against the virtually black background, one sees the fat as virtually white and the lean portion as a shade of grey. Figure 14.30b shows the first step of the algorithm. The image has been converted to a binary image, where the threshold (grey level at which grey is set to black) is set below the grey of the lean areas. As a result the entire slice appears as a white silhouette. The area of the total slice is calculated and is displayed at the bottom of the screen.

The image is now modified by changing the threshold so that the lean areas appear black. This is illustrated in Fig. 14.30c. Now only the fatty areas appear in white silhouette. Again the area is calculated and the ratio determined. This ratio is related to the percentage of fat, which is clearly related to the density of the slice. In an automatic control scheme this information and the total area is transmitted back to a controller, which issues commands to the slicer to change the target slice thickness.

In theory, the system can examine the fresh surface created after each movement of the slicer blade, so that the thickness is controlled on every slice.

Example: Detection of dented cans.   Dented cans, which arise from poor handling of cans, present a significant problem to the canning industry. It is desirable to reject dented cans before they are filled or labeled. Machine vision provides a straightforward method for performing this task. Figure 14.31a shows a can that has some dented areas, which are not always easy to see.

If the can is viewed in a dark environment, illuminated only by a line source of light, which can be obtained with a scanning laser or a focused incandescent filament, the damaged areas are readily identifiable. Figure 14.31b–d illustrate what the computer will "see" when the undented or dented cans are illuminated under these conditions. It is readily apparent that the dented are distinguished by either area analysis or direct comparison of the test image to a reference image.

The practical realization of this application would require considerable development in order to view all sides of the can at high speeds. The simplest method appears to be high speed rotation of the can while it is within the lense's field of view.

We hope, by now, that the reader is adequately introduced to this rapidly emerging computer vision technology. Clearly, many food industry inspection problems lend themselves to computerized automation.

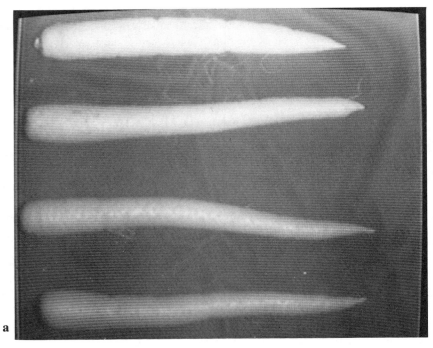

a

Figure 14.28 Computer vision measurements of carrots: (a) Original image; (b) Image converted into a binary form; (c) Area/moment calculations.

PROCESS CONTROL

After we have developed and discussed modeling, simulation, and optimization of industrial processes, let us see how we can use this information for process control.

Systems and Control Systems

A system can be defined as a set of components related in such a manner as to form and/or act as a single unit. This system concept establishes a framework that allows for the redefinition of components relative to their relationship to other components. Furthermore, it provides an understanding of the behavior of the output relative to the interrelationship of the components. Just because a specific set of components are designed to work as a single unit does not guarantee that

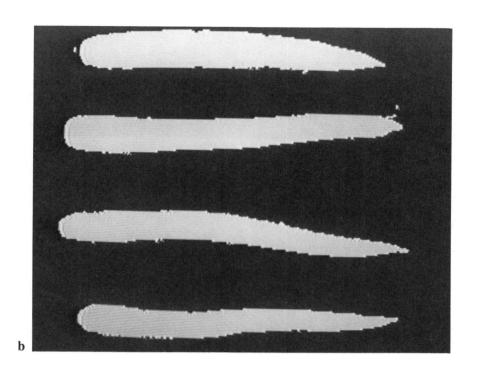

b

AREA = 3747, DIA = 41.1, LEN = 2347

AREA = 3742, 3742, DIA = 32.4, LEN = 3217

AREA = 3521, DIA = 34.7, LEN = 3254

c   AREA = 3521, DIA=34.5, LEN = 2474

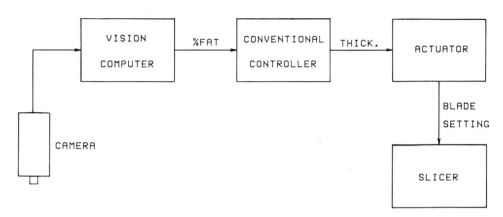

Figure 14.29   Block diagram for controlling the thickness of a bacon slice.

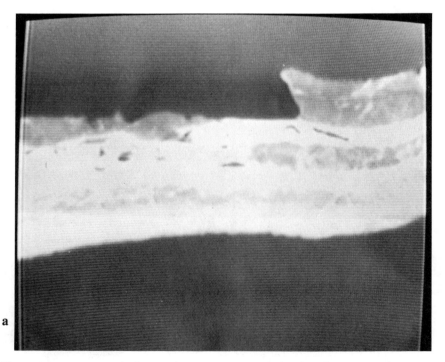

Figure 14.30   Image of a bacon slice: (a) Against a dark colored conveyer; (b) Binary image with a threshold level set below the grey level of the lean area; (c) Binary image of the fatty area only.

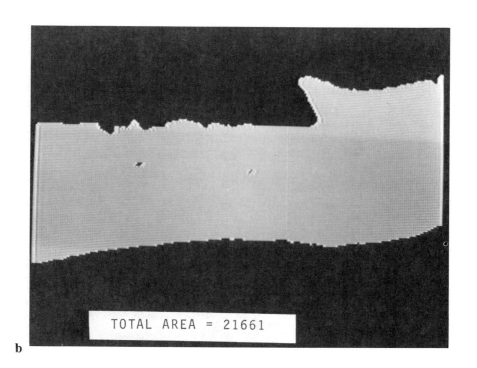

TOTAL AREA = 21661

b

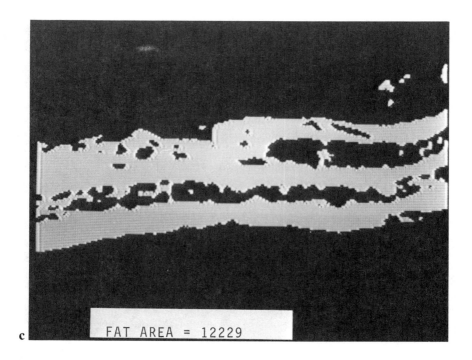

FAT AREA = 12229

c

a

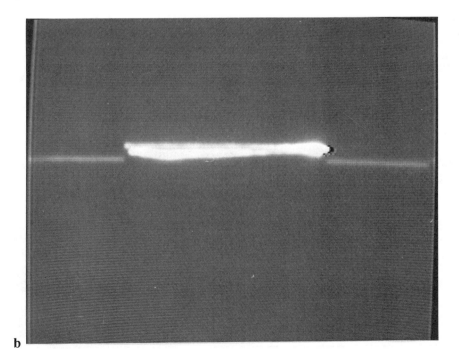

b

Figure 14.31 Image of a dented can: (a) Original can with a dented area; (b) Digitized image of undented area illuminated with a line source;

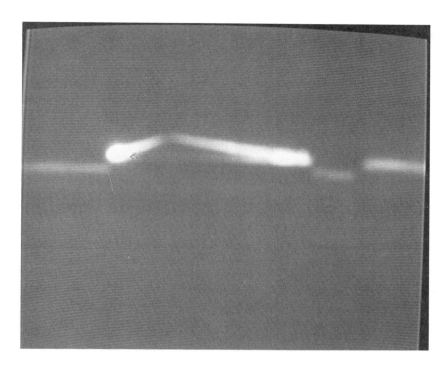

c

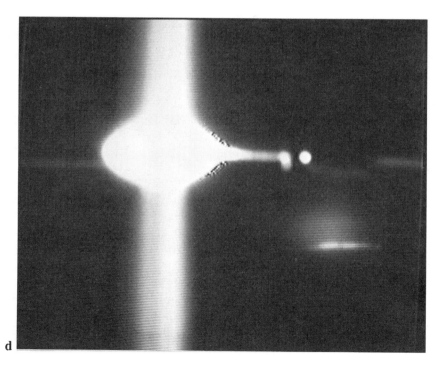

d

(c) Digitized image of dented area illuminated with a line source; (d) Digitized image of dented area illuminated with a line source.

the output from the system will be stable or predictable (e.g., the weather system). While this fact may be acceptable for mother nature, in a modern food manufacturing facility, control of the output is an absolute requirement to ensure food safety, quality, and the cost-effectiveness of the system.

A control system can be defined as a set of components related in such a manner as to modify its own behavior or the behavior of some other system. The desired attribute of a control system is an output that is stable and predictable.

Control systems can be classified into two general categories: open-loop and closed-loop. In an open-loop control system, the output is independent of the control action. The attributes of an open-loop control system are they are very stable and they are predictable only if care is taken in the determination of the controller settings.

In a closed-loop system, the control action is dependent on the system output. The attributes of a closed-loop control system are they can be very predictable, there is an increased range of acceptable input variation with acceptable output, nonlinear systems can be controlled, and a system can be unstable if poorly designed.

To demonstrate the differences in an open- and closed-loop control, let us consider the following examples:

Open-loop example: If the temperature control device for a water heater is an on/off timer attached to the heat source, after some experimentation a specific on/off timer cycle could be found to maintain water temperature.

Closed-loop example: If a thermocouple is used to measure temperature of the water in a water heater, and the information from the thermocouple is fed back to a device designed to turn off or on the heat source, based on a desired water temperature, the system would be able to heat and maintain water temperature following a simple setting (desired water temperature) versus a set of experiments to establish a new on/off cycle as in the open-loop control system in the previous example.

Batch Systems

Sixty to 80% of all processes in the food industry are estimated to be batch processes. Therefore, batch processing is a good representation of the current state of applied process control technology and critically important to the continued improvement in productivity (Russell, 1988). The attributes of a typical batch system are:

1. The batch process is controlled by an open-loop control system, and the set points for the system assume the inputs are both stable and predictable.
2. The inputs to the batch system are controlled by numerous open- and closed-loop controllers.

3. The systems to supply inputs to the batch process are in general independent stand-alone systems.
4. The state of the next unit process is unknown to the current process.

In general these systems work well, but there are a set of fundamental problems which limit the effectiveness of batch systems in the food industry. These problems may be summarized as follows:

1. The input to the system is generally some form of naturally occurring ingredient with properties that are not always stable or predictable, resulting in the frequent recalibration of the batch controllers' set-points.
2. The output of the batch frequently requires further processing within specific time constraints; if the next process is not available the entire batch may be lost or the performance of the next process may adversely be affected by the lower than normal quality.
3. If the output of one of the batch input systems malfunctions, the malfunction may not be detected until the batch is completed and all the value of the product is lost.
4. Frequent recipe changes require manual readjustment of control settings, resulting in possible errors in the control setting.
5. Without computer control strategies, all postproduction information related to the batch is lost. This makes subsequent troubleshooting difficult, if not impossible.

Computer Integrated Process Control

An effective first step in computer integrated process control is to incorporate centralized computer data acquisition functions into the current system without making fundamental changes in the batch control strategy. The benefit of this first step is the acquisition of real time data that currently is difficult or impossible to obtain and assimilate. In general, the initial data acquisition system is accompanied with the introduction of a simple recipe management system.

Once the system performance data is readily available from the computer data acquisition system, one is afforded an opportunity that is not generally taken advantage of. The acquired data may be used to understand, model, and simulate the system behavior and performance. This new understanding may then be used to incorporate novel and/or advanced control strategies that are only available through the computation power that the computer provides.

Examples of some of the advance control strategies that now become available are:

1. Expert control systems, which infer the proper control action based on "fuzzy" logic and the experience gained from the data acquisition system.

2. Via interaction with other plant operations, the timing of batch initiation may be adjusted to obtain the "best" (most economical) operation given the current plant performance status.
3. Through the use of multivariate, statistical models, one might be able to adjust system performance levels for quality, even though a direct sensor for this quality measurement is not available.
4. The computer provides the user with the opportunity to use predictive and adaptive control. The former uses a model of system performance to adjust controller actions for significant dead times that may occur in the process. The latter automatically optimizes the values of control parameters to maximize the system performance.
5. The use of the computer's computational ability to statistically analyze data provides the opportunity to perform on-line model identification. This is particularly useful in the food industry where significant year-to-year, season-to-season, and field-to-field raw material variations occur. These variations usually require a yearly relearning of control strategies. On-line model identification eliminates the relearning.

Concluding Comments

The subject of process control and future directions of process control is far too complex to cover in this limited format. While traditionally the subject of process control has been limited to advances in sensor technology and simple batch (recipe) control strategies, the authors believe the low cost availability of computers offers the food industry the opportunity for great advances in productivity through the application of automatic data acquisition, modeling, analysis, simulation and advance control strategies.

FUTURE NEEDS

Although predicting the future is usually a futile exercise, it seems appropriate to point out a number of future needs.

In modeling, simulation, and optimization we do not suffer a scarcity of mathematical tools. The field is well developed, and the number of procedures, including available software, is more than enough to accomplish many of the applications we envision. However, there is a bottleneck in the development of dependable models for the food industry, which implies fundamental knowledge in transport properties, equilibrium data, etc. Therefore, we perceive that future work will need to focus on providing this knowledge and to take advantage of the existing tools.

In the field of sensors and biosensors we need to continue their development for on-line sensing of food quality attributes such as color, texture, and composition. This must be done with high speed capabilities, so that we may address applications at the high production rates and the natural variability of the material typical in normal industrial production.

In the world of machine vision, we need processors containing both neural network (parallel processing) and Von Neumann architectures operating four times faster than the current computers. We believe the advances in fractal geometry may aid in breaking the machine vision processing bottleneck associated with Euclidean geometry.

In the area of process control, the major bottleneck is moving the theory to the production floor. Traditionally, major capital expenses are required to implement a new control model; hopefully, with the current advances in computer integrated process control, a new control strategy can be quickly and easily implemented using strictly software.

The best use of computer technology results not from the unbridled acquisition of leading-edge hardware, but from innovative and disciplined management and from the basic understanding of the processes and the technology. It is critical to look beyond short-range cost/benefit analysis only.

Finally, scientific and basic understanding are paramount steps towards computer integration. To address the special requirements of computer integration for food applications, enhanced computer curricula must be developed.

ACKNOWLEDGMENTS

The authors express their appreciation to The Pillsbury Company for the opportunity to participate in this project and to Professor Israel Goldberg for his critical comments and suggestion of biosensors. Special thanks are due to Leila Johnson for her dedication in typing and editing this manuscript.

REFERENCES

Anonymous. 1989a. Is the computer business maturing? *Business Week,* March 6: 68.

Anonymous. 1989b. System Integration. *Financial World,* Feb 21: 49.

Anonymous. 1989c. The personal computer finds its missing link. *Business Week,* June 5: 120.

Anonymous, 1988a. United States computers & auxiliary expenditure by food and beverage industry. *Datamation* 4(1): 83.

Anonymous. 1988b. Biosensor market to grow almost 100% yearly. *Biotechnology News* 8(1): 7.

Anonymous. 1986. Automated fish inspection system, *Robotic Eng.* 8(6): 19.

Anonymous. 1985. Sales Brochure for Loctronic Graders, Loctronic Graders Ltd., Danbury, Chelmsford, Essex, England.

Babbit, S. 1988. Machine vision: poultry quality control by computer. *Poultry International* 27(7): 27.

Balland, D. H., and Brown, C. M. 1982. *Computer Vision.* Prentice-Hall, Englewood Cliffs, NJ.

Bandoni, J. A., Rotstein, E., and Romagnoli, J. A. 1988. Simulation modelling and optimal operation of an apple juice concentrated plant. *J. Food Eng.* 7: 159.

Baxes, G. A. 1984. *Digital Image Process: A Practical Primer.* Prentice-Hall, Englewood Cliffs, NJ.

Berler, H. M. 1988. U.S. Market for smart sensors. Report A1909J. Frost and Sullivan, Inc., New York.

Brody, H. 1988. America's technology champions. *High Technology Business* 8(6): 22.

Casansent, D. 1987. Optical pattern recognition and artificial intelligence, a preview. *Proceedings of SPIE* p. 2–11. Society of Optical Engineering, Bellingham, WA.

Clark, P. J., Valentas, K. J., and Lund, D. B. 1985. Food process engineering/automation. Proceeding of the IFT Workshop on Research Needs. *Food Technol.* 39(6): 21R.

Deming, E. W. 1982. *Out of the Crisis.* MIT Press, Cambridge, MA.

Diamond, M. D., and Ganapathy, S. 1982. Cooperative solutions to the continuous graph labeling problem: review and reformulation. *Proceeding of IEEE Computer Society Conference of Pattern Recognition and Image Processing* p. 64–71. IEEE, New York.

Downer, E. 1988. Sensors for the food industry. Soft Drinks Management International, Leatherhead Food Res. Assoc., Leatherhead, Surrey, UK.

Draper, N., and Smith, H. 1981. *Applied Regression Analysis,* 2nd ed. John Wiley & Sons, New York.

Fraser, R. T. 1989. Personal communication. Pegasus Biotechnology, Ontario, Canada.

Gagliardi, G. R., Sullivan, D., and Smith, N. F. 1984. Computer-aided video inspection. *Food Technol.* 38(4): 53.

Gagliardi, G. R., and Sullivan, D. 1983. Computerized inspection for quality control. *Cereal Foods World* 28(9): 513.

Gagliardi, G. R. 1983. Personal correspondence with L. Levine, June.

Gerber, G. L., Holmes, Q. A., and Calhan, R. 1985. Industrial machine vision with x-ray sensor for on-line food processing inspection. Paper presented at the National Meeting of the Society of Manufacturing Engineers, Detroit, 1985.

Green, R. J. 1987. Introduction: The importance of manufacturing research. In *Manufacturing Research Perspectives USA Japan.* (Gertenfeld et al., Ed.) Elsvier, Amsterdam.

Guilbault, G. G., and Luong, J. H. T. 1989. Immobilization methods for piezoelectric biosensors. *Biotechnology* 7(4): 349.

Gunasekaran, S., Cooper, T. M., and Berlage, A. S. 1988. Evaluating quality factors of corn and soybeans using a computer vision system. *Trans. Am. Soc. Ag. Eng.* 31(4): 1264.

Hudson, D. L. 1984. Food for thought and appearance. *Proceedings of the 3rd Annual Applied Machine Vision Conference* p. 3–1. SME, Dearborn, MI.

Janata, J., and Bezegh, A. 1988. Chemical sensors. *Anal. Chem.* 60(12): 62R.

Kanegif, G. 1987. Sorting of fruits and vegetables with unique visions systems. Paper presented at the National Meeting of the American Institute of Chemical Engineers, Minneapolis, MN.

Karube, I., Matsuoka, H., Suzuki, S., Watanabe, E., and Toyama, K. 1984. Determination of fish freshness with an enzyme sensor system. *Japan J. Agric. Food Chem.* 32(2): 314.

Kelly, P. T., and Richardson, P. S. 1987. Computer modelling for the control of sterilization processes. Tech. Memorandum, Campden Food Preservation, Res. Note #459.

Kozempel, M. F., Sullivan, J. F., and Craig, J. C., Jr. 1981. Model for blanching potatoes and other vegetables. *Lebensm.-Wiss. U. Technol.* 14: 331.

Kozempel, M., Craig, J. C., Jr., Sullivan, J. F., and Damert, W. 1988. Computer simulation of potato processing. *Biotechnology Progress* 4(2): 63.

Lathi, B. P. 1974. *Signal Systems and Controls.* Harper and Row, New York.

Levine, L., Symes, S., and Weimer, J. 1987a. Simulation of the effects of formula variations on the transient output of a single screw extruder. *Biotech. Prog.* 3(4): 212.

Levine, L., Symes, S., and Weimer, J. 1987b. A simulation of the effects of formula and feed rate variations on the transient behavior of starved extrusion screws. *Biotech. Prog.* 3(4): 221.

Levine, L., Symes, S., and Weimer, J. 1986. Automatic control of moisture in food extruders. *J. Food Proc. Eng.* 8(2): 97.

Li, X.-M., Liang, B. S., and Wang, H. Y. 1988. Computer-aided analysis for biosensing and screening. *Biotechnol. Bioeng.* 31(3): 250.

Liska, B. J., and Marion, W. W. 1985. Executive summary. Proceedings of the IFT Workshop on Research Needs. *Food Technol.* 39(6): 4R.

Luong, J. H. T., Male, K. B., and Ngyen, A. L. 1988. Development of a fish freshness. *American Biotech. Lab.* 6(8): 38.

Malone, R. 1988. Modeling requires sound management. *Managing Automation* 3(8): 40.

McCormick, R. 1988. On-line scanner homes in on quality deviations. *Prepared Foods* 157(7): 143.

Mishkin, M., Karel, M., and Saguy, I. 1982. Application of optimization in food dehydration. *Food Technol.* 36(7): 101.

Mishkin, M., Saguy, I., and Karel, M. 1984. Optimization of nutrient retention during processing: Ascorbic acid in potato dehydration. *J. Food Sci.* 49(5): 1262.

Mishkin, M., Saguy, I., and Karel, M. 1983. Dynamic optimization of dehydration of processes: Minimizing browning in dehydration of potatoes. *J. Food Sci.* 48(6): 1617.

Patel, N. P., and Scott, A. O. 1988. An immunochemical approach to sensing food components and contaminants. Technical Memorandum, No. 476, Campden Food Preservation Res. Assoc., Campden Food Preservation Res. Assoc., Chipping Campden, UK.

Rand, W. M. 1983. Development and analysis of empirical mathematical models pertinent to food processing and storage. In *Computer-Aided Techniques in Food Techology* (Ed.) Saguy, I., Ch. 3. Marcel Dekker, New York.

Rechnitz, G. A. 1988. Biosensors. *CSEN,* Sept: 24.

Rechnitz, G. A. 1987. Biosensors, an overview. *J. Clinical Lab. Analysis* 1: 308.

Rehkugler, G. E., and Throop, J. A. 1985. Apple sorting by machine vision. Paper presented at the Winter Meeting of the American Society of Agricultural Engineers, Chicago, IL.

Roberts, S. A., and Guy, R. C. E. 1986. Instabilities in an extrusion-cooker: A simple model. *J. Food Eng.* 5: 7.

Russell, M. J. 1988. Process control: emphasis on batch. *Food Engineering* 60(7): 53.

Saguy, I. 1988. Constraints to quality optimization in aseptic processing. *J. Food Sci.* 53(1): 306.

Saguy, I. (Ed.) 1983. *Computer-Aided Techniques in Food Technology.* Marcel Dekker, New York.

Saguy, I., and Karel, M. 1980. Modeling of quality deterioration during food processing and storage. *Food Technol.* 37(2): 78.

Saguy, I., and Karel, M. 1987. Index of deterioration and simulation of quality losses. In *Objective Methods in Quality Assessment.* (Ed.) Kapsalis, J. G., Ch. 9. CRC Press, Boca Raton, FL.

Sarker, N., and Wolfe, R. R. 1985. Computer vision based system for quality separation of fresh market tomatoes. *Trans. Am. Soc. Ag. Eng.* 28(5): 1714.

Schaertel, B. J., and Firstenberg-Eden, R. 1988. Biosensors in the food industry. *J. Food Protection* 51: 811.

Schmidt, R. D. 1988. Trends in biosensors. *Biofuture* 37.

Schrage, L. 1981. *LP Models with LINDO* (Linear Interactive Discrete Optimizer). Scientific Press, Palo Alto, CA.

Shimizu, Y., Takao, Y., and Egashira, M. 1988. Detection of freshness of fish by a semiconductive Ru/TiO$_2$ sensor. *J. Electrochemical Soc.,* Oct.: 2539.

Suetens, P., and Oosterlinck, A. 1985. Critical review of visual inspection. *Proceedings of SPIE,* pp. 240–254. Society of Optical Engineering, Bellingham, WA.

Turner, N. P. F., Karube, I., and Wilson, G. S. (Ed.), 1987. *Biosensors: Fundamentals and Applications.* Oxford University Press, UK.

Watanabe, E., Endo, H., Takeuchi, N., Hayashi, T., and Toyama, K. 1986. Determination of fish freshness with a multielectrode enzyme sensor system. *Bulletin of the Japanese Society of Scientific Fisheries* (Nihon Suisan Gakkai-shi) 52(3): 489.

Watanabe, E., Nagumo, A., Hoshi, M., Konagaya, S., and Tanaka, M. 1987. Microbial sensors for the detection of fish freshness. *J. Food Sci.* 52(3): 592.

Wong, K., Bartlett, F., and Gill, T. A. 1988. A diagnostic test strip for the semiquantitative determination of trimethylamine in fish. *J. Food Sci.* 53(6): 165.

Wren, J. J. 1986. What the food industry needs from probes. In *Bio '86 Probes for Detection and Measurement in Biotechnology. Int. Conf. p. 13–1. London.*

Index